Success Stories in Satellite Systems

Success Stories in Satellite Systems

Edited by
D. K. Sachdev
SpaceTel Consultancy LLC, Vienna, Virginia and
George Mason University, Fairfax, Virginia

LIBRARY
OF FLIGHT

Ned Allen, Editor-in-Chief
Lockheed Martin Corporation
Palmdale, California

Published by
American Institute of Aeronautics and Astronautics, Inc.
1801 Alexander Bell Drive, Reston, VA 20191-4344

American Institute of Aeronautics and Astronautics, Inc., Reston, Virginia

1 2 3 4 5

Library of Congress Cataloging-in-Publication Data

Success stories in satellite systems / edited by D. K. Sachdev.
    p.   cm. -- (Library of flight)
    ISBN 978-1-56347-966-3
    1. Artificial satellities--History.   I. Sachdev, D. K.
    TL796.S83   2009
    629.43'4--dc22   2008055024

Cover design by Gayle Machey

FOREWORD

Perhaps no element of our culture is more expressive of the civilizing role of aerospace engineering than our space-based information, communication, and navigation infrastructure. In many ways, it was and is the central enabler of the worldwide broadening of human outlooks that has come to be called "globalization." And nothing has been a more nourishing source of cross-cultural engagement and international collaboration than the enterprise of building a space-based infrastructure shared across every continent and the entire human race.

Today, we take these space-based systems for granted; but, of course, they were conceived, engineered, and implemented at great effort by at least three generations of visionary and ambitious aerospace people. As we became and are becoming still more dependent on these satellite systems for our day-to-day existence, these systems became more reliable and robust as well as more capable—again, the result of a creative and dynamic aerospace community. The mere presence of this infrastructure system and of the visionary dreaming it seems to spark in each new generation continues to generate entirely new human enterprises in new areas of endeavor that were not foreseen: the GPS system was not glimpsed by anyone I knew growing up and yet today we are captivated by it and captive of it. Today, we read of identity theft and criminal enterprises carried out across the satellite data links and hear talk of innovative key distribution from the same satellites to defeat those enterprises—none of this was even imagined when the first satellites went up in the 1950s.

This collection of original essays, assembled and edited by D.K. Sachdev, documents the movement that established our now-crucial infrastructure systems and highlights the exceptional people that drove the creation and the international organizations they spawned. Many of the authors, including our editor, Professor Sachdev, were and are themselves key leaders in the moment and, thus, the book carries the internationally diverse flavor of the community whose story it tells.

The Library of Flight series is part of the growing portfolio of information services from the American Institute of Aeronautics and Astronautics. It extends the Institute's publications with the best from a growing variety of

v

topics in aerospace from aviation policy to histories, studies of aerospace law, management, and beyond. The Library of Flight documents the crucial role of aerospace in enabling, facilitating, and accelerating global commerce, communication, and defense, and so, this volume occupies a significant position on the library shelf.

Ned Allen
Editor-in-Chief
Library of Flight

TABLE OF CONTENTS

PREFACE

For the past eight years, I have taught a graduate class on systems engineering, using satellite-based systems as its basis, at George Mason University, Fairfax, Virginia. The students carry out group projects simulating as much as possible real-life environments. As part of this course, we discuss the applicable criteria for success that often go beyond financial success. Almost invariably the questions in the class gravitate to a discussion on what factors make a project or an organization truly successful and to what extent the leadership at the top counts. I try to answer these questions with examples of ventures that have been successful and those that have not. However, often after such discussions, I felt that I had not really addressed the questions in full and that there was still a genuine need for a few well-documented stories of success at least in our field, hence this book.

I approached several friends and contemporaries about such a compilation, and the response was uniformly encouraging. However, the common advice was that such a compendium should not be yet another survey book but should instead be based on first-person accounts as much as possible. I heartily agreed but was not sure that this could indeed be achieved in all cases because most of such authors are either really busy persons or have already moved on and might not be willing to put in the effort.

I took up the task in earnest. Identifying the organizations and topics was not that difficult. What indeed took much longer than I had allowed for was to locate the right authors and to persuade them to put down on paper events going back sometimes as much as 25 years or even more. After several fits and starts, we are finally here. The response from a number of industry leaders and their willingness to contribute has indeed been gratifying to me; however, in spite of my best efforts, I just could not cover all of the topics that I had originally embarked upon because of my inability to locate the right (and willing) authors. Nevertheless, I believe the book you are about to read is indeed a unique collection of many fascinating and instructive stories. It is also my expectation that these stories would interest a fairly broad audience of managers, leaders, and students, hopefully even beyond the satellite communications field as well.

I have tried to fulfill my duties of editor very carefully and often gingerly. Throughout this process, my approach was to locate and persuade the right authors for a particular topic, have extensive discussions with them about the book objectives and the contents of their chapter, and from then on to let them narrate their stories as they thought fit. This accounts for the somewhat different styles adopted by practically each author. My general guidelines were that the emphasis should be on how success was achieved and to name as many contributors as possible. The authors were also encouraged to dig out as many pictures of historical importance as possible. Thanks to the support and cooperation of the publishers, the end result is that the pictures interposed throughout the book are perhaps as memorable as the accompanying text.

I wish to record my thanks and appreciation for all those who have helped me make this book a reality. Above all are of course the authors themselves who have written these stories, each remarkable in its own way. Although I had known many of them in the past, some of them I came to know for the first time through this book effort. All of them showed remarkable patience and not once expressed annoyance at my frequent reminders. Overall, it has been my privilege and a matter of great professional satisfaction to work with many illustrious persons in our industry. They have taken the time and effort to document some really unique stories that in many ways only they could write. Not just me, but may I say that the whole industry owe them our sincere thanks.

Several other friends and professional colleagues pointed me in the right direction and also provided me valuable historical milestones and perspectives. They are just too numerous to list by their names.

On behalf of all of the authors and certainly myself, we would like to record our appreciation of several executives at AIAA, the publisher. Rodger Williams got the ball rolling and paved the way for the book to be on its publishing calendar. Dan Cafaro piloted the program once all of the manuscripts were in. Pat DuMoulin deserves our special thanks in providing valuable professional advice on the structure of the book and its manuscripts and managing the whole process in a true professional manner. Janice Saylor deserves our appreciation for the unique book cover and for the ongoing marketing campaigns for this book. Thank you all.

The whole process of putting together this book took a bit over two years. It was inevitable that some of the early manuscripts would be rendered obsolete in certain parts because of later organizational and other changes in the marketplace. I thank all those who have helped me update the information as much as possible. One notable event in 2008 was the merger of XM Radio with Sirius Radio. Chapter 13 as written tells the story of the evolution of XM Radio only, and all of the key points are still valid except for some executive changes in the interim.

Several authors, including me, have highlighted contributions of Arthur Clarke in one form or other. Despite his advancing years, Sir Arthur was alive and active while this book was being put together. Unfortunately, in March 2008, he passed away shortly after celebrating his 90th birthday. Ever since his now famous paper in 1945, Arthur Clarke has contributed to the advancement of science in general and satellite communications in particular. The frequent references in this book to his vision and contributions are our collective homage to this true legend of humanity. But he is not quite done yet because long after he has passed away, his predictions and projections through his writings will continue to benefit all of us.

Thank you, Sir Arthur.

D. K. Sachdev
April 2009

Chapter 1

OVERVIEW

D. K. SACHDEV*

It is my privilege to briefly introduce the chapters in the order they are presented in this book. Wherever relevant, I have tried to fill the gaps if any to highlight events of historical significance; once in awhile I have also added some personal memories when I felt they added some value to the stories.

The concept of space communications as we know it today started with the now famous article by Arthur Clarke in 1945 wherein he conceptualized a worldwide radio system consisting of three geostationary satellites spaced equally around the globe. This article is perhaps the most quoted publication in our industry; indeed, that is also the case in several chapters in this book itself. It is fascinating to see the variety of ways through which individual authors have recognized Clarke's unique position and contributions to the industry. From my perspective, what is most impressive is that this paper was written well ahead of the emergence of any type of microwave systems or launch vehicles. That this was no accidental one-time musing by an under-worked person at a military base in a war that was winding down is borne out by the series of extensive writings, books, movies, and speeches that Sir Arthur continued to create right up to his death on March 19, 2008.

The cessation of World War II in 1945 opened up several technologies for commerce and business. Specific areas of interest to telecommunication system designers were the totally new capability to deploy communication equipment beyond Earth's atmosphere and extension of the frequency range of communication devices well into the microwave bands. Several programs were initiated to exploit these advances. The most fascinating of these were related to attempts to launch a payload beyond the shackles of the Earth's gravity. However, these were still relatively uncoordinated small programs in

*President, SpaceTel Consultancy LLC, Vienna, Virginia; Adjunct Professor, Electrical Communication Engineering, George Mason University, Fairfax, Virginia.

different budgets, and any concerted efforts at a national scale had not yet
emerged. The successful launch of the first-ever satellite Sputnik by the Soviet
Union in October 1957 changed all of that, and all of a sudden a global competition for space supremacy was underway. This was particularly challenging for the United States in the context of the failures of two successive launch
attempts by the Vanguard program. As described in more detail in Chapter 16
on the Defense Advanced Research Projects Agency (DARPA), many programs got accelerated and better coordinated, and soon the United States
assumed the mantle of leadership in this nascent industry with both commercial and military potential. The first communication satellite, SCORE,
was launched in December 1958, but its operating life was only 12 days as its
batteries failed. Bell Laboratories launched their Telstar I and II satellites
during 1962–1963. These satellites were tested extensively over the United
States and across the Atlantic for telephony and television.

All of the programs just mentioned were in what we now call low Earth
orbit (LEO) or inclined orbits. Although they did have acceptable time delays,
they did require expensive tracking Earth stations and multiple satellites to
provide continuous service. Nevertheless, they did create a lot of interest
around the world in this exciting new field. Soon there was a fairly active
debate underway in the telecommunication community between LEO and
geostationary-orbit (GEO) configurations. Bell Labs, who had developed the
Telstar satellites, argued against GEO satellites on grounds of excessive time
delays. In addition, it was indeed a challenge to launch a reasonable payload
in that orbit. There were a few voices arguing for the GEO satellites, one of
them being that now famous Harold Rosen of Hughes Aircraft.

Before I talk about the opening story in this book by Harold Rosen himself
on the first geostationary satellite, I seek the indulgence of the reader to talk
a little about my first meeting with Harold almost 40 years ago. About a year
before I first met Harold, I was introduced to satellite communications in a
professional sense by Vikram Sarabhai, the legendary and charismatic founder
of space technology efforts in India. Soon after that, I was a member of an
Indian government group sent to the United States in late 1968 to develop
viable system concepts for India for its own domestic satellite system. I was
in the group deputed to Hughes Aircraft at El Segundo, California. I arrived
at the Los Angeles airport on a Saturday evening and was a bit puzzled not to
find any Hughes person there to pick me up and take me to the hotel that they
had reserved for us. I was obviously ill-prepared for my trip because the only
name at Hughes I knew was that of Harold Rosen, who had successfully
deployed Syncom a few years earlier. Luckily he responded to the number in
the phone book at the airport and promptly came to pick me up. Not knowing
where I had been booked by his company, he took me to another hotel nearby
until Monday. While at my hotel, I realized that I had made a mistake about
my itinerary and had not taken into account the crossing of the International

Dateline while coming via Tokyo! This embarrassing incident was in a way the start of my friendship with Harold. During my three-month stay at Hughes Aircraft in 1968–1969, not only did Harold assist us in many ways in our system study but also on more than one occasion took me in his car for long rides and shared with me the struggles and travails of getting Early Bird approved and launched a few years earlier. These conversations have stayed in my mind ever since, and I always thought I knew the Syncom story pretty well, that is, until I received Harold's draft for his chapter for this book.

In Chapter 2, Harold takes us down memory lane and beautifully records all of the issues and hurdles that he and his team had to resolve or overcome before the first truly geostationary satellite could be ready just a few weeks before the 1964 Olympics in Tokyo. While reading through this account, one is reminded of many other firsts within the spacecraft itself, ranging from stabilization techniques to a workable travelling tube amplifier in space. Throughout this narrative, Harold is keen to give ample credit to others whenever it is appropriate.

I have read this chapter several times and have talked to Harold and his very capable wife Deborah Castleman several times about its contents and photographs. What comes across to me above all is the clear vision and tenacity with which Harold pursued his mission. Any failure or setback was never a dead stop but simply a hurdle for which a solution had to be found either by himself or by approaching the person or persons more capable of solving the problem. In terms of such a clear-headed approach, Harold and associates were several decades ahead in terms of leadership skills. Lastly, long after the Syncom series of satellites were just some milestones in the distant past, Harold continued to be active in innovation for future satellites for many of the Hughes customers, including of course Intelsat. However, in his typical professional courtesy, Harold allows others to take the bow for what followed his monumental project, which was nothing short of the very beginning of the global satellite communication industry.

The successful demonstration of Syncom and Early Bird in early 1960s was not just a technical or engineering achievement for the first time. It also initiated what will remain perhaps forever the successful formulation of a powerful, successful, and equitable structure for international cooperation. The first embodiment of this structure was Intelsat, and that takes us to the next chapter (Chapter 3).

It is always somewhat harder to maintain a balanced perspective when it comes to writing about an organization where you have yourself worked for a long time. That indeed is the case for me. A major part of my nearly four decades' association with the satellite industry was spent at Intelsat. Whenever an opportunity arises, I do say publicly that much of what I know about satellites is largely what I learned at Intelsat. Therefore, I was very happy when Conny Kullman, former CEO of Intelsat, accepted my invitation to write a first-person account of the Intelsat story. Conny joined Intelsat a few years after I did, and it was a matter of pride for all of the engineers to see him rise through the professional cadre through his remarkable sense of duty and hard work. His story about Intelsat is as personal as it can be and at the same time very accurate. One can see his perspective broadening as he rose through the ranks; what has not changed even today, however, is his strong, unflinching professionalism.

In Chapter 3, Conny takes us from Arthur Clarke to the current Intelsat, fully privatized and now the largest operator in the world after its merger with Panamsat. He very appropriately provides several insights into the recent events that led to the transformation from an inter-government organization to a fully privatized entity. This process was by no means easy or smooth, and Conny had at many instances to apply his firm leadership skills to stay focused on the objectives and achieve them despite all of the distractions.

I spent 18 years at Intelsat and all but the first five years concurrently with Conny. Like Conny, I had come to Intelsat for a short tenure with a leave of absence from my service career in the Indian PTT, but decided to stay on (like Conny and many others). Until 1979, Intelsat was largely run by Comsat. As Intelsat gradually assumed its own responsibilities, many managers like me came onboard and helped in this transition. Intelsat was literally born via the Early Bird, and for over two decades it had the implicit responsibility of helping the satellite industry grow. This was achieved through two overlapping methods: funding of near-term and long-term R & D around the world and through the development and procurement of larger and larger generations of spacecraft. I had the opportunity to lead both of these efforts to a degree. I have added this reservation not necessarily from a sense of modesty but also to recognize that international organizations do not always empower the managers fully and like to oversee and contribute to almost all of the processes at all levels. (Conny has touched upon this aspect as well.) Although at certain times such "owner vigilance" at all levels was an irritant, overall I will be the first to admit that it invariably led to a better decision and product. Given the time frame when Intelsat was not seriously challenged in the marketplace, this deliberative consensus-based process did not require excessive time, provided one had the foresight to start well ahead of the need date.

The competition to Intelsat came slowly but surely and from several directions. First, several regional and domestic organizations came into existence, thus providing competition in their respective domains. Although this challenge was not always serious, what was irksome and difficult was that some of the very owners who scrutinized and approved Intelsat's plans, programs, and tariffs were often also heading organizations back home that competed against Intelsat in their respective countries. Talk of multiple conflicts of interest!

The second and far more serious competition came, of course, from optical fiber cables. Before the late 1980s, the undersea copper cables had capacities much smaller than those of Intelsat satellites. However, all of that changed in 1988 when the first cross-Atlantic fiber cable, TAT-8, was commissioned. Several years before this deadline, several commentators began writing about the imminent end of satellite communications literally. Fortunately, Intelsat had the foresight to introduce digital technology in 1983 and sign long-term deals with its users with attractive rates compared to the past. Although it temporarily disappointed the satellite designers who feared that the days of bigger and bigger satellites were over, it was a sound business decision both in the technological and the marketing sense. In my mind there is no question that TAT-8 was a wake-up call for Intelsat and inculcated the very necessary discipline of being ready to compete, irrespective of the administrative structures at the top.

The third type of competition originated from what came to be known in the 1980s as "separate systems." These were not just domestic operators who wished to co-exist with the inter-government format of Intelsat, nor were they other such bodies like Inmarsat and Eutelsat, which focused on specific regions or markets. Rather the "separate systems" were private entities who wanted to operate internationally in direct challenge to Intelsat. They argued that the special landing rights granted to Intelsat in all its member countries' marketplaces were not equitable as they did not provide fair opportunity for others to compete. What was particularly annoying to them was that any such separate system had to meet the "economic harm" criterion, the arbiter of which ironically was the top body of Intelsat itself! The leader of this separate systems effort was unquestionably Panamsat, founded by the one and only Rene Anselmo.

Rene Anselmo's story would have been the heading of one chapter that I desperately wanted to add to the book, but did not succeed in finding a willing author with firsthand knowledge. (The irony of the whole process is that while I was busy trying to find an author to talk about Panamsat and how it competed aggressively against Intelsat, the "new" Intelsat had successfully merged with Panamsat!) Although I was unsuccessful in finding a willing author, I could collect some of the essential elements of his achievements. Many aspects of Rene's achievements were in fact known to me as an Intelsat executive, but at that time we saw it more as a threat rather than with the objectivity that I can have now as a book editor.

Rene had made his money through Mexican broadcasting stations and was determined to start a satellite company of his own. He literally spent a good part of his personal fortune for a spacecraft already built for somebody else and got it modified to serve Latin America. He went further and accepted an attractive discount price on a new yet-to-be-proven Ariane launcher. To top it all, according to one of his associates at that time, he overruled his team and did not insure the launch! In other words, a failure of the launcher—a real possibility for a maiden flight of a new type of launcher—would have wiped out not only his dream of a competitor to Intelsat but also his own personal fortune. Fortunately, the launch was a success, and he immediately went about establishing a market economy for transponders by initially offering them at almost no cost! Although this would not make too many waves today, it was literally a revolution in the early days of 1980s when a uniform price for all services all over the world was the cornerstone of Intelsat policy.

Panamsat continued to grow, and Rene continued to attack Intelsat's "privileged position" in the marketplace, sometimes through his inimitable cartoons of Dog Spot doing his thing in public. There is no question that he started the wave that less than two decades later led to a fully privatized Intelsat.

Following a chapter on Intelsat, it first appeared logical to talk about other international systems that, at their inception at least, emulated the framework established by Intelsat for equitable cooperation among nations. However, from considerations of historical precedence, it also appeared logical to first place a chapter on the long and successful story of systems in Canada, a country with many "firsts" to its credit in this field. I have opted for the latter.

Canada has, of course, a long history of innovation in many fields. Given its vast area and relatively small population compared to its southern neighbor, it also seems to have perfected the art of focusing on what it is capable of and then doing it very well. The Canadian space efforts fit this definition very well.

Deciding that there should be a chapter on Canada space programs was easy. However, locating the right and willing author turned out to be such a difficult process that at one point I came to close to giving it up. However, through some very helpful friends I got introduced to Harry Kowalik, a veteran of Canadian space programs who had recently retired from service after a lifetime of contributions. Although I have to yet meet him in person, through phone calls I have come to know him quite well, including his impressive biking adventures all over the world despite being in his seventies. Once Harry agreed to undertake this task, he went about it very diligently indeed.

For each and every facet of his story in Chapter 4, Harry located the right executives, both former and current, to gather backgrounds and to corroborate his narratives. The final story is not only exhaustive and fair, but it also includes many memorable pictures of historical significance. One of the notable parts of his narrative is his very fair summary of how governments formulate policy on new issues and directions and how adjustments often become necessary in midcourse. He has addressed not only all of the key historical milestones but also highlighted Canada's contributions to the technology as a whole.

Harry has also documented many Canadian "firsts," some of which might not be that well known. Intelsat's system was explicitly put together for international communications, and most members were careful to keep domestic services within their boundaries outside Intelsat's domain of business. (Eventually, of course Intelsat did decide to provide domestic leases, starting initially with the in-orbit "spare" capacity.) Concurrently, many nations, especially those with large landmasses were giving consideration to satellites for purely domestic applications. These included the United States, Canada, India, and many others. However, on 11 January 1973, Canada had the unique honor of making the first purely domestic call over its own domestic satellite, Anik A1. Another first for Canada was the very first launch of commercial satellite, Anik C3, on the shuttle in November 1982. Since this chapter was written, Telesat has changed hands, and as a result in late 2007 was the fourth largest network in the world.

An interesting piece of information that Harry provides is that in the early 1980s the shuttle launch was much cheaper than an equivalent expendable launcher. Of course, that is vastly different from the status today when the shuttle does not offer commercial launches at all and the cost of each shuttle launch is well out of the reach of any commercial customer. Lastly, Harry reminds us that the very first Canadian spacecraft, Alouette 1, launched as early as September 1962 had a very large deployable dipole antenna as big as 45.7 meters long and 22.9 meters wide. This early development led to a long-term business for Canada not only for many other programs that followed all over the world but has also given Canada a leadership role on deployable arms used in a whole range of space missions.

We now revert back to a number of international and regional systems. The first one is Chapter 5 on Inmarsat. Here also, it was somewhat of a struggle to get the right author, but eventually my long-time friend and professional colleague, Ahmad Ghais, came to my rescue. At short notice and in spite of a temporarily partially disabled right hand, he has produced a remarkable chapter on the story of Inmarsat. Ahmad, whether as a committee chairman at Intelsat, or at a conference, has always been an extremely articulate person and invariably demonstrates a clear comprehension of the key issues involved. That is also the case for the chapter he has produced on Inmarsat.

Inmarsat was born several years later than Intelsat and thus had the benefit of adapting what worked and what did not. Right from the start it fulfilled a long-standing need for providing reliable and dependable communications with ships in deep seas. Ahmad's personal knowledge (as a senior executive of both Comsat and Inmarsat) provides a valuable insight into the interplay of national interests with international imperatives, a drama that was not unique to these two organizations but can be seen to be underway on the global arenas in many ways. In this chapter, Ahmad provides a very succinct history of how the whole field of mobile satellites has progressed. In many ways this chronicle is more fascinating than that of fixed communications because it is this very same field that saw not only many successes but also three major bankruptcies in the 1990s. Looking forward, as we will talk more in the very last chapter, instant mobile access (with Internet access) is once again spawning potential progress and in some ways might well lead the whole field of satellite communications in the future.

Dr. Ghais' narrative, not unlike other chapters in the book, highlights many interesting pieces of information that often remain buried either in the minds of the dwindling cadre of industry "white-beards" or at the fading archives of organizations. Thus, it is fascinating to know that aeronautical communication needs were addressed even well before Inmarsat took shape and one of the MARISAT satellites, launched in the 1970s, is still providing some kind of service over Antarctica!

Inmarsat has been essentially a story of steady growth and progress for several decades, much like Intelsat. During the 1990s, however, there were several attempts to successfully develop direct-to-user satellite mobile systems. The three major among these were Iridium, Globalstar, and an offshoot of Inmarsat, ICO. Unfortunately all of these failed for different reasons. Iridium and Globalstar were both the victims of a delayed business plan implementation that led to their financial failure against the rapidly growing terrestrial cellular systems, whereas ICO did not complete its 12-satellite medium-Earth-orbit (MEO) architecture at all because of financial issues. However, there was one such system that did start a bit later in the 1990s and succeeded. This is the Thuraya geostationary mobile satellite system, the subject of Chapter 6, following the chapter on Inmarsat.

Thuraya was a venture of entrepreneurs from the Middle East and targeted that market. They were quick to learn the perils of developing right away a global system and chose a progressive growth approach, something that geostationary satellites are eminently suited for. Such a plan had its own challenges, one of them being to have a satellite powerful enough to reach handheld phones on the ground. Recognizing the close interaction required between the space and ground segment, they had the wisdom to entrust the whole system implementation task to one team that could optimize the interfaces, etc. The team chosen was Hughes, which at that time included companies capable of making very powerful satellites (Hughes Aircraft, now Boeing) and Hughes Network Systems with the largest experience in VSAT networks and other user devices for satellites. This chapter is written by the leaders of the two teams: Ali Saeed Al Mazrooei of Thuraya and Adrian J. Morris of Hughes. Its style is typical of those who have spent many sleepless nights solving the many challenges that they had to encounter. The chapter takes pains to recognize some notable engineers who have made—and continue to do so even today—very impressive contributions to the mobile technology in general, a true success story in both the technical and business sense.

Next we have two very interesting stories, both of which had their origin in Europe. Interestingly, the two systems started quite differently from a marketing perspective, but today are strong competitors in satellite broadcasting. Both are expanding their service mix, and both are also spreading their network outside Europe as well. These are, of course, the Eutelsat and SES-Astra networks. Until Intelsat merged with Panamsat, SES-Astra was the biggest satellite-based system in the world.

Chapter 7 on Eutelsat is written by another old friend whom I have admired ever since I came in contact with him in early 1980s when he was at ESA and I was at Intelsat. Although their personalities differ quite a bit, both Conny Kullman and Giuliano Berretta through their respective careers have demonstrated that in spite of all of the hurdles and obstacles of international organizations, it is possible to reach the very top on the strength of their abilities, hard work, and demonstrated contributions. Interestingly, Conny and Giuliano achieved such pinnacles more or less contemporaneously, and both successfully took their organizations through the complex privatization processes.

Eutelsat started in 1977 initially as the "European" response to Intelsat, which was seen as an "American" organization despite its global charter. I joined Intelsat just a year later, and at first it appeared that the creation of Eutelsat was somewhat redundant because in principle the entire Eutelsat market was also within Intelsat's domain. However, that was perhaps a biased view of an Intelsat executive, and, as subsequent events have demonstrated, Eutelsat has played a very crucial role in developing appropriate spacecraft and services for the landmass-centric market of Europe, whereas Intelsat spacecraft were traditionally designed until then almost exclusively to serve from the midocean orbital locations.

Giuliano's style in presenting the Eutelsat story is quite different than that of Conny, but he in his own way does ample justice to the continuing success that this system has achieved. Early on in his chapter, he emphasizes that one of the basic reasons for the continuing success of satellites is that they adapt very well to the relevant sector and are willing and able to assume their due role in the larger missions of that sector. He also highlights that Eutelsat early on decided to shift the focus towards television broadcasting and away from telecommunications. In hindsight this decision was very timely as the use of satellites for trunk telecommunications in developed landmasses such as Europe declined quite rapidly once the fiber networks were in place. The early challenge to satellite broadcasting was of course the cable penetration in several European countries. Eutelsat and many similar systems have always taken advantage of the visibility of satellites to capture markets where terrestrial system do not or cannot penetrate easily. Satellites also went digital much earlier than cables all over the world, thus gaining significant economic leverage by multiplying the satellite capacity many times over. The latest in this arena is, of course, the rapid expansion of HDTV that once again allowed the satellite medium the first mover advantage in many markets. As he ends his chapter, Giuliano also recognizes that the satellite medium does have an upper bound because of its finite radio spectrum. However, that is the boundary that technology has been pushing forward for several decades.

The evolution of SES Global in the 1980s followed a totally different path than that of Intelsat, Inmarsat, and Eutelsat, the three systems we have discussed in preceding chapters. SES was not the creation of countries or large operators—well-financed by their respective government in most cases, but rather purely the result of entrepreneurial spirit of a few pioneers. In some ways, the story of SES's origin during the 1980s reminds us of how Panamsat was created on the other side of the Atlantic only a few years earlier. The parallel goes even further. Although the creation of Panamsat was opposed by Intelsat, in Europe the predecessor organizations of SES had similar opposition from Eutelsat!

Chapter 8 on SES Global is interestingly titled by the author as "The Mouse That Roared." SES originated in one of the smallest countries in the world, Luxembourg, and it also started very small indeed. In 1986, according to one of the founding executives, it had "no money, no frequencies, no regulatory approval, no satellites, no rocket, no TV channels ...". However, one by one through sheer grit and determination everything fell in place, and today it has global presence with only a few satellites less than the leading operator, Intelsat.

Every so often, we come across a vindication of the "small is beautiful" principle, and the origin of SES also confirms that. As this chapter so carefully puts together, it required a less powerful transponder and not a bigger one to open the doors of success! While "technology-based" satellites backed by large organizations on both sides of Atlantic struggled with their reliability and a business plan based on just three to five transponders, here came a company from nowhere that made a cheaper satellite with as many as 16 transponders that could be directly accessed by the users. SES also got around the cumbersome regime of broadcast frequencies by using the so-called FSS frequencies, a regulatory "license" (pun intended) that has been taken countless times since then. That was not all; SES also demonstrated that one can grow with the business without building risky huge satellites with capacity levels required perhaps a decade later. Instead, for a new business still exploring the size and strength of a new market segment, it makes much better business, and often technical, sense to first launch a smaller satellite using only part of the spectrum and then follow up with additional satellites in adjacent frequency segments and locate them at the same orbital location. This principle of colocation that SES established in a major business sense (and has continued to lead ever since), has since been emulated all over the world and is instrumental in enabling hundreds of TV channels through a single rooftop dish.

We now come to two chapters on domestic or national systems. Compared to North America and Europe, the story in Japan followed a somewhat different path. Given its much smaller geographical area, Japan did not provide the same kind of rationale for satellite communication as the other much larger areas, or the oceans, provided. Nevertheless, Japan has always been an important participant, both industrially and as an important user of satellite technology.

The story of satellite communication in Japan is told in Chapter 9 through the story of JSAT, currently the fifth largest network in the world (after Intelsat, SES, Eutelsat, and Telesat). The authors of this chapter put in a tremendous amount of effort and research to present to us a large amount of historical information with the names of the key players at different stages. There are some very interesting aspects that perhaps deserve to be highlighted. Japan, like many developing countries, was anxious to develop local industries and capabilities while being keen to provide the very best there was to its consumers. This led to some very interesting dialogues with the U.S. manufacturers, particularly Hughes Aircraft. In this story, Hughes comes across as a very active player that often did not hesitate in insisting on how things should be done. For a considerable part of the JSAT's early history, Hughes Aircraft was also an equity holder of this company. One interesting aspect is how Hughes persuaded the authorities to use Ku-band instead of Ka-band despite an earlier policy to reserve the Ku-band to the extensive terrestrial users in this geographically small nation. This episode reminds me one of a somewhat similar situation in the United States in the 1960s when Hughes (with Harold Rosen leading the effort) persuaded the authorities that C-band could be used even in urban areas by taking proper care in shielding any interference pathways.

Another interesting historical item in the fascinating JSAT story is the one on the debate about 27- vs 36-MHz transponders from two competitors. Once again, such a story also played out in the United States. Today all over the world, long after analog transmissions have become history, the whole industry still measures the capacity in equivalent 36-MHz transponders.

While introducing Chapter 2, I touched briefly on my first meeting with Harold Rosen in the late1960s when I had supported a study about a possible satellite system in India. Chapter 10 presents the story of this very system that has today grown to be another major domestic system. The story of INSAT gives me an opportunity to say a few words about two other pioneers in this field. The first one is the great visionary, Vikram Sarabhai, briefly mentioned in the beginning of this chapter, who is universally acknowledged as the "Father of Indian Space." I had an opportunity to interact with him during the 1960s with an interesting twist. My very first meeting with Sarabhai was where I was asked by my director to *oppose* satellite systems because they were seen as a competitor to microwave systems that I was involved in developing at the Indian Telecommunication Research Center. That meeting in a curious way opened the door for me in the space technology where I have been now for over several decades now. The other pioneer is U. R. Rao, the lead author of this chapter. He worked for Vikram Sarabhai and had the responsibility to start the development of satellite systems in India. During the 1970s, while I was busy expanding telecommunication equipment development at Bangalore, Rao was building a small satellite for the first time across town in a small industrial hanger. We came to know each other and have kept in touch ever since. Unlike me, Rao continued his quest for several decades in Bangalore and was largely responsible for the systems described in this chapter.

INSAT started in the early 1980s as an operational system. Before that there were two unique field trials, each with a new system of its own kind. The SITE experiment in 1976 was probably the first experimental broadcasting system of its kind anywhere. It used a NASA satellite, ATS-6, that was moved over India just for this experiment. I recall dazzling my colleagues in my telecom campus in Banglalore with a direct reception of television, when at that time there was no commercial TV in India of any kind! The second experiment was with the world's very first three-axis spacecraft, Symphonie, developed in Europe.

Today, the INSAT network is fairly large; it is still managed by ISRO, which also designs and develops satellites and launch vehicles as well, a business model that is not duplicated anywhere else!

We now come to the perhaps the most "visible" part of the satellite industry, namely the direct broadcast of television programs. Many of the chapters before this one have already talked about it in the context of the systems they are describing. In Chapter 11, we look at the evolution of this largest sector of the satellite industry (in terms of revenues and not spectrum or number of satellites) in the United States. The authors, Jimmy Shaeffler and Lloyd Covens, have both lived through this fascinating piece of history, and they have presented the story almost exclusively by giving credit to its leaders from the very first page. They also highlight the basic tenets of success that perhaps could apply to many other chapters as well. Throughout this chapter, there are very interesting and valuable nuggets of information about early systems and personalities.

It might be pertinent to touch upon one important part that the authors have decided not to describe in any great depth. As mentioned in the context of the SES chapter, DBS had some major failures in the beginning, both technological as well as business-wise. The early satellites pushed the technology too hard and/or aimed at a marketplace that just did not have the capacity to garner adequate revenues for the business plans to succeed. Only when the industry adopted digital compression and could compete against cables in urban areas did this industry flourish.

In today's day and age, global positioning system (GPS) is probably the most well-known application of satellite technology, especially among the younger age groups all over the world. Newer satellites and applications keep on emerging frequently, many of them leading to whole new segments of the consumer electronic industry. Probably the most important application is to get directions from anywhere you might be to another place, close by or far away. Cellular phones are in the middle of incorporating GPS function in order to dramatically improve emergency services; management of commercial aircraft in crowded skies is the next major frontier this exciting technology is about to get into its fold.

How did all this happen? Did it start like many other high-tech breakthroughs by some bright college undergrads dreaming away in the dorm? On the contrary, GPS was the result of old-fashioned investigations around the 1970s by several U.S. government agencies all addressing different aspects of the then common challenge of improving navigation and position determination capabilities of ships, aircraft, and other tactical units. A large number of investigators were involved spread among government agencies—both commercial and U.S. Department of Defense (DoD)—and also private sector and academic institutions.

The story for the early days of the evolution of this perhaps the most successful application of satellite technology is told in Chapter 12 by Keith McDonald, who has first-hand knowledge of this phase because he was an integral part of many of the key activities. Keith got involved as early as 1963 and by the mid-1970s was playing a central role in the formulation of the system architecture of this technology.

Unlike other chapters, Chapter 12 is full of detailed and specific pieces of information without which it is difficult to do full justice to all the efforts that went into it. Broadly, we can envision three overlapping phases over about 10- to 15-year period starting in the mid-1960s. In his chapter, Keith documents with details the various efforts that were undertaken by various agencies to address several individual technological challenges. Once it began to appear that there was an adequate body of knowledge to focus on how the overall system will look like, a formal centralized coordinating activity was started in what we could call now the second phase. Keith had an important official role in coordinating this activity through different executive positions.

In addition to a lot of details, Keith also provides several interesting snippets of information that only a first-hand involvement can provide. Thus, initially, the system came to be known as "Navstar," which stood for navigation system (or satellites) for timing and ranging. However, sensing the then prevailing image of satellite technology as being often unreliable and costly, the senior authorities decided to choose GPS because it avoided the word *satellite*! Not a very flattering justification for our industry, but it did lead to the needed funding to move forward!

Chapter 13 on XM radio is written by John F. Dealy who has had much more to do with its success than is known beyond a small circle of his associates. This chapter also gives me a chance to say a few words about my own involvement with this project.

During 1996–1997 as we were developing the WorldSpace L-band digital radio system, the possibility of starting similar systems in the United States was an attractive arena to apply our expertise in a different market. I would often push Noah Samara that WorldSpace should also venture into the U.S. market, but he would always push me back largely because he was afraid that my team would get diverted away from the task of building the pioneering WorldSpace system in Africa and Asia. However, we felt that we could do both systems in parallel by creating separate teams. Eventually, Noah agreed and sent me across the street to meet one John F. Dealy for the first time. Over the course of my career, I have made several new professional friendships; however none of them has been as fast to happen as the one with John F. Dealy. In one single meeting, we were both on the same page, and we worked with Noah and Gary Parsons, just as John describes in this chapter. It was particularly satisfying for me to get at least two bids for the spacecraft, one of them from a partnership of Boeing and Alcatel, which required some effort from us to happen. To manage this program, I hired several new engineers from the industry, most of whom are even now at XM Radio including the versatile Derrick DeBastos mentioned by John. Understandably, it was a moment of both pride and pain to let them go from WorldSpace to the new company, shortly to be called XM Radio (and no longer AMRC, its "birth name"). In parallel with the spacecraft, we also developed the terrestrial technology, which was licensed to XM Radio.

John's chapter on XM Radio is, as I expected, a short treatise on how to achieve success in new ventures. He highlights what I thought was the common feature in the otherwise different management styles of Gary Parsons and Hugh Penaro. They made absolutely sure that all pieces required for going into operations got their due importance at the right time and in correct proportion. This sounds like common sense but does not always happen and has led to subcritical success in the nascent years of many new organizations. As the system moved from the engineering phase to the operations phase, one could see a smooth transition from technical focus to content and marketing. Another notable reason for XM's continued success is the setting up of an extremely efficient and competent group in Florida headed by Stell Patsiokas to develop the user equipment. They are behind the succession of newer devices that continue to come on the marketing shelves and as OEM in newer cars.

This story on XM Radio has been overtaken by subsequent events as foreshadowed by John F. Dealy in his narrative. After a long regulatory and review process, Sirius Radio and XM Radio have merged into a single system, SiriusXM Radio. As a result, there have been many personnel changes in the combined company, and some of the names mentioned are no longer with the company. Nevertheless, the story as told by John is very much valid from the perspective of documenting the evolution of a brand new system.

We now come now to a different but unique success story in Chapter 14. Unlike other stories, this one does not extend over decades, but is concentrated to just a period of about two years (1990–1992). This is a story of how a large satellite got stranded in space, was recovered by some real cooperation among diverse organizations, and then lived on in the geostationary orbit to provide revenue services for a long time. The narrator of this gripping story is Leonard R. Dest, who at that time had a critical and important role in this whole effort. Len's story is thorough and complete and is based on painstaking effort put in by him to get all of the facts right. Complementing this narrative is Conny Kullman's own perspective of the dramatic culmination of this story in his chapter on Intelsat.

Here I would add to this absorbing chapter a few personal memories that could be relevant. Just before the Titan launch of Intelsat VI (F3), I had assumed the additional responsibility of overseeing the Satellite Control Center and launch missions. I was looking forward to a quiet and routine start of this role when I walked in that afternoon to the control center to listen in to the teams about to manage the launch. It was unnerving, to the say the least, to see the launch go wrong. After the perigee separation, we made a quick trip to Houston to discuss with NASA experts about the best means of rescuing the spacecraft. One of the outcomes of this trip was that the orbit of the spacecraft had to be raised to save it from atomic oxygen. However, that step would cost more fuel later to rendezvous with the shuttle for any rescue. The decision had to be made immediately, and I had to make the call before we boarded the plane back to Washington, D.C. Fortunately, that turned out to be the correct thing to do.

The second piece of memory related to this historic spacecraft rescue that I would like to share is related to the decision-making process of the then Intelsat. Once all of the pieces were put together for the rescue, we had to get Board approval for nearly $200 million cost (over and above the already sunk cost of the most expensive spacecraft until then). With active support and guidance from John Hampton, I had persuaded the Board (meeting in Bahamas!) to approve the whole rescue when a phone call from Washington, D.C. almost derailed the whole project. Hughes was not prepared to assume the liability of any patent infringement involved in the rescue; the Board wouldn't take that undefined exposure either. With a time-proven means of breaking a logjam, John Hampton asked for a coffee break, and we made the DG of Intelsat, Dean Burch, talk to the Hughes CEO. Hughes agreed to take patent liability insurance if Intelsat accepted the cost of the $4 million premium. The Board approved the project, and it was finally underway. Only then could we enjoy the meeting in Bahamas!

When I was putting together the plan and order of chapters for this book, I struggled a bit to decide where and how to cover the progress in user equipment for satellite systems. One option was to just focus on systems hoping that each author would do justice to his ground equipment. The other option was to have a separate chapter to give an overview of the ground systems. Finally, I opted for both provided I could get the right author for such a chapter that would survey the full landscape from the perspective of ground equipment.

I did succeed in this particular case, and the legendary Mark Dankberg in Chapter 15 takes us through a very interesting journey through the industry while doing so.

Chapter 15 starts off with a quick look at the various drivers that have influenced the size, complexity, and cost of ground equipment. These include satellite, system, and network technologies and of course the technology that goes in the user equipment itself. Mark has a real insight into how and why certain historic milestones happened. He highlights a not so well-known fact that the cable TV started originally as a rural system to cover areas too far from VHF/UHF TV transmitters. However, when pioneers like Ted Turner added satellite receiver antennas at cable TV head-ends, the availability of large number of channels made cable TV systems grow very rapidly in urban areas. The satellite DBS went through a somewhat similar evolution although its initial efforts were costly and unsuccessful. As we also discuss in the SES chapter, initially such satellites targeted only a few channels for rural areas but using analog techniques and overconservative parameters that ignored the advances in ground equipment. However, once digital techniques were introduced, the choices multiplied, and user equipment became small, and it became a global success for satellite systems.

Mark traces the evolution of all types of ground equipment, from giant Standard A antennas for the early Intelsat system all of the way to handheld systems. Not surprisingly, he comes down heavily on the poor system economics of the LEO systems that failed in the marketplace. His chapter ends with a very interesting system economics review of broadband systems. At the time of writing this chapter, his company has also announced a broadband satellite to match his ground equipment!

The final story in the book in some ways takes us to the very beginning of our industry. By now, the name DARPA (or strictly speaking its earlier name ARPA) is quite well known for its pioneering role in the birth of the Internet. What is not so well known is how DARPA was created and its rather unique management style that has been its hallmark from the very beginning. Chapter 16 by three experienced executives at DARPA takes us through its history and achievements in space through a very well-researched narrative supported by a series of historic pictures, references, and quotes.

DARPA was formed in February 1958 by President Eisenhower to bring some structure and order in the rather duplicative efforts following the launch of two Sputnik spacecraft in late 1957. Within a matter of months, the new DARPA proved its value through a series of successes as well documented in this chapter. These programs laid the foundation of many applications we know today that include communications, weather forecasting, early warning, reconnaissance, and geo-location.

The authors explain from first-hand knowledge the key to DARPA's success while operating somewhat in the background. From the very beginning, DARPA has had an "agile management style" through which it makes projects happen by arranging for the necessary funding and then assigning it to the group judged most competent to deliver the results. They highlight, in the context of one of the early programs, the agency "concentrated on developing and fostering innovative concepts, as well as making key project funding decisions, while leaving the burden of facility and workforce maintenance, system development and production, and operations to industry, research labs, and the military services."

DARPA's involvement in space had its peak in the very early days soon after Sputnik launched, and then it "expanded its mission to one of preventing technological surprise in domains other than space." However it has remained active in space technology development and demonstration to this day. During the 1980s, it developed a low-orbit data relay system. It has also nurtured development of new small launchers that include Taurus, Pegasus, and more recently the ongoing Falcon program. The latest noteworthy achievement has been the very successful demonstrations of orbital servicing and refueling through the Orbital Express program.

In Chapter 17, the final chapter, I do some crystal-ball gazing in terms of predicting future success stories to come. It is a risky thing to do, but it does give me an opportunity to put my thoughts in front of the reader for whatever that is worth.

Chapter 2

SYNCOM: WORLD'S FIRST GEOSTATIONARY SATELLITE

HAROLD A. ROSEN*

The story of Syncom begins with the convergence of two events in the late 1950s related to the cold war and the beginning of the space age. These events affected the viability of the radar department at Hughes Aircraft Company, my employer at that time. One of the department's biggest projects—an advanced radar for an interceptor being designed to counter a fleet of high-speed Soviet bombers—was abruptly cancelled when it was learned that the bombers themselves were to be replaced with intercontinental ballistic missiles (ICBM). Shortly thereafter, the advanced state of the USSR rocketry was dramatically emphasized by their launch of Sputnik, the world's first artificial satellite.

My department head, Frank Carver, challenged me to find a new project that would use some of our radar technology to keep our staff gainfully employed. After conferring with my colleagues Tom Hudspeth, a brilliant communications engineer and avid radio amateur, and John Mendel, who was leading the development of advanced traveling wave tubes for our radars, both independently suggested to me that the new project be a communication satellite. Both pointed out the then sad state of international communications: telephony was hard to schedule and very expensive, and transoceanic television was impossible. These problems, they said, could be overcome with a properly designed communication satellite system. This excited me. I began to learn all I could about what appeared to be an important and relevant field.

Don Williams, another brilliant colleague whom I had lured into my small advanced development group with the promise of working on space projects, had been thinking independently about satellites and was already developing a concept for a navigation satellite. I wasn't thrilled with that objective, but was intrigued with several features of his design: first, it was in a geostationary orbit; and second, he had analyzed the mechanics of the orbit sufficiently

*Former Vice President, Hughes Aircraft Company, El Segundo, California; Consultant, Boeing.

to show how little impulse was needed to change from one orbit location to another in a reasonable time.

My most valuable reading was of a seminal paper in the March 1959 IRE Journal titled "Transoceanic Communications Via Satellites," written by John Pierce and Rudy Kompfner of Bell Telephone Laboratories. I knew of John Pierce's reputation as one of the world's finest communication scientists, who among many other accomplishments had tamed the traveling wave tube's propensity to oscillate, thus making it a valuable wideband amplifier. The paper, which was very instructive, contained a link budget that showed how little satellite transmitter power was needed to provide a useful communication link to a ground terminal. I was struck by this: compared to our department's radars, which had to transmit powerful signals in order to detect weak reflections from the target, transmitting a signal from a satellite seemed relatively easy. The paper discussed many different possibilities for satellites: active repeaters, passive reflectors, fleets of satellites in low-altitude orbits, and a system using geostationary satellites. Of these options, the authors favored the low-altitude fleet equipped with active repeaters. To them, the geostationary solution was deemed too complex to be practical. (Unbeknownst to me at that time, there was an ongoing government program, Advent, to design a geostationary satellite that was proceeding as smoothly as the biblical Tower of Babel, and I now believe John had based his negative assessment of the geostationary solution at least partly on that program's many problems.)

To have the reliability and long useful lifetime for a good communication system, their proposed low-altitude satellites had neither attitude nor orbit controls to perform their mission. To do this, their antenna patterns were nearly omnidirectional, and the orbital spacing between satellites was not maintained. I could see that this system design had drawbacks. It needed a large number of satellites to prevent frequent gaps in service and relatively high power built into the link to accommodate the low gain antenna. On the other hand, the geostationary satellite requires a number of active attitude and orbit control elements that must match the lifetime of the communication system. This, of course, would make the satellite heavier. Not only that, but a geostationary satellite would need to be launched into an orbit that needed an additional 4000-meters-per-second velocity increment, which would require more fuel, over that of a low-altitude satellite. Given the small launch vehicles then available, it seemed to me that the authors considered this to be too difficult a challenge.

FIRST GEOSTATIONARY SATELLITE CONCEPT

I, however, reached a different conclusion, despite my respect for the authors. They were communication experts; I had spent the first part of my career (at Raytheon) in the design of high-performance antiaircraft guided missiles and felt I was a better judge than they of the difficulty of guidance and control for a geostationary satellite. I was inspired to try to find a practical design for a geostationary communication satellite with the technology that was then available or that could be readily developed. I knew that both the basic communication and control elements would have to be as light as possible.

While pondering various possibilities in August 1959, I remembered back to a Caltech classroom discussion from years earlier on the dynamics of rotating bodies. The physics course was being taught by Carl D. Anderson, a renowned but modest Nobel laureate. I asked if he could explain in simple terms the powerful stabilizing effect against external influences that spin had on objects such as footballs or artillery shells. He couldn't do so offhand, but together we worked out a comparison of the effects of external torques on similar bodies, one spinning and one not, over a period of time. The effect of the spin was to reduce the angular disturbance by a factor approximately equal to the angle the body had spun in that period, which could easily be millions of radians.

Using this earlier knowledge, I made a few calculations that led me to an estimate that the spin-axis attitude of a spinning satellite could be maintained for a useful period with no active attitude control. In an epiphany, I realized that if the spinning configuration were adopted, with the spin axis parallel to the Earth's axis, it would enable spin phased impulses to permit a single thruster to control both the period and eccentricity of the orbit, a major simplification. An obvious disadvantage of spinning was the difficulty of incorporating a high-gain antenna for the communication system. However, even with a beam pattern that was a figure of revolution around the spin axis, significantly more gain than omnidirectional could be achieved by narrowing the beam pattern in a plane containing the spin axis, and valuable communication bandwidth resulted despite this less than optimum pattern. I felt that in the future an electronic or mechanical beam despin system could be employed to provide still more gain, but that it was best to start with the simplest possible approach. A second disadvantage was the relative ineffectiveness of the spinning solar panel compared to a flat, sun-oriented array. However, adequate power for the lightweight, efficient payload could be obtained with a cylindrical array.

When I conveyed this concept to Don Williams, he enthusiastically jumped aboard, and together we started filling in some of the details, such as how to achieve the initial orbit and orientation. Because we were unaware of the Thor Delta booster development program then underway, we selected the low-cost Scout sounding rocket as the launch vehicle. We added a small fifth solid rocket to the Scout's four solid stages to achieve the high energy transfer orbit and incorporated an even smaller sixth stage (an apogee kick rocket) inside the satellite to change from the transfer orbit to the synchronous orbit when it reached apogee. For our launch site we selected a small equatorial island, Jarvis, so that no plane change (removal of inclination) would be needed to achieve the equatorial orbit. Because the upper stage of the Scout was spin-stabilized, we retained the spin to stabilize the added stages. Because the booster's spin axis was in the equatorial plane, we required a 90-degree reorientation to get to the operational spin-axis attitude. This was done in two steps: the initial spin was killed by en electric motor reacting against the spent fifth stage, followed by separation and spin in the desired attitude affected by tiny spin rockets.

For attitude sensing, Don invented (and built in his garage!) a V-beam solar sensor that determined both the spin phase and the angle between the spin axis and the sun line. The second attitude sensor that was required to unambiguously determine the spin-axis attitude was the polarization of the communication signal.

While Don and I were working these issues, I asked Tom to design a communication payload that would also provide the necessary telemetry and command links. Because ITU frequencies had not yet been allocated for satellite use, we picked our own: UHF (470 MHz) for the uplink and 2 GHz for the downlink. Microwave transistors were not yet available, and so frequency multipliers were used for the local oscillator signal and a traveling wave tube amplifier (TWTA) for the transmitter. Tom's design turned out to be remarkable for its low weight and excellent performance. He also designed the collinear dipole array antenna that generated the desired communication beam pattern.

For the satellite transmitter, I knew that we would need a lightweight (lighter than any then available) traveling wave tube. For this design, I approached the expert in this area, John Mendel, who worked in a different area of the company. TWTAs were then emerging as the preferred microwave amplifiers for many applications, and high-power versions were being developed for use in radar transmitters. This ingenious device, invented by Kompfner and made practical by Pierce, was even at the relatively low power of our satellite transmitter simply too heavy for use in our design. The weight was mostly in the focusing magnet used to confine the beam radially. John decided that he could make the focusing structure much lighter by using a field of alternating sign along the beam axis, affected by a multiplicity of very small permanent magnets of alternating field direction replacing the much larger single magnet design then in vogue. To compensate for the less effective focusing than that was provided by a uniform field, he made the envelope of metal rather than glass, with ceramic windows for the input and output leads and the power connections. This construction permitted the tube to be baked out during manufacturing at a temperature high enough to handle the higher beam interception than that occurring in the heavier tube. This metal ceramic, periodic permanent magnet TWT was considered a difficult but doable development whose advantages far outweighed the risks. It has proven to be an extremely valuable product that is the transmitter of choice in most communication satellites even today, nearly 50 years later.

With a paper design of the communication payload now available, Don and I worked together to improve the aforementioned control system. Don was a gun fancier and had used bullets in his navigation satellite design to provide the control impulses. When we learned of the availability of fast-acting pneumatic valves, we changed to bursts of compressed nitrogen for providing orbit control impulses. I realized that we could now replace the relatively cumbersome despin-respin initial orientation with a reaction-controlled reorientation maneuver. I thought that four additional thrusters would be required to minimize the nutation during this reorientation maneuver; Don showed me that the maneuver could be achieved more simply with just a single additional thruster. This orientation thruster would also enable us to adjust the spin-axis orientation over the lifetime of the satellite if solar radiation pressure unbalance or other perturbations caused it to precess excessively.

SEARCH FOR PROJECT SPONSORSHIP

I was now convinced that we had a practical design that could provide substantial transoceanic bandwidth at a relatively low cost. At the going rates for transoceanic telephony, it could be very profitable. There was no established rate for television because transoceanic service was not available. With the support of Frank Carver, we presented our plan for a communication satellite business based on our satellite design to Allen Puckett and subsequently to Pat Hyland, our general manager. (Howard Hughes, the company founder, had by then become a recluse but retained the title of president.)

The presentations led to a formal study by Sam Lutz, who directed engineering for the Hughes communications division. The resulting Lutz report enthusiastically made the case for the communication satellite venture that would be funded and operated by Hughes. But Hughes upper management was more cautious, and the project did not receive the support needed for the next steps—developing and demonstrating a prototype of the satellite. Instead, we were encouraged to seek government support for the project. To that end, in January 1960, Don and I prepared a descriptive brochure summarizing the technical features of the system. We made presentations to high levels at the newly formed NASA, but they dismissed the concept with no sign of further interest. As time passed without this hoped-for support, Tom, Don, and I decided to try to do it ourselves. We would pool our resources to the extent of $10,000 each as seed money and try to find outside investors. We thought that an additional million dollars or so would be needed to get to the project off the ground, literally.

This was not the most auspicious time to propose a commercial space program—the most vivid impression most people then had of space-related activities was of rockets blowing up at Cape Canaveral. The thriving venture capital businesses we have today had not yet been formed—it would be another 20 years before my brother Ben would become a noted venture capitalist in the emerging personal computer sector. Faced with universal skepticism, we quickly ran out of contacts. In March 1960, in desperation, I called a former colleague at Raytheon, Tom Phillips, who had by then risen through the ranks to become its executive vice president, and outlined our plan to him. He invited the three of us to brief him and the rest of the upper management of Raytheon.

This briefing occurred shortly thereafter at Raytheon headquarters in Waltham, Massachusetts. Among those attending were Raytheon's president, Charles Francis Adams, and the vice president of engineering, Ivan Getting. Tom was sold on the merits of the satellite system, but not on the private-venture aspect of our proposal. He offered instead to pursue the system as a Raytheon venture, with the three of us becoming Raytheon employees in Massachusetts. It wasn't all that we wanted, but I felt it was the best we could do.

Upon returning to Hughes the next day, I told Frank Carver I was resigning to accept Raytheon's offer. He wouldn't accept my resignation without a fight, and arranged a meeting with Allen Puckett. Allen told me that he and Pat Hyland had had further thoughts about the project and had decided to invest company funds in a prototype development. When Pat Hyland himself personally assured me of this decision, I felt elated. Even though we would not have a direct financial stake in the enterprise, we could now proceed with my main objective, the development and demonstration of the system.

Frank Carver provided laboratory space for the design of all of the communications electronics (except the for the traveling wave tube, which would take place where John Mendel worked in the microwave tube lab). The skills we had honed in designing state-of-the-art airborne radars were well suited to those needed for our satellite: a greater sensitivity to weight and efficiency issues than those involved in the design of ground equipment. Before long, we were able to demonstrate a communication transponder that relayed a television signal or multiplexed voice signals in the laboratory, using flight-like electronics minus the traveling wave tube whose development, as expected, took longer. The telemetry and command functions, whose requirements were quite modest, were integrated into the communication system in order to save weight. The antenna that would provide the flattened-doughnut-shaped beam pattern during the satellite's operational phase consisted of a coaxially fed linear array of dipoles. To handle telemetry and command signals while the satellite was still attached to the launch vehicle, the antenna was split into three identical sticks fed in parallel, which were spread out to lay on a conical surface during the launch phase, in order to provide continuous telemetry and command access.

Don demonstrated the smooth precession of a spinning wheel obtained by pulsing a single thruster in synchronism with its spin. He used a toroidal tube to store the gas and provide the spin inertia, the assembly being mounted on a spherical bearing. (Motion pictures taken of this precession later proved to be an important part of subsequent patent infringement litigation, which was eventually settled in Hughes' favor.)

Feeling pleased that most of the major elements of the satellite had been demonstrated, we once again tried to obtain government support. Allen Puckett was now an enthusiastic advocate and presented the concept to the Space Science Panel of the President's Science Advisory Committee. He arranged for Don and me to brief many of the defense agencies within which he had contacts. But the Department of Defense (DoD) had its own project, the previously mentioned Advent satellite, and wasn't interested in hearing of a practical alternative. Unfortunately, NASA had signed an agreement with DoD that prevented it from supporting a competing synchronous satellite system. So these overtures to the government were all doomed from the start. We also tried to form a venture with the General Telephone and Electronics Corporation (GTE), whose West coast technical staff endorsed our concept only to have their recommendation rejected by their management. We also presented our case to Bell Telephone Laboratories, but we were unable to persuade them to reconsider their objections to the geostationary system. (John Pierce has since written that he regretted his negative assessment at this meeting. We later became friends and shared the 1995 Draper Prize for our work in communication satellites.)

The year 1961 began with neither a government program nor a commercial partner, and the outlook was bleak. When NASA issued a request for proposals for a low-altitude satellite, Hughes decided to bid even though no part of our design was appropriate to that orbit. The low-altitude system, to be called Relay, was NASA's attempt not to be left out of the communication satellite field—AT&T/Bell Lab's low-altitude system was beginning to move from concept to active program—and, as previously mentioned, a geostationary orbit system was precluded by its agreement with the DoD. I was neither surprised nor unhappy when we lost the Relay competition to RCA. I felt that RCA could become a formidable competitor in this field and was glad to see their able technical staff tied up in a program that I was certain had no future.

The situation was still bleak.

In the spring of 1961, Hughes decided to display our satellite at the Paris Air Show, which had become the world's most important trade show for air- and space-related products. Tom and I were sent to demonstrate the relay of television signals via our satellite repeater across the display booth. For us, it was an exciting time. The demonstration was quite popular and generated much interest and press coverage. Company Francais Thompson Houston (CFTH) hosted a demonstration of the satellite on the Eiffel Tower, revealing

Fig. 2.1   Tom Hudspeth and Harold Rosen are seen with the Hughes prototype satellite on the Eiffel Tower (May 1961).

the system to an even wider audience (Fig. 2.1). Alas, in the end no sales resulted. But things were about to change.

John Rubel, a former Hughes executive who had become assistant director of Defense Research and Engineering, had been given a briefing on our satellite program during an official visit to Hughes. At that time, I could tell from his reaction to our demonstration of the communication and control systems that he was extraordinarily excited, but I didn't know why. I learned later that he had become convinced that the troubled Advent program that he had inherited would never work and saw our program as a possible substitute. In the spring of 1961, working behind the scenes in Washington with both DoD and NASA personnel, he managed to annul the agreement restricting NASA from the geostationary orbit and helped create a joint DoD–NASA program that became known as Syncom. The Syncom satellite development would be contracted for by NASA. NASA would also supply the launch vehicles, whereas the ground stations, which had been developed as part of the Advent program, would be supplied by DoD. A contract was awarded to Hughes in

August 1961 to construct three satellites based on our design. We were, to say the least, exuberant. (I have remained in contact to this day with John Rubel, a vibrant and lively man in his 80s.)

FIRST SYNCOM

Although Syncom was based on our system design, significant changes were nonetheless required. The launch vehicle chosen by NASA was the new Delta, which launched from Cape Canaveral. The higher payload capability of the Delta allowed the use of redundancy in all of the communication and control elements, which was, of course, desirable. The 28-degree latitude of the launch site meant that the satellite could not initially be injected into an equatorial orbit. This resulted in considerable north–south motion as seen by the ground stations, so that tracking would be necessary. But planned improvements in the Delta would make this only a temporary impairment because later augmented Deltas would permit launch trajectories from which the inclination could be removed by the apogee kick motor. Another significant deviation from our original design was a change in the uplink frequency, required for compatibility with the Advent Earth stations. This required another exotic multiplier development to provide the local oscillator signal, not an easy task with the components then available.

Our team grew quickly as we set about building the three satellites. On the NASA side, overseeing the technical compliance was Bob Darcy, who proved to be particularly helpful in resolving technical issues as they arose. Under John Mendel's supervision, completion of the development of the traveling wave tube progressed smoothly because adequate funds were at last available for this essential task.

Another improvement resulted from a visit to the launch site in the fall of 1961. I observed a small rocket on a stand emitting puffs of a white gas and was told that it was steam resulting from a test of its hydrogen-peroxide control thrusters. Up to this time, I had not been aware that hydrogen peroxide could be used in a pulsed mode. Because this liquid propellant provided a much higher specific impulse than compressed nitrogen, I convinced NASA to let us use hydrogen peroxide for our redundant control system, which substantially increased our control capability.

By the end of 1962, the first flight model of Syncom was nearing completion (Figs. 2.2 and 2.3), and a launch was scheduled for 13 February 1963. After three years of struggle, the moment of truth had arrived.

The launch preparations proceeded smoothly, and the Delta rocket lifted off at the appointed time. The Delta performed flawlessly, placing Syncom 1 in the desired transfer orbit during which normal telemetry was received. Five hours later its solid fueled apogee kick rocket was ignited for an expected

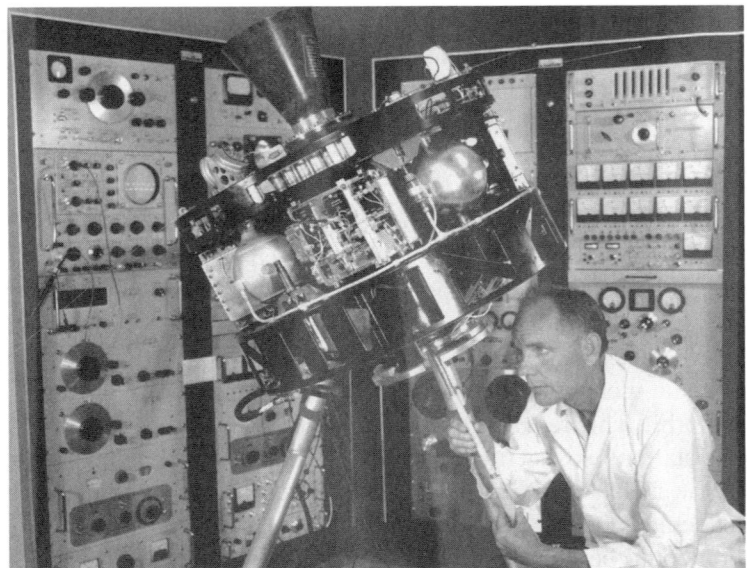

Fig. 2.2   Tom Hudspeth is peering into a Syncom satellite with its solar panel removed. Visible are the nozzle of the apogee kick rocket, the V-beam solar sensors, the hydrogen-peroxide and compressed-nitrogen fuel tanks, the communication electronics, and the colinear array communication antenna.

22-second burn. But one second before burnout, all signals disappeared, never to be heard again. The disappointment was almost unbearable.

But I knew that we could waste no time on mourning. We had a second flight model to launch. We quickly identified three possible causes of the disaster: an apogee rocket explosion, a burst of one of the nitrogen tanks, or a wiring failure. To be safe, corrective measures were taken to all three areas, and Syncom 2 was shipped to the Cape to prepare for a July 26 launch.

The launch preparations this time were extremely difficult because of heightened concern about the consequences of a second failure. Many imaginary problems were dealt with by our team, and just when it appeared that we were good to go, a real problem arose, or so it seemed—the communication repeater broke into a hard oscillation. The satellite was by now fueled and mounted atop Delta's third stage, and bringing it down for troubleshooting at that point was almost unthinkable. Instead, we lugged a heavy spectrum analyzer up to the 13th floor of the gantry, the last three without benefit of an elevator, so that we could troubleshoot in situ. Tom Hudspeth then performed some more of his magic: by placing his fingers between the folded-down communication antenna and the receiver, he could suppress the oscillation, thus determining that the feedback path was between the antenna and the receiver. Because we had never before tested the communication

Fig. 2.3 Don Williams, Tom Hudspeth, and Harold Rosen are with the first Syncom, ready for its shipment to Cape Canaveral (January 1963).

system with the antenna folded down, which was the launch but not the operational configuration, we felt the system was probably okay in its operational mode, but we had to find a way to prove it. When we reproduced the oscillation on the prototype down below and showed the NASA folks that it disappeared when the antenna was erected, we were finally given a launch go ahead.

Once again the Delta performed flawlessly, and the telemetry was normal during the transfer orbit. The apogee rocket was fired on time, we held our breath, and to our great relief, this time we survived the burn. We were in a 24-hour orbit!

NEW SPACE COMMUNICATION ERA

A few days later, the reorientation maneuver involving thousands of spin-synchronized pulses was performed for the first time on a satellite. With the spin axis now orbit normal, we could communicate with the Earth stations 24 hours a day. The new era of continuous space communications had begun.

Fig. 2.4   President Kennedy is at the inaugurating service over Syncom, speaking to the Prime Minister of Nigeria (August 1963).

President Kennedy used Syncom in a historic conversation with the Prime Minister of Nigeria, in August 1963 (Fig. 2.4), beginning its five years of service to the U.S. government.

These five years clearly demonstrated that, despite many predictions that it wouldn't be reliable enough, Syncom proved that a geosynchronous satellite could have the necessary reliability for communications.

With most of the technology for geostationary satellites proven, we started planning modifications in the last of the three Syncoms to achieve the remaining objective, the demonstration of a geostationary orbit. Taking advantage of planned improvements in the Delta, we designed a mission in which some inclination reduction would be performed by the Delta at perigee injection into the transfer orbit. The remaining inclination would then be removed by the improved apogee rocket (already used for Syncom 2), which would finally result in an equatorial (hence, geostationary) orbit.

Syncom 3 was scheduled for launch in the summer of 1964, shortly before the Tokyo Olympics. We received a request to provide television coverage of these events. What an opportunity—never before had the Olympics been seen in real time overseas. To improve the quality of the television signal, we

Fig. 2.5   An image of the first-ever live transoceanic television of the Olympics (Tokyo, 1964).

increased the bandwidth of one of the satellite transponders and installed a liquid-helium-cooled maser as an ultra-low-noise receiver for a large ground antenna at Point Mugu, a naval facility 50 miles to the north of our Hughes plant (Fig. 2.5). (This antenna had originally been installed for the Advent program.)

We and the Delta team were getting good at this. The launch and subsequent operations proceeded flawlessly, and the first geostationary orbit was achieved in August, just a few weeks before the opening ceremonies. Even though the effective radiated power of satellite was only about seven watts, the sensitive receiving system allowed continuous transoceanic television of the events.

The Communication Satellite Corporation (Comsat) had been chartered by an act of Congress to provide commercial satellite service. Noting the success of Syncom 2, Comsat had contracted with Hughes in April 1964 for delivery in one year of a commercial derivative of Syncom for its first satellite. This new satellite would later be called Early Bird. The era of commercial satellite communications was about to begin.

Chapter 3

INTELSAT

CONNY KULLMAN*

MY PERSONAL INTRODUCTION

I will tell Intelsat's story from my perspective of growing up in the 1950s and 1960s with space-related business taking off, and eventually seeing myself starting to work at the company in 1983. The phrases "It is never boring to work at Intelsat" and "Time flies when you're having fun" pretty much summarize my 23 years at Intelsat.[†]

Before I came to Intelsat, I worked at SAAB Space in Gothenburg, Sweden. There I learned all aspects of how satellites and rockets are designed and built. My first job coming out of university was to design a piece of the attitude control computer for the first Ariane Launcher. So I can with certainty claim to be a "rocket scientist!" This was a fact that Intelsat's corporate communications liked to use from time to time. It seemed to somewhat impress the financial markets. In this industry, it can be good to be able to say that you are a "rocket scientist"; perhaps because it adds a little bit of credibility and might give the impression that you know what you are talking about—as long as you also understand the other aspects of the business.

Something that was said to me by an engineer at Intelsat in early 1983 when I visited for the first time has stuck in my memory: "Intelsat is a good place to work; we have a simple mission that everyone here understands—to connect people everywhere around the world so they can communicate. Compared with other places where I have worked, I really feel I am making a positive contribution here."

When my family and I came to the United States in 1983, we were looking for two years of adventure; I was at Intelsat on unpaid leave from my Swedish

Copyright © 2008 by Conny Kullman. Published by the American Institute of Aeronautics and Astronautics, Inc., with permission.

*Former Chairman and Chief Executive Officer, Intelsat, Washington D.C.

[†]Before privatization in 2001, Intelsat was known as INTELSAT. The current name is used throughout this chapter.

company—we expected to go back after two years. But, I must have taken a wrong turn somewhere, because what was supposed to be two years became 23 when I left the company in late 2006.

I have had the honor and privilege of working in the company as it has gone through enormous, challenging, and stimulating change. I have held positions in pretty much all parts of the company except in legal and finance (thankfully for us all). I have had the opportunity to change jobs without having to change workplaces.

It has certainly been an incredible journey for me personally—one I could never have imagined when I first showed up at Intelsat's L'Enfant Plaza offices in 1983. Back then, we were just months away from launching a new, highly sophisticated satellite, 507, and it was the year after celebrating our 20th anniversary.

I'll never forget how proud and excited I was to join the Intelsat family when I stood at Intelsat's doorstep—and my pride only grew stronger since that day. I am proud of our heritage, our employees, and our commercial success. Most of all, I am immensely proud to have worked with all Intelsat staff, helping to create the leading satellite company in the world.

It has truly been an honor to serve as the company's CEO during this exceptional period in its history. When I took over as the CEO in 1998, I had three objectives: 1) privatize the company; 2) create shareholder liquidity by making an initial public offering within a year of privatizing; and 3) expand our services beyond selling space segment only. Two out of three is not too bad under the circumstances of the collapsing telecom market in 2001/2002. It took until January 2005 before I could deliver liquidity to our shareholders by selling the company to a group of private equity funds. In the end, the shareholders were very happy with the outcome. Our new owners also appear to be quite pleased with the returns on their investment.

On 1 April 2005, I handed my CEO responsibilities to David McGlade. I was very happy that Dave was inheriting a company well positioned to meet the strategic and market challenges that Intelsat faces. For the five years since privatization, we had worked to transform the company and position it for continued leadership in our industry and commercial success. We had weathered the storm of a sector-wide downturn and remained strong, with a highly diversified business, an excellent mix of customers, and the most skilled and respected staff in the business.

From April 2005 until I left the company in 2006, I served Intelsat as board chairman. It was a new and exciting challenge for me. I got to do it in Bermuda, which was an added benefit. The business continues to be very dynamic and interesting; the PanAmSat acquisition that closed on 3 July 2006 resulted in a dramatically changed operating profile for the company. With over $2.1B in annual revenues and a diversified business mix with strategic presence serving the video sector, it was rewarding to see how far Intelsat had come.

THE BEGINNING

Arthur C. Clarke has a rich history of developing "crazy," implausible, or economically impossible ideas, and this is probably why he has been so successful as a science fiction writer. At one point, a global satellite communication system was one of those fantastic ideas. In 1945, he got his famous "Extra-Terrestrial Relays" article published in *Wireless World*—a speculation that was realized a little less than 25 years later when the Intelsat system in 1969 achieved true global coverage by placing the Intelsat III F-4 over the last ocean region—the Indian Ocean, just weeks before our historic broadcast of the first Apollo moon walk.

My own first memory of Intelsat was the worldwide transmission with live music from all the continents of the world. My strongest memory was the live participation of the Beatles singing "All You Need Is Love." This was in 1969, so they were in their "flower power" attire, and they were surrounded by a horde of young people in a large living-room setting (that is the way I remember it; someone might tell me that I am slightly off). It was a great emotional and memorable experience for me. At that time to even ponder that I, one day, would be the CEO and chairman of the company that made this possible was not in my wildest dreams. But as all kids who grew up in the 1950s and 1960s, I was mesmerized by Sputnik, by the space dog Laika, Yuri Gagarin, and eventually the race by the Americans to reach the moon before the end of the 1960s. I knew I wanted to be an engineer, even if I did not realize at the time that I would be working in the space industry.

Many chapters in this book, I am sure, must have at least a small section on the given company's first days of history. So, here is a timeline for the early years of Intelsat:

20 December 1961	The United Nations General Assembly adopts Resolution 1721, stating that global satellite communications should be made available on a nondiscriminatory basis.
31 August 1962	President John F. Kennedy signs the Communications Satellite Act, with the goal of establishing a satellite system in cooperation with other nations.
20 August 1964	The International Telecommunications Satellite Consortium (INTELSAT) is established on the basis of agreements signed by governments and operating entities.
6 April 1965	Early Bird (Intelsat I) is launched into synchronous orbit. This is the world's first commercial communications satellite, and "live via satellite" is born.
28 June 1965	Early Bird begins providing television and voice services. Officials in the United States and Europe exchange greetings in a transatlantic ceremony introducing the new service.
26 January 1967	Commercial satellite service between the United States and Japan is established, with live television coverage in both countries.

1 July 1969	The world's first global satellite communications system is complete with the Intelsat III satellite covering the Indian Ocean Region.
20 July 1969	Intelsat transmits television images of the moon landing around the world—a record 500 million television viewers worldwide see Neil Armstrong's first steps on the moon "Live via Intelsat."

I will stop with the history lesson there. Apart from what has happened in the last 6 to 7 years, I believe these events are the most significant for Intelsat; in fact, I regard them as the most important years and events for the entire satellite communication industry. In the beginning, the cost and risk associated with developing and launching satellites were high enough to bring countries together in cooperation for the creation of Intelsat. Intelsat was the pioneer of this industry and showed the commercial potential of satellite communications to the world.

What is not mentioned in the history lesson is some of the very strong reasons why the company that became known as Intelsat really was created. The idea was to create a global coverage system available to all people of the world on a nondiscriminatory basis. As an example, until the Intelsat system was created, the developing countries often had to route their communication traffic through the telephony systems of the old colonial powers in Europe. For example, it was not unusual for a telephone call from East Africa to West Africa to be routed through either France or the United Kingdom. With our system, these African countries could set up independent and direct links of communication. The advantage spread to all parts of the world as soon as the Intelsat system had global coverage, and it was an idea that appealed to national and political interests around the world. So whereas television broadcast was what people saw and knew as the evidence of the Intelsat system, basic telecommunication or telephony was what it really was about in the beginning. In 1995, telephony type of services accounted for 53% of Intelsat's revenues as compared to the single digits as a percent of revenue today.

THE BUSINESS

Intelsat is the pioneer and an innovator in the satellite communication industry. Starting from a telephony customer base, Intelsat's business has grown to be the most diversified in the industry, supporting telecom, data, video, and government service applications.

Many are surprised by the continuing use of satellite for traditional voice connectivity. However, the telecommunication carriers still need satellite for telephony for a number of their connections and applications. For example, in early 2007 British Telecom and AT&T used Intelsat's satellites to reach approximately 50 remote countries in their respective global networks.

Whereas sea fiber has rightfully taken over the transmission of much of the world's point-to-point traffic, the demand for satellite capacity has been quickly replaced by "new telecom" such as wireless network extensions and voice-over IP.

Data and private network applications generate a greater and greater proportion of Intelsat's revenues. This includes small-antenna (VSAT) networks, corporate wide-area networks, intranets/extranets, and video conferencing for corporate service providers, system integrators, and multinational corporations. Intelsat also provides Internet trunking to tier 1 and 2 Internet service providers, especially in developing regions of the world. We also sell broadband access to residential service providers, for example, in the Middle East and in North America through our 30% investment in WildBlue.

Intelsat has a leading market position serving the media industry, especially after the acquisition of the Loral North American business in 2004 and now with the addition of PanAmSat. Video applications fall into three categories. First is video contribution—feeds from, for example, sports events or news events to the broadcasters' central facilities. Second is video distribution—programming distribution from the broadcasters' central facilities to the affiliates around a nation or region or the cable programmers' distribution of programming via our satellites to the cable head-ends. Third is DTH (direct to home) video distribution—direct distribution of large programming packages to single family houses or apartment buildings using very small antennas; already before merging with PanAmSat Intelsat served as the platform for 11 DTH systems around the world, including the largest ethnic programming network in the United States. Intelsat has also launched an innovative MPEG4-based platform to serve the burgeoning IPTV (Internet Protocol TV) market.

Intelsat provides all types of communications for the military and other government entities: Internet, voice, data, command and control, protected services for government agencies, etc. We have created a strong government business because of our excellent network availability record and customer support. A four-star U.S. general told us that they had been trying for more than a day without success to get an urgent requirement satisfied by one of our competitors; they came to us, and the service was up in less than an hour. This is the performance that has earned Intelsat a leading position serving the U.S. government.

The Intelsat satellites provide connection between ships (both cruise ships and military vessels), and shore. Some cruise ships have mobile cell structures onboard as well as Internet "cafes." Our satellites then carry the signals from the ship to shore for connection to the terrestrial network. We also resell satellite mobile from other operators to the government.

In addition to service diversity, Intelsat enjoys regional diversity as well. Our business is spread across the different regions of the world. This is a key strength of Intelsat and differentiates us from our competitors. Intelsat is strong in the developed markets, but also very well positioned in the growth markets in Asia and Africa. In fact, Intelsat's business in Africa is growing faster than in any other part of the world.

As the company exited privatization, my management team executed a strategic initiative to expand our portfolio away from capacity services and into hybrid "managed solutions," which embraced fiber and other terrestrial services as part of the overall Intelsat system. This was a service Intelsat could not provide as an intergovernmental organization (IGO), but which my management team recognized as essential to capture the growth from the "new telecom" players following the privatizations that were happening around the globe. From 2002 to 2005, the growth of managed services has largely offset the decline we have seen in the channel/telephony product, and it has been one of the most successful new product launches in the recent history of the industry. Managed solutions have seen a fast revenue growth from only a few million in 2002 to around $115M in 2005 (excluding PanAmSat).

Intelsat enjoys a leading market position in each of the segments it serves, and now with the PanAmSat acquisition Intelsat has a leading market share in video as well. While I was CEO, a main element in the company's strategy was to further diversify away from telephony and to strengthen the video business. We accomplished our strategy through successfully introducing the managed solutions business and through acquiring the North American satellite assets of Loral. The merger with PanAmSat further diversifies Intelsat into the mainstream of video applications, and Intelsat now operates many of the world's top video neighborhoods.

Moving forward, Intelsat remains committed to leveraging and building on our strengths—the size and diversity of our satellite fleet, 40 years of knowledge and experience, a talented and multinational staff—to carry us into the future and maintain our competitive position as the industry and market leader.

THE HARDWARE

We had to deal with the misconceptions about the modern, space-age look of our Van Ness building (see Fig. 3.1). For those who have not seen it, it has enclosed round, staircases on the exterior. As the building was coming up in the early 1980s, we held meetings with the neighborhood community organization. At one of those meetings, a lady raised her hand and asked if we were going to use those round staircases, or silos, of the building to launch rockets. We said that we had not thought of that before, but that we would take it under advisement.

Out of the 290 or so communication satellites in operation in early 2006, the Intelsat system had 27 of them; none of them launched from our staircase silos. After the PanAmSat merger, we now have 51. Intelsat has more satellite bandwidth in orbit than anyone else in the industry. In September 2006,

Fig. 3.1   The Intelsat Van Ness building.

we operated approximately 2250×36 MHz equivalent station-kept transponder units.

The traditional Intelsat satellites have been designed for flexible connectivity. This has major advantages, but also a disadvantage. The disadvantage is that we do not always have "beams" or coverage that is tailored for maximum power for a particular geography. However, our flexibility more often positively differentiates us from our competition. We have movable beams that easily can be repointed to a new area of the world as the business is changing. For example, when a television operator in Kazakhstan was looking for a DTH platform, we could instantly repoint one of our satellite antennas to provide the services that the customer was looking for. Our competitors generally cannot do what we could do for this Kazakh customer.

As opposed to satellites that have been tailor-designed for a particular region, most of our traditional satellites can be moved from one region of the world to another. We had a lightly loaded satellite over the Pacific Ocean that we could off-load to neighboring satellites and move over Africa and the Middle East as traffic requirements grew there. The flexibility of our satellites and the resilience of our system allowed us to retain virtually all revenue when we had problems with two of our satellites in late 2004 and early 2005.

In addition to the C- and Ku-band payloads, Intelsat also flew L-band on the 505–508 satellites launched between 1982 and 1984. These L-band payloads were leased by Inmarsat. Before Inmarsat had its own satellites in orbit, Intelsat provided them with all their in-orbit capacity. At the time, the major Intelsat and Inmarsat owners were largely the same, and this arrangement made perfect sense when the Inmarsat business was started up. Inmarsat continued to use these satellites until they were all deorbited, the last one in August 1999.

Since we privatized in 2001, we have built out a terrestrial infrastructure to complement our satellites. During my tenure, we procured Earth stations on the West and East Coast of the United States, a very large station in Germany, and leased Earth station capabilities in Hong Kong. We also lease extensive fiber-optic cable facilities (over 45,000 km/28,000 miles during a count in early 2007) to provide cost-effective hybrid network solutions to our customers. As a result of the PanAmSat merger, we add significantly to these resources, and Intelsat now owns seven teleports. We use this hybrid network to provide mainly video services and managed services such as broadband and Internet trunking services.

THE PEOPLE

When I joined the company in 1983, I was struck by how well people worked together despite all of the different national backgrounds. I have always sensed and felt that this diversity was a great advantage for the company. We always have someone around who understands a particular market or business situation, a certain culture, a language, or behavior. We had a bit of a debate in the company when we started to offer health insurance coverage to same-sex domestic partners; otherwise, we really have had no noticeable conflicts or discussions because of the different backgrounds of our staff.

I believe that our cultural diversity differentiates Intelsat from our competition. The diversity becomes routine to us in the company, but we are often reminded of our unique characteristic. One time I was in China with a group of staff. We were visiting Beijing to meet with customers and representatives from the control and monitoring station we had there at the time. This was a year or so before we got the company privatized, and I tried to meet as many political decision-makers as possible to make sure we had support for privatization. I had gotten an audience with Minister Wu who was heading up the Ministry of Information Industry in China at the time. He was well read on the company and my delegation. He started out by noting that I was bringing a very international group of people: a Sri Lankan from our contracts department, a Chinese woman who was responsible for sales in China, an Egyptian responsible for technical matters, a British person responsible for control and monitoring operations, and myself from Sweden. Until Minister Wu made these comments, I had not reflected on the fact that we all came from so many different places.

In the early days of the company, the Indian community at Intelsat invited all staff one afternoon to celebrate Diwali, the Hindu festival of lights, and one of the most celebrated and grandest festivals in India. Diwali marks the beginning of the New Year for the traditional business community. Our Indian staff prepared Indian food for the entire organization; everyone celebrated and enjoyed. As the staff grew, it became burdensome for Indian staff to cater for everyone, and the celebration changed to an international festival with food and cultural treats from all over the world. Everyone enjoys an afternoon of sharing each others' food, drink, and cultural arts such as dance and song.

Normally, the cultural entertainment at the festival was of high quality and both entertaining and educational. However, there was an infamous exception. At a festival not too long after I had become CEO, to my surprise my male direct-reports took the stage—all dressed in drag, masquerading as the all-female singing group the Supremes. They lip-synced to the song "Stop! In the Name of Love." Even though some of their female accomplices from staff had tried to dress them up properly, it was an ugly sight! Their less than perfect control of high heels turned almost fatal for the general counsel when he fell off the back of the stage. I was honored with a dance on stage with the lead singer portraying Diana Ross. Those participating in this forgettable stunt shall remain nameless. I would not want to destroy their reputations.

Our staff has not only been dedicated to the company and its mission, but they have also shown a great interest in supporting the society in which they work. A prime example of this comes from one of our principal satellite attitude control engineers. Earl Main is a dedicated marathon runner. Not only has he run in numerous marathon events, he also ran from the U.S. West Coast to the U.S. East Coast to raise money for the Children's Hospital in Washington, D.C. He also annually mobilizes staff to run in the Marine Corps Marathon in Washington, D.C., of course with the ulterior motive of raising more money for Children's. In the last few years, Intelsat has had as many as 50 staff participating in the event. I doubt that any other main company in the region (or country) has had a higher percentage of staff participating in such a grueling event to raise money for charity.

THE PRIVATIZATION

In the early 1980s, when the construction of the grand new Intelsat building was in progress and the occupancy date was getting nearer, there used to be references made to Intelsat by the tour guides in Washington, D.C. When they went up north on Connective Avenue, they used to announce, "Ladies and Gentlemen, please see the National Zoo on your right side." When they moments later were passing by the new Intelsat building, they said, "Ladies and Gentlemen, now you see on your left side the International Zoo."

Jokes aside, the IGO cooperative structure worked well in the beginning, but over time it became both a commercial and political burden. The Intelsat IGO was a public intergovernmental organization created on an interim basis by its initial member states in 1964 and formally established in February 1973 upon entry into force of the Intelsat intergovernmental agreement.

The member states that were party to the treaty governing the IGO designated certain entities, known as the *signatories*, to market and use Intelsat's communications system within their territories and to hold investment share in Intelsat. Signatories were either private telecommunications entities or governmental agencies of the country or territory. Some signatories authorized certain other entities located within their territories that used Intelsat's satellite system, known as the *investing entities*, to invest in Intelsat as well. Both signatories and investing entities made capital contributions to Intelsat and received capital repayments in proportion to their investment share. Signatories and investing entities were also Intelsat's principal customers. Each signatory's and investing entity's investment share in Intelsat was based on its percentage level of use of the satellite system.

As a public IGO, it was exempt from various taxes and enjoyed privileges, exemptions, and immunities in many of its member states. However, because of its status as an IGO, Intelsat's business was subject to certain operating restrictions. For example, Intelsat could not own or operate its own Earth stations anywhere or provide retail services directly to end users in most countries. It also could not set market-based pricing for its services or engage in business relationships with non-signatories without first obtaining signatory approval.

Although the governments really did not get involved in running the company, just having them there as part of the structure led to some suspicion of favorable treatment for Intelsat compared with our competitors. This increased the political pressure primarily from the Western world to change Intelsat into a regular corporation.

We used to have board meetings that lasted for a week with a board of 27 to 28 directors, or governors, as they were called. The board consisted of governors who generally were staff members from the investing entities or companies. It was not unusual for the governors to also have operational and commercial responsibilities within their own organizations. It happened that this led to conflicts of interest. As an example, when transponder prices in the late 1990s in the market were $1.7–1.9M per 36 MHz, our prices were around $1.3M. This is perhaps somewhat of an apples-to-oranges comparison because of the tailored, high-powered satellites some of our competitors had. But it is an example of how we were differently run and focused than our competition.

The board also had a preference for arriving at decisions through consensus. This meant that decisions sometimes were delayed. For instance, if a governor or group of governors disliked a proposal from management, they would ask for more studies and examples. As a result the decision was delayed for another three months until the next board meeting. Having owners, who at the same time were our main customers, so close to us could be an advantage because we generally heard first hand what they were looking for. But, as the owners of the company also became more invested in fiber-optical cables, other satellite operators, and other entities that competed directly with us, the nature of our board and the speed with which it worked started to become a burden. It was necessary to change not only for political but also for very strong commercial reasons.

The first phase of privatization was the creation of New Skies in 1997. It came after years of painful negotiations among the member governments and owner operators. Some governments wanted a complete break-up and sell off of the company's assets. There was also some discussion to break up Intelsat into three or four smaller companies. Thankfully, in the end only six satellites were spun off in the creation of New Skies. Unfortunately, with these satellites went a significant part of the then existing Intelsat video business. In my view, the spin-off of these assets was an unfortunate and unnecessary event. Over the preceding years, our competitors had grown larger especially by focusing on the television business. Their satellite fleets and businesses were at the time of the creation of New Skies on par in size with Intelsat's.

When I became CEO in 1998, completely privatizing the company was my number one priority. I made a quick, back-of-the-envelope calculation and said that we could get it done in around two years, from fourth quarter 1998 to fourth quarter 2000. But, I have to confess that I had no real clue how long it would really take. What followed was a period of excruciating, but necessary, pain for the management team. It involved convincing the governments and owners that this was the right thing to do.

First, because Intelsat was an IGO, we had to follow accepted international law and principles to get it done. What did it really mean to change a large unwieldy IGO to a private company? With the help of a few specialists on international law, we concluded that we needed to get a unanimous decision by all of the member states that had signed the Intelsat Treaty or at least get a consensus decision where no member country explicitly expressed dissent. No disrespect intended, but the expression "herding cats" came to mind. We had to find a privatization solution that satisfied the concerns of 144 governments. Also, we had to devise a solution satisfying the signatories and investing entities in all of those countries. We had to do this while making sure that the resulting company was not crippled by the resulting construct, making it impossible for the company to function as a true commercial organization.

Simultaneously, the U.S. government was developing the Open-Market Reorganization for the Betterment of International Telecommunications (ORBIT) legislation. U.S. Senator Conrad R. Burns of Montana was the main driver. The legislation, which became law in 2000, stipulated that Intelsat must become a private company by 2001 and hold an IPO by the end of 2002. The draft of the legislation originally had a later deadline, but the U.S. Congress advanced the deadline when they saw my aggressive two-year schedule for getting the work done. The selection of these arbitrary deadlines created significant pressure on the management team.

For a two-year period, we had at least quarterly work group meetings with the member states, separate quarterly meetings with the signatories, and also a legal working group drafting the necessary documents. In effect, we held week-long meetings every month. All of this work, although necessary, obviously took a lot of management time away from actually running the company.

In November 2000, we had moved all of this work forward to a point where we were ready for the final meeting with the member states. The main issues for the governments were as follows: 1) the orbital filings, that is, the rights we had to the orbital locations and frequencies that we were using; 2) the continued landing rights or market access in all of the countries where Intelsat did business; 3) the public service obligations including the nondiscriminatory access to the Intelsat system, and the continued global coverage; 4) the protection of the service contracts and prices of the countries that did not have any alternatives to Intelsat for their communications, the so-called Life Line Connectivity Obligation (LCO); and 5) the country of incorporation of the new, private Intelsat.

In November 2000, at the very end of the last day of a grueling week-long government meeting with representatives from approximately 120 member states present in Washington, D.C., we had an agreement. Intelsat would become a Bermuda-based corporation with the operational center still in Washington, D.C. The United States and the United Kingdom would hold our orbital and frequency filings; we had been guaranteed continued market access in all places where we were doing business, and the governments were satisfied that we had made the right contractually binding commitments to the public service requirements and that the LCO contracts were satisfactory for the protection of the countries relying on Intelsat for their communications.

Privatization was an enormous team effort by everyone in the company. Those who did not directly work with privatization had to perform double duty making sure that the business continued to function well—and they did. Our business continued to be very strong in 1999 and 2000. At the time that we privatized, only the 603 Reboost could compete in intensity and effort with what the company had achieved by the way in which it privatized. On 18 July 2001 came the proud moment when we shed our intergovernmental past and became a regular private company with shareholders and a new board.

CHANGE OF CULTURE

In the earlier days, Intelsat was built on a culture of consensus. The governments, or Intelsat parties, and the investing operators, or Intelsat signatories, always tried to resolve an issue by accommodating each other and eventually reaching a consensus. True commercial values took a second seat to agreement and understanding. It was very rare for the board to take a vote on any issue. Although a culture of consensus was meaningful and important in the earlier days of the company, it led to delays in making decisions, and it led to inefficiency.

As we privatized, I made it clear that I wanted to fight the consensus culture and instead move towards speed in making decisions, encourage risk taking, independence, and creativity. We added new staff with new capabilities and also replaced staff who left the company, either because of their own free will or because they had difficulty adapting to the commercial realities after privatization. Even as we added a large portion of new staff from commercial industry, I was somewhat surprised to see how quickly they adopted our traditional consensus culture. When I left the company, five years after privatization, I believe the consensus culture was gone, but it had been a hard battle.

A major task for the Intelsat management team after the PanAmSat acquisition is to merge the two company cultures. Over the last few years, the cultures have moved closer, which should help smooth the way for the complete merger of the two businesses.

MERGER AND ACQUISITION ACTIVITIES

As a private company, Intelsat could now look at any acquisitions or mergers that made sense for the business. We started reviewing potential deals across several strategically important categories: 1) acquisition of other communication satellite operators to broaden our geographical and customer category reach—primarily to get into North America domestic services and build our video business, markets that had been closed to us by regulation before we privatized; 2) terrestrial infrastructure complementing and adding to the value of our satellite network; 3) acquisitions that would strengthen our customer relationships; and 4) "big bet" acquisitions and relationships that could add to our satellite business, but also expand our business into new territory. All of this, of course, was done with shareholder value as the key driver.

The first priority was to look at ways of diversifying our customer base. We had a very strong network services and telecommunication business. This business is characterized by shorter contracts and contracts for smaller amounts of capacity. It was a priority for us to complement this business with a stronger media segment. The media sector contracts for full transponders of capacity, and often for the entire life of a satellite, creating a larger backlog of future revenues.

As a warm-up for our acquisition activities, we bought the COMSAT World Systems operation from Lockheed Martin (LMT) in 2002. This gave us direct relationships and access to the U.S. customers, including some of the largest users of our system, such as AT&T and MCI. Through this transaction we also got Earth stations in Maryland, California, and Hawaii. Later, in 2004 we closed on the purchase of the remaining COMSAT business from LMT when we bought COMGEN. This business gave us some additional Earth station resources and direct relationships with a number of U.S. government customers. We could then provide ground services, consulting services, and hardware integration directly to our government customers. These additional capabilities allowed Intelsat to bid on government contracts that we could not compete on previously.

We set our sights on Loral in 2003. Loral had a strong video business and like PanAmSat very good coverage of North America. Their business was an excellent combination with ours. After the ill-fated investments Loral made in Globalstar, they were in financial difficulty and needed to raise cash to survive as a company.

We initially held discussions to take over Loral's entire satellite communications business. But, we soon realized that the mesh of cross-liabilities in the company made it a very difficult proposition to execute without incurring the risk for drawn-out negotiations and the potential for lawsuits from different Loral stakeholders. Instead, we came to the conclusion that we could buy only part of the satellite assets and this only if the company declared bankruptcy. This was a very difficult decision for the Loral management team and board, and it took time for them to accept that this was the only way out.

Eventually, in July 2003 we had all of the basics of a negotiated deal for the Telstar 4, 6, 7, 8, and 13 satellites and all of the necessary support conditions that we needed to take over the operation of the satellites and the Loral U.S. satellite communication business. What then followed was a seven-month review by a long list of U.S. authorities: the Department of Justice (in both its law enforcement role and antitrust role), the Department of Homeland Security, the Department of Defense, the Federal Bureau of Investigation, the Federal Communication Committee, and the Committee on Foreign Investment in the United States.

In the middle of this process, the oldest of the Loral satellites, Telstar 4, failed in orbit and was declared a total loss. Under our agreement with Loral, the proceeds from the Telstar-4 insurance came to us. Although we would have preferred to keep the satellite flying, the insurance settlement well covered for the loss of business on this aging satellite. Then eventually, in March 2004, everyone had approved, and the assets were finally ours. We renamed the satellites to Intelsat Americas. Our team then very quickly integrated both the business operations and the satellite control operations into our organization. Our integration was completed ahead of schedule, in just about a year's time.

My philosophy was always to fully integrate the assets that we bought with our existing operations. By using our operational and technical strength, we could make this happen seamlessly and quite fast. I set the objective that the customers should experience the result of this integration as an improvement compared with their past experience working with the acquired company.

We operate satellites from all major manufacturers from our single primary satellite control center. We deal with customers from all over the world from our single primary communication control center (the Intelsat Operations Center, better known as the IOC). This philosophy set up Intelsat to be the "power consolidator" enabling tremendous value creation, particularly from the later PanAmSat acquisition. I believe that this value creation potential was a major factor in later making the Intelsat business attractive to the private equity funds.

We also embarked on a few smaller more speculative investments. WildBlue, which now is up and running successfully, is an example of this. WildBlue delivers affordable two-way wireless broadband services via satellite, direct to homes and small offices, throughout the United States. WildBlue is the first company to launch Ka-band spot beam satellite-based services to lower the cost of providing consumers high-speed Internet access via satellite in the United States. WildBlue leverages proven terrestrial cable modem technology (following the DOCSIS standard), resulting in lower customer equipment and installation costs, an essential requirement in satellite-based consumer services. WildBlue's service is especially appealing to the millions of homes and small offices in the United States that lack access to DSL or cable modem service.

WildBlue's customer uptake has exceeded expectations. Intelsat is together with Liberty Satellite & Technology, Inc., and the National Rural Telecommunications Cooperative (NRTC) the largest owners of the WildBlue venture. We have gained significant knowledge about this type of service from our investment in WildBlue and also from another high-speed Internet access service that we have built with Orbit in the Middle East based on the same DOCSIS technology. We can apply this experience in other markets where broadband access shows business promise.

THE SELLING OF THE COMPANY—IPO OR PRIVATE EQUITY

My biggest disappointment as CEO was that I did not get the IPO done within one year of privatization as I had set as my target. Our shareholders were looking for the liquidity that an IPO provides. We started to plan for an IPO immediately after privatization in 2001. However, in early 2002, the IPO climate worsened, especially for anything related to telecommunications. Then in May 2002, Teleglobe went bankrupt. Weeks later MCI also went bankrupt. Both were among our top five customers. At that time, an IPO became impossible for us.

The business continued to perform reasonably well under these circumstances. Although we saw our channel and carrier business rightfully move to fiber-optic cable, we had a healthy growth in our GlobalConnex managed services that started to offset the business that we lost to cable. We also saw a healthy growth in the business on the Intelsat Americas satellites, the business we had bought from Loral.

In late 2003 and early 2004, as the stock markets improved, we again were contemplating an IPO. At the same time, private equity firms had started to show an increasing interest in satellite communication assets. Private equity entities had made significant investments in Eutelsat starting already in 2002. Funds advised by Apax and Permira agreed to buy Inmarsat in October 2003. In April 2004, KKR agreed to buy PanAmSat. KKR later brought in two other buyout firms, Providence Equity and Carlyle Group, to join their team. Blackstone agreed to buy our New Skies spin-off in June 2004. The trend was clear. The private equity firms saw the valuations of the satellite communication assets as favorable and the strong cash flows supporting higher leverage. The financial markets were also flooded with cash that could be had for very reasonable interest rates.

We had filed with the U.S. Securities Exchange Commission for an IPO in April 2002; we kept the filing active and updated it again in March 2004. At the same time we held discussions with interested buyers. The IPO had been filed for a price range of $11–$13. We were unhappy with the range; our owners were expecting more. In April 2004, the discussions with private equity firms picked up speed. Two groups showed the greatest interest: Apax and Permira on one side and Madison Dearborn and Apollo on the other. Eventually, these four firms joined forces, similar to what had happened around PanAmSat with KKR, Providence, and Carlyle.

On 16 August 2004, we had a deal, a deal where funds advised by Apax, Apollo, Madison Dearborn, and Permira would buy all shares of Intelsat for $18.75 per share. Our shareholders were very happy when they compared this price with the $11–13 price that had been discussed for the IPO. Actually, the CEO of one of our larger shareholder companies called one person on my executive team to congratulate us for a job well done. He said, "First, I thought I heard $8.75 per share; now I understand it is $18.75 per share—God bless you, my son!" Another feature of the deal that our old owners liked was that it was for the entire company. Most of our owners were traditional telecommunication companies, and many of them needed all of the cash they could lay their hands on for investment in their own core businesses.

After the deal was signed in August of 2004, the review by the authorities ensued. We estimated that the review would take around six months. Then one dark day, 28 November 2004, we lost control and all contact with our IA-7 satellite. We thought that the satellite was a total loss, but our engineers did not give up. They kept looking for it and sending commands to it in the blind to try to recover the satellite. After a heroic effort, they eventually brought IA-7 back from the dead. Our team managed to get commands in that reconfigured the satellite and contact could be reestablished. The satellite was tested, and roughly half of the transponders could be brought back into full operation. By 1 December, virtually all customers on the satellite were up and running again, either on IA-7 or on alternate Intelsat capacity. We lost almost no revenue or backlog as a result of this event.

After this incident, because the deal had not closed there was of course some discussion initiated by the buyers on lowering the price. In the end, because we had lost virtually no business as a result of the event, the price was not changing, and we continued to wait for the authorities to complete their review of the deal. In December, we started to get indications that the final approvals would come in mid to late January 2005. By 10 January, we were out on the "road" in the United States to finance the deal. We needed to raise around $3B. We started out on the East Coast that week and finished on the West Coast. All went very well; the investors were responding well to our story.

We were just boarding the plane in Los Angeles on Friday afternoon (Fig. 3.2), at the midpoint of our financial road show, when we got the phone call we absolutely did not want—shock and horror, even Stephen King could not have scripted the story better: Our operations group called—we had lost contact with our IS-804 satellite.

This time we were not as fortunate as with IA-7. After several days of attempted recovery, we had to write the satellite off as a total loss. Again, our staff had to move traffic around to other satellites. This time, a larger part of the business had to be put on a competitor's satellite, but the majority of the

Fig. 3.2   The team in front of the road-show airplane, minutes before we got the IS-804 news. (See also color figure section at the back of the book.)

revenue and backlog stayed with us. Our buyers remained committed to the deal with no price adjustment.

The Monday after the event was a holiday, so we had time to regroup and assess the situation before we continued the road show for the financing of the deal. We did a second stretch on the East Coast and also headed south to Texas. Despite the road bumps on the way, the road show turned out to be a success. We got good financing, and approximately 80% of the investors we met with or talked with on the phone ended up buying into the financing deal. We had gotten all of the necessary approvals from the authorities, we had the financing, and we eventually closed the deal on 28 January 2005.

The IA-7 and IS-804 problems were the first serious in-orbit problems that Intelsat has experienced for over 30 years. Despite the (fortunate) lack of real experience with this type of event, the customer-facing and technical teams at Intelsat did an outstanding job in resolving the customer issues following these two in-orbit problems. Services were migrated to another capacity and in some cases to our competitors' satellites. When an anomaly like this occurs, the players in the industry try to help each other out. In the end, what is important for us all is that the customers can rely on the services that the industry provides. It is in everyone's interest to ensure that the customers are satisfied. The fact that we managed to retain almost all revenue and backlog after these two events is also a testament to the resilience and flexibility of the Intelsat system.

PRIVATE EQUITY OWNERSHIP AND THE BIG DEAL

We had traveled a long road to accomplish this final, important step in the transition of Intelsat. I have talked about my three objectives when I became CEO in 1998. The sale of the company was a surrogate for an IPO, but in the end it had the same effect of delivering value and complete liquidity to our traditional mainly telecommunication company owners.

The fact that the deal closed at our initially agreed terms, despite the IS-804 and IA-7 issues, underscores the strength of our business. The private equity group bought this company for many reasons. We were a company with highly attractive assets and a diverse business. We had a large and stable backlog, with a drastically improved "book-to-bill" ratio over the last 12 months before the deal closed. We had proven our ability to integrate outside assets and customers—a strong positive in a consolidating market where more acquisitions were likely. We had changed our internal culture to be much more commercially focused. Unexpectedly, we had also proven the true value of our flexibility and ability to cope in crisis, over the last eight weeks before closing the deal.

Most importantly, we remained very well positioned to capitalize on the changes and advancements that we had made to move Intelsat forward and generate strong cash flows for our new investors. With the acquisitions that we had done, we had a full set of assets and customer relationships to succeed in the market. Our major capital investment cycle was over for the time being, so the future was all about reaping what we had sown.

The private equity team does not run the company directly, but the new, smaller board of directors consists of their representatives. The directors spend a lot of their time on the company. They brought fresh air and new focus to the work of the board. They have crystal clear objectives and a very sharp focus on company performance. The company and especially the managers in the company have learned a lot from their presence.

The new owners appointed me as the chairman of the board of Intelsat, Ltd., and Dave McGlade joined the company as the new CEO on 1 April 2005. For a long time, I have viewed continued industry consolidation as inevitable. The old, traditional owners, boards, and sometimes also management teams were not always conducive to making it happen. However, with more than 50% of this industry owned and controlled by private equity funds in the beginning of 2005, the conditions had changed drastically. Before taking control, the new owners looked at how they either can sell the asset with good profit or how it can be combined with other businesses to enhance value. This new dynamic led to two important consolidations. First, SES closed on the acquisition of New Skies. Then on 3 July 2006, Intelsat closed the acquisition of PanAmSat after almost a year of review by the authorities in the United States and in other parts of the world. I was very happy to see that the merger of the two companies eventually took place. It is no doubt the right thing to do for both companies.

With the addition of PanAmSat's video market expertise, the advanced satellite fleet, and blue-chip media customer base to Intelsat's portfolio, the new Intelsat is now the largest provider of fixed satellite services (FSS) worldwide to each of the media, network services/telecom, and government customer sectors. Intelsat acquired all of the outstanding common shares of PanAmSat for approximately $3.2B. As a result of the merger, PanAmSat is now a wholly owned subsidiary of Intelsat. The total value of the transaction, including PanAmSat debt, was approximately $6.4B. For the 12-month period ending 30 September 2006, pro forma revenues for the combined company total more than $2.1B, and EBITDA on a pro forma combined basis was $1.6B. By September 2006, pro forma combined revenue backlog, which is based on long-term customer commitments of up to 15 years, was approximately $8.0B. These are impressive numbers.

Using optimized capacity on the combined fleet of 51 satellites and a large, complementary terrestrial infrastructure including eight owned teleports, fiber connectivity, and over 50 points of presence in almost 40 cities, the new Intelsat carries one out of every four television channels transmitted over fixed satellites; supports 27 DTH platforms worldwide; operates 16 satellites that are part of video neighborhoods around the world; is the number one provider of transponders for video programming worldwide; carries more high-definition (HD) programming than any other FSS carrier; is the largest provider of commercial satellite services to the government sector; and is the leading provider of services to enterprise, Internet, and mobile network operators.

The acquisition of PanAmSat is an indication of Intelsat's strength. It gives Intelsat an unparalleled telecommunications network, which enhances our ability to meet the needs of all customers, wherever they might be. In particular, the addition of PanAmSat gives Intelsat greater redundancy to deal with satellite anomalies; enhanced flexibility to respond to humanitarian needs, such as earthquakes, tsunamis, and hurricanes; and a wider geographic reach to all corners of the globe. The combined assets of our company provide the highest level of service and network reliability to existing customers, while opening doors to new business opportunities in key communications growth markets such as HD, IPTV, and applications resulting from the convergence of video, voice, data, and mobility.

The integration process, which was already underway when I left the company, will ensure a smooth and seamless transition for all customers. The objective is to give the customers a better service and customer experience than they received from the individual companies. Intelsat intends to adopt a "one-company" operating philosophy and expects to fully integrate PanAmSat's assets and operations. Since the merger was announced in August 2005, the companies have conducted disciplined integration planning and execution in order to drive the benefits of greater scale and complementary service offerings to customers and to deliver strong operational synergies to the company's stakeholders. By making key functional and systems decisions ahead of time, the new Intelsat is positioned for an accelerated start.

THE 603 REBOOST MISSION

Editor's Comment: Chapter 14 provides a more detailed story on this mission. This is Conny Kullman's recollection and perspective of this momentous event. The editor's own recollections are briefly touched upon in Chapter 1. Although all of these may not agree in detail, they all share the common sense of achievement and pride. As these are all first-person accounts, the editor has restrained himself from trying to reconcile small differences in detail or technical parameters.

Although I have many exciting experiences from my years at Intelsat, the most memorable event without a doubt is the 603 Reboost Mission. It all started when Intelsat 603 was launched on 14 March 1990. Because of an error in the wiring of the Lockheed Martin Titan 3 rocket, the satellite did not separate properly from the upper stage and was unable to reach its geostationary orbit.

To prevent reentry into the atmosphere and a complete loss of the satellite, the mission management team had to react quickly to move the satellite to a safe orbit. The engineers in the Intelsat Launch Control Center working with tracking and commanding stations around the world managed to command the satellite to separate from its unused perigee kick motor (PKM), which took with it the burned-out Titan stage. With additional maneuvers, they raised the orbit to around 560 km [300 nautical (n) miles] to avoid reentry and to prevent the damaging effects of atomic oxygen at lower altitude. The bombardment of the atomic oxygen could have destroyed the thin silver interconnects that are an integral part of the spacecraft's solar-power generators. There the satellite was left in a stable, but unusable orbit, while the engineers considered how best to raise 603 to its appropriate place.

After studying several options, we decided that the best solution was to bring a new motor to the satellite rather than to bring it back to Earth, attach the new motor, and re-launch the satellite. NASA was willing to take on the mission, partly because it would be an excellent opportunity to train for the later assembly of the space station. The maiden voyage of the new Space Shuttle *Endeavour*, or STS-49, was selected by NASA.

A new PKM was built. Intelsat worked closely with Hughes and NASA in planning the recovery mission. Among other things, Hughes built a docking adapter and a grapple fixture or capture bar, which the astronauts would latch to the satellite before bringing it into the shuttle's cargo space where it would be mounted on top of the new motor. This capture bar would dramatically increase the excitement of the recovery mission.

We had a very involved preparation and training program getting ready for the recovery. NASA's safety requirements for manned missions are substantial. We had to work out specific mission rules that ensured the crew's safety at all times. We had to track the satellite and get telemetry at least every 90 minutes. To achieve this, we had to build a new Earth station on the west coast of Africa. The choice made was to use the location of Gandoul in Senegal. A new station was built in record time.

We had numerous internal rehearsals and also eight joint rehearsals with NASA. The most exciting one was with the shuttle crew working in the water tank in Houston with real life-size models of all of the equipment to be used during the mission. In some cases, the joint rehearsals stretched over three to four days. The rehearsals included a simulation of problems. At one rehearsal, the "referees" handed out "red cards" to a large number of mission team members. It said something like, "You had contaminated food for lunch; you are disabled and can no longer perform your work." The remaining "healthy" team had to prove that they could continue the mission without the "sick" team members.

We had to fly the satellite at 560 km altitude, but the shuttle could only reach around 350 km (190 n miles) altitude. So for the first time in the history of the exploration of space, two spacecraft had to meet each other in orbit as part of a "dual active rendezvous." This maneuver required substantial expertise and confidence in the mission team's abilities.

Then, eventually, the day came when the shuttle took off, and the mission was underway. Immediately after *Endeavour*'s successful liftoff on 7 May 1992, the Intelsat Mission Control in Washington, D.C., issued the commands to lower the satellite from its 560-km orbit to meet the shuttle at 350 km. The shuttle and 603 met in what NASA called a "control box." This is like a moving window in the sky extending over 6 deg of arc at the rendezvous altitude. The satellite reached this box about 26 hours after shuttle liftoff. It had to reach the box within 46 hours after liftoff; otherwise, the mission would risk failure. NASA was impressed with how fast we got the satellite in the box. This was a good start to the mission.

Our excitement got even higher when we could see the satellite on video for the first time since March 1990 as the shuttle approached it in orbit. On 10 May 1992, one of the astronauts, Pierre Thuot, performed an extravehicular activity, or EVA, to catch the satellite with the capture bar; Thuot's feet were strapped to the robotic arm from the shuttle's cargo bay. The Intelsat 6 satellite looks like a large coffee can. To keep it stabilized, it rotates around its central axis. To enable the astronaut to catch it, we had spun down the satellite from 12 rpm (revolutions per minute) to 0.65 rpm so that it could safely be grabbed by the astronaut. At this time, the satellite had a mass of 4045 kg (8909 lbs). In space it is weightless, but it still has its inertia. Pierre did several gentle attempts to catch 603, but the $5M capture bar would not work as intended. Instead the satellite got pushed into a flat spin, almost perpendicular to its normal axis of rotation. At the end of the day, there was a sense of despair in the mission team. So much work and it seemed it all was going to waste.

The following day we felt a bit better. NASA had agreed to send Pierre out for a second attempt. Overnight, we had spun up and stabilized the satellite, again placed it in a location where the shuttle safely could approach it, and spun it down so that Pierre could grab it. In suspense, we again watched Pierre approach the satellite. His attempts this time were even gentler than the day before. He tried to grab the satellite with the capture bar. After a couple of attempts, it looked like he had it, and cheering got underway at mission control. But, after a few seconds we realized that we had cheered too early. The satellite tumbled away in space. The night that followed nobody slept. Our team had to rethink the approach.

Our mission management team in Washington D.C. got together and concluded that the only way to make this work was to send three astronauts out at the same time and catch the satellite with their "bare gloved" hands. Now, all we had to do was to convince NASA. They said no: they'll try with two astronauts, but not three, as the shuttle's airlock is only designed to house two astronauts. NASA did not believe it could be done with three, and in any case it was definitely outside of what was allowed by the mission rules.

Then the morning came for the astronauts. They had also spent a mostly sleepless night trying to figure out what to do. They had done this in the dark, with their microphones switched off. We listened in on the morning discussion between the crew and NASA mission control in Houston. We silently smiled and cheered in Washington when we heard the mission commander, Dan Brandenstein, say that the only way was to send three astronauts outside and grab the satellite by hand. That helped push the three-man solution through. The NASA rule book was thrown out the window and replaced with some true risk taking and ingenuity.

The following day, 12 May 1992, was a day of rest for the shuttle crew and for our team in Washington. However, in Houston some heavy-duty planning took place. A group of astronauts was sent into the training water tank. They simulated what needed to happen to get the three astronauts out and how they would be positioned in the cargo bay and on the robotic arm for the right configuration to catch 603.

So came the dawn of 13 May 1992. Rick Hieb, Pierre Thuot, and Tom Akers got ready to go outside and catch the satellite. Because only two of them could be hooked up to cooling and air supply in the airlock, one of them had to do all preparations before getting out. Crammed into the airlock like sardines, head to toe, the astronauts got their next surprise—a computer glitch onboard forced them to stay there a full 92-minute orbit before they could exit. Pierre remarked, "Wish we had brought some M&Ms."

The three safely got out from the air lock. Then Dan flew the shuttle slowly and gently under 603 inch by inch. This time, we had slowed the satellite's rotation to less than 0.3 rpm, which of course made it unstable, and it started to nutate ("wobble") lightly. But, eventually the astronauts decided it was time for the big grab. Six hands took hold of the bottom ring of 603. True and lasting cheering and relief spread throughout the mission management team in Washington. Shuttle Commander Dan Brandenstein reported: "Houston, I think we got a satellite."

Then what might have been the longest 90 minutes of my life followed. The astronauts held onto the satellite and felt the fuel inside sloshing around. Eventually, they felt comfortable enough to attach the misbehaving capture bar to the bottom of the satellite. This time Pierre got it to dock properly, and the 603 could safely be moved into the cargo bay with the robotic arm and mated on top of the new motor. When all was said and done, the astronauts returned to the shuttle after what was the longest and most complicated EVA in history. The astronauts spent eight hours and 29 minutes in the cargo bay, breaking the 20-year old spacewalking record set on the final Apollo moon flight.

At this point we all felt pretty good, but the ordeal was not over yet. Payload specialist Kathy Thornton was responsible for taking the actions from inside the shuttle necessary to redeploy the spacecraft. She went through the process of activating power time switches on the perigee stage, and she eventually flipped the switch that would release the spacecraft with its new motor from the cradle in the cargo bay. We watched in disbelief; nothing happened. The satellite would not move. It still stubbornly sat on top of the motor and the cradle in the cargo bay. Kathy spoke words that I will never forget: "No joy on deploy." It was another harrowing experience. Was it never going to end?

One of our contingency scenarios was to deploy the satellite and the motor with the cradle and fire through it. But, this would have meant losing five years of satellite lifetime. The mission specialists in Houston went through the switch sequences every which way. Finally, they figured out that the procedure was wrong. The switches had been thrown in the wrong sequence. The procedure was quickly updated, and Kathy went through it again. This time we saw the satellite slowly leave the cargo bay, gently spinning. NASA's work for us was done; there was a communal sigh of relief in all of the mission control and management teams.

Now it was up to our mission management team to get the satellite into the right operational orbit. It involved the first ever supersynchronous transfer orbit for launching a synchronous orbit satellite, violating several mission constraints, and a few other extraordinary events. Compared with what we had been through the preceding week, it all seemed fairly pedestrian. We got the satellite to the right place, without any further hiccups.

Kitty Horer, one of our staff, always wrote a small poem to our launch missions. This is what she wrote after the 603 Reboost:

> *Ode to the Reboost*
> With bar at the bottom, gloved hand at the top
> somehow or another, this bird we will stop!
> No longer will 603 wander through space!
> The free fall is over; it's the end of the chase.

As of second quarter 2007, 603 is operated in inclined orbit. We believe we can continue to operate the satellite towards the middle of 2009. The satellite has been one of the best behaved in the Intelsat 6 series. We have lost some of the payload redundancies, but otherwise the spacecraft is in very good shape. The "shake-and-bake" it got during the two years before the reboost appears only to have improved it!

THE WRAP-UP

When my wife and I were about to leave the Washington D.C. area to go to Bermuda, we met with some close friends. I said that I had a great career at

Intelsat, but that I was ready to leave the United States and get out of the high-pressure job of being the CEO. One of our friends said that he was surprised, "But, Conny, you have lived the American Dream." When I think about it, he was right. I got to do what most people only dream about. I made it to the top job in the company, but what really made the dream come true was to have been part of the team that made the recovery of 603 possible, part of the team that privatized the company, part of the team that transformed the company through acquisitions, and eventually part of the team that made the sale of the company possible at a very favorable value for our owners.

At the end of August 2006, I left the company. Joe Wright, the previous CEO of PanAmSat, took over as chairman. I had worked 23 years for Intelsat, and as I said in the beginning, "It is never boring to work at Intelsat," and "Time flies when you're having fun." I had fun at Intelsat; my experiences with the company and its people will always be among my fondest memories.

I would be remiss if I did not mention my main mentor and supporter, John Hampton. John came from Australia and was our Chief Operating Officer (COO) from the mid-1980s to the mid-1990s. He challenged me by moving me from one job to the next larger one. I did not always like what he asked me to do; he made me sweat, but I kept saying "Yes, sir" whenever he had a new job for me, and I am glad I did. It allowed me to have many different jobs and take on challenges without having to move between companies.

One day in 1997, as the previous CEO was getting ready to leave office, John came to me and said, "You should go for the CEO job!" This time I did not say "Yes, sir." My reaction instead was, "What the ... have you been smoking?" Nobody from inside the company had made it to the top. It was even quite unusual for someone to reach a vice president post from inside the company. The senior management appointments tended to be a bit political. But, I thought about it, considered who my competitors for the job could be, and decided to go for it. The rest is history, as they say.

I started out by talking about Arthur C. Clarke's brilliant idea for "Extra-Terrestrial Relays," which could be used for communicating around the globe. As I was preparing for a presentation to a group of people from the aeronautical industry in 2004, I thought the audience might be interested in what Clarke would be asking the industry if he were standing in front of them. I had, in fact, the opportunity of asking him before my presentation, and this is what he told me he wanted:

> First, I would ask the telecommunications folks in the crowd for the creation of a flat videophone rate for services around the world. This is entirely possible because, thanks largely to satellites, distance has become irrelevant. This would eliminate legions of auditors and accountants. I say this with feeling, as I once was an auditor myself.

"My main interest now," he wrote in an email message to me,

> is in the space elevator, which I first heard about in the early '60s. This technology could assure easy and efficient access to space at a very low cost, and would enable not only satellites to be launched very cheaply, but also would allow spacecraft in orbit to be checked and serviced. It would move the focus of satellite operators off of pyrotechnics and onto productivity. In fact, the space elevator would make access to space so inexpensive that the main cost of getting there would be for movies and in-flight catering. If I were the CEO of Intelsat, this is what I would ask of the world's space community.... Of course, budget is no consideration!

At the time, I was the actual CEO of Intelsat—I can tell you that budget is in fact always a consideration. But as Clarke knew better than anyone, inspiration is still where it all begins.

Chapter 4

TELESAT: THE FIRST DOMESTIC SATELLITE SYSTEM

HARRY KOWALIK*

On 11 January 1973, the Minister of the Department of Communications of the Canadian federal government in Ottawa, placed a telephone call, via Anik satellite, to a settlement manager in his office in Resolute, an isolated community in Canada's High Arctic at 74.4° N. This call marked the world's first service via a domestic communications satellite. It was a giant leap forward in the quality of telecommunications services for isolated points in Canada, particularly for its northern communities. For the first time, reliable, interference-free, dial-up service was suddenly available between points anywhere in Canada, at a quality equivalent to that among major cities in the south. It was an enormous step forward in telecommunications services as is the nature of the introduction of satellite communications anywhere. But it took several years of studies, political decisions, engineering, and construction to get to this historic milestone.

HISTORICAL BACKGROUND

The development of satellite communications in Canada began in the research laboratories and various departments of the federal government. The government controlled the system and financial studies, including the technical services to be provided. This program culminated in an Act of Parliament, the Telesat Canada Act, which established Telesat Canada, with the mandate to provide satellite communications for Canada. The company was incorporated on 1 September 1969. The initial satellite technical studies in Telesat were very much an extension of the Canadian space program of the federal government. Thus, a historical review of space activities in Canada prior to Telesat is considered appropriate.

---

*Former Vice President, Space Systems, Telesat, Ottawa, Canada.

The Canadian space program began in the government research laboratories in Ottawa in 1958–1959 in the Defense Research Telecommunications Establishment (DRTE) of the Defense Research Board (DRB). The launch of Sputnik 1 on 4 October 1957 was the dawn of the space age, and it was not long before excitement was sparked among Canadian upper atmosphere and space research scientists; it was a new scientific frontier in which to get involved. To convince the politicians of the merits of entering into research using satellite technology, tangible reasons had to be identified to justify funding for this newly emerging scientific field.

During this period, the electronics laboratory of DRTE was engaged in measurements of the ionosphere from the groundside, and it was thought that if a polar orbiting satellite could also map the topside of the ionosphere then a much better evaluation and understanding of the ionosphere would be achieved. The scientific objective would be to measure the electron density distribution in the ionosphere at altitudes between 300 and 1000 km. It was believed that this knowledge would be of immense value to enhance the capabilities of the Ballistic Missile Early Warning System (BMEWS) and other tracking systems concerned with the Soviet intercontinental ballistic missile (ICBM) threat. The DRTE scientists convinced the politicians of the usefulness of such a satellite, and the program was approved.

DRB then approached NASA regarding the merits of this program, and NASA and its scientists agreed. NASA would provide the launch services in exchange for the scientific data from the satellite. Thus was born the Canadian space program. This initial agreement with NASA was the forerunner of Canadian participation in future NASA space programs, up to and including the Canadian astronaut program on the space shuttle and the Canadarms on the shuttles and on the International Space Station.

The first Canadian research satellite to carry out measurements of the ionosphere from the topside, Alouette I (Fig. 4.1), was launched on 28 September 1962 into a 1000-km circular, polar orbit. The scientific results were considered so successful that U.S. scientists invited the Canadian scientists to continue with follow-on programs, and Alouette II was launched in 1965. These successes led further to the International Satellites for Ionospheric Studies (ISIS) program. ISIS I and II were launched in 1969 and 1970, respectively.

The Alouette I spacecraft was designed, developed, and constructed in the DRTE laboratories, with some parts such as structure, antennas, and batteries procured from outside. Novel characteristics of the design were the 45.7- and 22.9-meter crossed-dipole antennas for the ionospheric sounder. This was the first space application of the Storable Tubular Extendible Member (STEM), an invention by a scientist at the National Research Council of Canada. Stored in a spool much like a carpenter's tape, its initially intended application was for extendable antennas on military tanks in World War II. Developed by Spar Aerospace Limited for Alouette I, the four antenna elements were deployed

Fig. 4.1   Alouette 1, the first Canadian satellite.

once the satellite was safely in orbit in a weightless environment. With no structural strength in terms of their own weight, for prelaunch ground testing the elements were suspended on guy wires from a ceiling. The STEM concept is particularly suited for space hardware applications because of its storability properties during the spacecraft launch phase. Since Alouette, the STEM has been developed extensively, into various and complex configurations and is among the great success stories in sustained space structure design. The STEM has been employed on many spacecraft of various configurations, including Voyager I and II, GPS, Hubble Space Telescope, Mars Pathfinder, Solar Maximum Mission, Nova, Dynamic Explorer, SEASAT, INSAT, Radarsat I, and others.

It was government policy to transfer the design and construction technology for future programs to Canadian industry. Therefore, the ISIS satellites were contracted to RCA in Montreal. These satellite programs were vital to the development of engineering experience for the upcoming geostationary communications satellites for Telesat Canada.

While the Alouette program was underway in the early 1960s, the communication satellites' concepts being explored in the United States were being closely followed by the Canadians. Initially, low-orbiting satellites such as Telstar and the Echo balloons were believed to be the way to success. But it was the Hughes Aircraft Company that set the stage for satellite communications when it pioneered Arthur C. Clarke's [1] 1945 concept of a global geostationary satellite system, capable of providing continuous real-time global coverage. NASA launched Syncom I and Syncom II in 1963 followed by Syncom III in 1964. The success of the geostationary orbit concept captured the imagination of the developed industrial nations very soon, and the International Satellite Corporation, Intelsat, was quickly inaugurated in 1964. Hughes was awarded a contract to deliver the first commercial communications satellite in 1965, Early Bird.

In light of the rapidly developing space technologies in the early 1960s, including satellite communications, the Canadian Science Secretariat commissioned a task force in 1966 to evaluate and make recommendations for various space programs for Canada. The task force presented its report titled "Upper Atmosphere and Space Programs in Canada" to the Science Council of Canada in January 1967. The report provided recommendations for various space programs and, in particular, a satellite communications system for Canada. It recommended a Canadian-owned system as opposed to leasing services from foreign suppliers. In particular, it identified an urgent need for Canada to apply, on a first-come-first-serve basis, for prime orbital positions in the North American corridor.

On the basis of this report, in July 1967 the Science Council submitted a summary paper to the federal government in the form of a report entitled "A Space Program for Canada." It identified social, economic, and industrial benefits of the proposed program and concluded that it would be highly prudent for Canada to develop its own capabilities in space hardware and services. The paper pointed out that Canada should press for international agreement on the use of the geostationary orbit and should identify and obtain agreement on its planned orbital slots. The adverse consequence of not obtaining the required orbital positions was one of the main motivators for an immediate go-ahead. The successes of the Alouette and ISIS programs also had a significant impact on the confidence for deciding to get on with it.

Hence, the next significant milestone was a white paper by the newly minted Department of Industry in March 1968, titled "A Domestic Satellite Communications System for Canada." The paper reviewed the world status on satellite communications development and the potential benefits of a satellite communication system for Canada. An early decision was considered to be of vital importance. There was a need to fulfill the federal government broadcast policy, which was to provide telecommunications in both official languages to all parts of Canada. Potential for satellite communications was most evident in the North, but it applied to all developing areas of the country; for example, to mining communities, for petroleum exploration and development, for stations monitoring pipelines, and to many other isolated locations in need of reliable and real-time communications. Government departments and agencies also required reliable communications to remote areas. East–west general communication requirements were also growing annually at about 20%. Satellite communications would be significantly beneficial to this expansion and would also provide route diversity. Although all of the benefits of satellite communications could not be identified, nevertheless there was a feeling in the technological, commercial, and government circles that this new technology for communications had enormous potential.

The paper included recommendations for a baseline system of two satellites and a variety of Earth stations. Most importantly the paper provided an outline for a corporation to be established by the government to provide satellite communications on a commercial basis. The paper also reinforced the Science Council's view that Canada be a leader in the development and manufacture of satellite communications equipment.

For Canada to become a leader as such, among other initiatives, the federal government set a policy objective of developing a Canadian prime contractor, capable of supplying geostationary communications satellites on a competitive and commercial basis. To have the capability of a prime contractor, an organization must have overall system and bus design and construction know-how and experience. Canadian industry did have the ability and experience for Alouette and ISIS type satellites, which were relatively simple, free-orbiting, electronic systems. However, geostationary satellite bus design, in addition to structure and electronics, requires control systems with sensors and control jets using toxic fuels, rocket engines, large solar arrays, and various deployment mechanisms. Also, the commercial and competitive business world requires highly efficient designs in terms of maximizing the number of channel years of service for a given size and weight of satellite and launch vehicle at a competitive price. These posed new challenges for the Canadian industry. Alouette and ISIS were experimental research satellites, sole source, and funded by the federal government on a cost-plus basis and where there were no financial impacts on late delivery. By the very definition of research programs, cost overruns and late deliveries are anticipated and allowed in order to facilitate research. It would be a huge step to develop the capability to compete in the open market and to deliver on time and at a competitive and fixed price.

There are also other factors that were not all that visible, especially at the policy-forming period. Bus development costs are extremely high and of long duration, construction programs are few, and only a few spacecraft are required each year throughout the world, distributed over already well-established suppliers. For some corporations it is not really profitable, but it provides great marketing publicity for them as a whole. In the United States, only very large corporations with several billions of dollars of annual revenue enter this high-technology field, and satellites are only a small part of their total business portfolio. Also, they invariably have military contracts to provide them with the extra technical and revenue base. In Europe, in order to compete with the United States suppliers, several companies formed consortiums to share the development costs. In addition, as in most other high technologies, the state of the art changes rapidly. Another important requirement for a satellite prime contractor to be sustainable is that the domestic market must be large enough to provide continuity of employment and development funds. International contracts, if any, are an extra bonus. In Canada, the domestic market is too small, and federal government satellite programs are few and far between. There are no military satellite programs.

Canadian government officials did not readily appreciate the financial depth and market that a supplier must have in order to exist on its own, without forever being subsidized. Many did not even understand that a prime contractor, by definition, must inherently have system design and satellite bus manufacturing capability. Canadian government policy makers did not readily recognize these factors but were eager to forge ahead in this very interesting, prestigious, and challenging technology.

The end result is that, even with substantial government subsidies over a period of some 25 years, a competitive Canadian prime contractor, for commercial geostationary communications satellites, had not materialized. However, that is not to say that Canadian participation in space technology is somewhat of a failure—far from it. The Canadian government and the Canadian space industry have made enormous strides. One perfect industrial example is Telesat Canada as a leading communications satellite operator, service provider, and the world's leading satellite consulting company. Renowned federal government Canadian Space Agency programs are the Canadarms on the NASA space shuttle and on the International Space Station. These robotic arms have required very high cutting-edge technologies and have been highly successful.

As a result of the recommendations of the white paper, the federal government decided to proceed with the program. It established a project office to prepare the framework for a satellite communications corporation and to continue with the technical program. Because the technical program was to be implemented by Canadian industry and not in government research laboratories, the project office was set up in the Department of Industry. However, the new Department of Industry had no engineers yet, so about six were seconded from various organizations for the start-up phase. In the technical program, defining a satellite in terms of its communications capacity, its service life, and cost was of prime importance. Beginning in May 1968, the small team of engineers was quite uncertain about what would be a reasonable specification for a satellite in terms of the weight and size capabilities of the designated NASA launch vehicle. Hence they estimated a set of performance requirements that were then used in two study contracts, one issued to RCA, Ltd., and the other to the Northern Electric Company, Ltd., both in Montreal. These two companies did not have bus technology, so they needed to team up with bus suppliers in the United States.

In early 1968 when the federal government founded the Department of Industry, it also developed a Department of Communications (DOC) to fulfill its broadcast policy. The DRTE, along with its Alouette/ISIS team, was trans-ferred to the DOC to provide it with the necessary research and development laboratories, an ideal fit for the DOC.

The reports for the study contracts were presented to the project office in January 1969, but the project office still did not have sufficient engineering personnel or office space to evaluate them. Hence, the Alouette/ISIS team was loaned to the project office, and the work was carried out at the newly acquired DOC facilities.

A program definition phase (PDP) was the next step in the technical pro-gram. In the spring of 1969, RCA, Montreal was invited to prepare a formal proposal for the delivery of two satellites. The satellites were to be optimized in terms of channel capacity, transmit power, eclipse capability, and satellite life, within the launch weight capabilities of the then current NASA Delta launch vehicle. RCA was required to submit its proposal, together with pric-ing and schedule in the first quarter of 1970. With no bus capability, RCA teamed up with TRW in the United States. RCA's parent company in the United States had not yet developed bus-manufacturing capability.

In drafting the legislation for the satellite communications corporation, the federal government scheduled it to be completed in good time for the corpo-ration to be ready for the RCA proposal. But what should be its financial and ownership structure? The government did not want yet another Crown Corporation, that is, a corporation wholly owned by the government, which might become a financial burden, forever subsidized by the government. Historically, there had been cancellations, for one reason or another, of major government open-ended development programs, such as the Avro Arrow and the Hydrofoil. For these kinds of reasons, the government decided to set up the new satellite corporation to operate on a commercial basis. However, the government desired some form of control over the satellite company so that it would be able to influence the company to buy Canadian manufactured space and ground segments. Hence ownership in the corporation was to be one-third each among the federal government, the telecommunications carri-ers, and publicly issued shares. Public shares were never issued, and the government ended up with 53%. After much consideration and deliberation the name Telesat Canada was chosen for the corporation, often simply referred to as Telesat.

A key reason for government ownership in Telesat was to enable it to promote Canadian industrial content in the supply of space and ground equipment and ultimately the development of a Canadian satellite prime contractor, particularly for geostationary communications satellites. This was a very important policy issue of the Canadian government. Notwithstanding the occasional government satellite program, Telesat would be the only Canadian commercial satellite procurer. Telesat procurements would therefore be required to have significant Canadian content. Telesat was allowed to procure its space segment in a competitive market for the best technical performance, quality, reliability, and price, but suppliers were required to maximize Canadian content in their proposals. Presumably subsidies would be provided for the additional costs of buying Canadian, and in that sense Telesat would be treated as a Crown Corporation, after all. On the other hand, in its products and services Telesat Canada would be expected to operate as a commercial enterprise and to compete directly with terrestrial common carriers, such as terrestrial microwave. It was under these kinds of constraints that Telesat Canada was inaugurated, and on 2 September 1969 it opened its doors for business.

For the first few years, initial financing for capital procurements and for operating costs of the company was provided through the issue of common shares, procured by the government of Canada in terms of its ownership and by certain, approved telecommunications common carriers. Future financing would be provided through debt and equity.

TELESAT SPACE SEGMENT

RCA, Montreal delivered its proposal to Telesat under the PDP in the first quarter of 1970. During this time frame, Telesat was also evaluating an offer from the Hughes Aircraft Company of California. Hughes was strategically developing a communications satellite bus designed for the domestic market that was soon to open up in the United States and elsewhere. However, it turned out that Telesat was the first potential customer on the horizon, and Hughes took advantage of the opportunity. After extensive evaluations of both proposals, Telesat negotiated a contract for delivery of, not two, but three satellites from Hughes. The Hughes satellites would have 12 channels each, twice the number proposed by RCA; their contractual service life would be seven years as compared to RCA's five years, and the unit price was substantially lower. However, in compliance with the government's policy objectives, substantial Canadian content was provided through two major Canadian subcontracts, one with the Northern Electric Company, Ltd., of Montreal and the other with Spar Aerospace Products, Ltd., of Toronto. These subcontracts raised Hughes' price to some extent from its initial price as a sole supplier. Although this fell short of a Canadian prime contract, nevertheless both the government and Telesat felt that this was an optimum solution for this initial procurement, all things considered. The signing of this contract was indeed a major milestone in the Telesat satellite program.

The Hughes Aircraft Company delivered the first Anik satellite in October 1972, a mere two years from date of contract. (The word *Anik*, which means "little brother" in Inuit, was the winning name in a publicly conducted contest by the federal DOC. The winner received an expenses-paid trip to the launch of Anik A1.) Delivery of the second and third flight models followed at four-month intervals. This timely delivery facilitated Telesat to maintain its initial schedule objective, even though signing the spacecraft contract had been delayed somewhat by all of the studies, study contracts, and evaluations in 1968 and 1969.

Anik A1 (Fig. 4.2) was launched on 9 November 1972, a historic milestone for Telesat, and it was the first geostationary domestic communications satellite in the world. Aniks A2 and A3 were launched in April 1973 and May 1975, respectively, to complete Telesat's first-generation fleet (Table 4.1). The Telesat satellite communications service was highly successful right from the beginning with no disasters or serious breaks in customer service. It was considered a great achievement in that this service was the application of a new technology with no previous experience to fall back on. The satellites functioned well throughout their design service life and then some. There

Fig. 4.2   Anik A1, the first geostationary *domestic* communications satellite in the world.

TABLE 4.1 MAIN CHARACTERISTICS OF TELESAT SATELLITES[a]

Parameter	Anik A	Anik B	Anik C	Anik D	Anik E	Anik F1	Anik F2	Anik F1R	Anik F3	Nimiq 1 & 2	MSAT M1
Prime contractor	Hughes	RCA	Hughes	Spar	Spar	Boeing	Boeing	EADS Astrium	EADS Astrium	LMC	Hughes Spar
Satellite bus type	HS333	RCA 3000	HS376	HS376	GE 5000	702	702	Eurostar E3000	Eurostar E3000	A2100 AX	HS601
Number of satellites	3	1	3	2	2	1	1	1	1	2	1[b]
Launch vehicle	Delta	Delta	STS	Delta/STS	Ariane 42P	Ariane 44L	Ariane 5G +	Proton B/M	Proton B/M	Proton D1e,B/M[c]	Ariane 42P
Launch year	1972–75	1978	1982–85	1982–84	1991	2000	2004	2005	2007	1999–02	1996
Launch weight, kg	560	920	1140	1217	2930	4711	5950	4470	4640	3590	2830
Array power, W, EOL SS[d]	235	620	800	800	3900	15,200	15,500	9900	10,500	8800	3200
Contractual service life, yr	7	7	10	10	12	15	15	15	15	15	10
Total number of channels	12	18	16	24	40	84	94	58	58	32	6F, 5R[e]
Number of active channel amplifiers per frequency band	12C	12C/ 6Ku	16Ku	24C	24C/ 16Ku	36C/ 48Ku[f]	24C/ 32Ku/ 38Ka	24C/ 32Ku/ 2L	24C/ 32Ku/ 2Ka	1.32Ku[g]/ 2.32Ku/ 2Ka	1Ku/ 16L
HPA power per channel, per frequency band, W	5	10/20	15	11	12/50	40/115	30/127/ 90	40/120/ 40	35/130/ 100	1.120[g]/ 2.120/92	120/35
Total RF power, W	60	240	240	264	1088	6900	6700	4840	5200	3840	600

[a]Data courtesy of Telesat Canada.
[b]One satellite was supplied to American Mobile Satellite Consortium (AMSC) and one to Telesat Mobile, Inc. (TMI), a wholly owned subsidiary of Telesat Canada.
[c]Proton D-1-e for Nimiq 1 and Proton Breeze M for Nimiq 2.
[d]End of service life at summer solstice.
[e]Six forward and five return channels.
[f]Geographical diversity allowed for additional channels per band in the orbital slot.
[g]The 1. and 2. indicate Nimiq 1 and 2, respectively.

was the expected degradation in their power systems for which allowances had been made. Travelling wave tube amplifiers (TWTAs) failed at a rate greater than expected at first, but adjustments in their operating levels reduced the failure rate considerably. The spare TWTAs pretty much filled the gaps and services were reconfigured as needed without incident.

In 1975, Telesat awarded a prime contract to RCA Astro Electronics Division of Hightstown, New Jersey, for the delivery of a second-generation satellite, Anik B (Fig. 4.3). RCA, Montreal was the main subcontractor. It was launched in December 1978, the first dual-band communications satellite in the world (Table 4.1). Its 12 C-band channels were a replacement for failed channels in the aging Aniks A1 and A2. The six Ku-band channels were for the federal DOC.

Following ISIS I and II and prior to Anik B, the DOC changed from upper atmosphere research satellites to geostationary communications satellite applications studies. Various government institutions were interested in experimenting with the application of satellite communications for isolated and remote communities in health care, education, community development,

Fig. 4.3   Anik B, the Telesat second-generation satellite.

and other services; these were only ideas at that time. For these experimental services the DOC, in partnership with NASA, in January 1976 launched Hermes, the high-powered direct broadcast Ku-band Communications Technology Satellite (CTS). To continue with these studies beyond the life of Hermes, the DOC contracted with Telesat for Ku-band capacity to be added to Anik B. This provided Telesat with an excellent opportunity to acquire Ku-band experience.

The next two generations of Telesat satellites were single band satellites in C- and Ku-band (Table 4.1). Both series were based on the Hughes spin-stabilized HS376 bus. Contracts were signed for three Anik Cs in early 1978 with the Hughes Aircraft Company and for two Anik Ds with Spar Aerospace, Ltd., in 1979. Each contract had approximately 50% Canadian content, which was very substantial, considering that there was very little component manufacturing in Canada. But it was sufficient to finally satisfy the DOC regarding the Canadian prime contractor and Canadian content policies of the federal government. Hughes delivered the two Anik D buses to Spar as its subcontractor, with satellite integration and test carried out at the David Florida Laboratories of the DOC, under guidance and training to Spar by Hughes.

Telesat had contracted with NASA to launch all five satellites on the space transportation system (STS), otherwise generally known as the space shuttle. But the shuttle production and test schedule was very tight, and Telesat required the Anik Cs and Ds to also be compatible with the Delta launcher just in case there was a schedule problem. Hence, the diameters of the Anik Cs and Ds were approximately the same as those of the Anik As, as constrained by the diameter of the Delta launch vehicle fairing. Hughes met the array power requirements by an extendable cylindrical solar-array section (Table 4.1). It turned out that the shuttle was delayed, and the first Anik D was launched on a Delta on 26 August 1982. On 11 November 1982, Anik C3 was launched from the *Columbia*, on the first commercial flight of the shuttle.

The Anik E series consisted of two dual-band three-axis-stabilized satellites with Spar as the prime contractor and GE Astro as the bus supplier (formerly RCA Astro Electronics Division, prime contractor for Anik B). The Anik E GE5000 bus was a derivative of the Anik B RCA3000 bus (Table 4.1). With the number of channels of a combined Anik C and D and over four times the array power of a C or D, a three-axis bus was the only viable design. The launch weight of an Anik E was more than twice that of an Anik C or D. They were launched on Ariane IVP launch vehicles. With the Ku-band experience gained in the Anik B experimental program of the federal government, Telesat engineering stepped into this service band with added confidence.

In the F series [2], all four satellites are different. Boeing supplied Aniks F1 and F2 from the satellite design and production facilities once owned by the Hughes Aircraft Company. These are very large satellites with 84 and

94 channels on Anik F1 and F2, respectively. The F2, with a tip-to-tip solar-array span of 48 meters is the highest-powered satellite in the Telesat fleet (Table 4.1). The massive 94-channel satellite was launched on an Ariane VG+ in July 2004. With Anik F2, Telesat introduced the first commercialized broadband service in the Ka-band via satellite, providing two-way high-speed data services for business and consumers in Canada and mainland United States. The next Telesat satellite was the F1R, supplied by EADS Astrium, on a Eurostar E3000 bus. With two L-band channels (Table 4.1) Telesat joined the North American GPS. The procurement of an Anik F3 from EADS Astrium followed soon thereafter. Launched in April 2007, on a Proton launcher, F3 meets Telesat's requirements for its expanding business in the C- and Ku-bands over North America. It also provides additional flexibility in customer service in general and additional backup capacity.

In parallel with the Anik fixed satellite services (FSS), Telesat embarked on direct broadcast services (DBS). Planning and implementation for significant new business such as this take time. In-house planning was followed by applications to government organizations, sometimes processes of long duration. Telesat applied to the Federal Communications Commission (FCC) in the United States for approval for U.S. customers to access the Telesat DBS services. At the same time it applied to the Canadian federal government for approval of its plans. After several iterations both initiatives were approved, and Telesat contracted with Lockheed Martin Corporation for the delivery of Nimiq 1. (*Nimiq* is an Inuit word for any object or force that unites things or binds things together.)

Traditionally, Telesat took ownership and responsibility for its satellites at liftoff. For Nimiq, Telesat, for the first time, took ownership of the satellite in orbit and fully tested. Nevertheless, the company monitored the construction of the satellite and launch services in its traditional manner. Under a separate contract a second direct broadcast satellite, Nimiq 2, was procured from Lockheed Martin. Launched on a Russian Proton (Table 4.1) from the Baikonur Cosmodrome in the Republic of Kazakhstan, the two DBS satellites joined the Telesat fleet in 1999 and 2002, respectively. Nimiq 3 was procured on orbit from another satellite operator and is colocated with Nimiq 2.

Satellite mobile services have been introduced in Canada as well, but the development was an exceedingly long and complex process. Experiments by the Canadian Department of National Defence (DND) began in trials with the United States as early as 1967 using UHF, whereas the launches of two MSATs did not occur until 1995 and 1996 (Table 4.1).

A Canadian military satellite mobile system was never implemented, and the DND MUSAT (Mobile UHF Satellite) study was taken over by the DOC as a civilian MSAT. Much of the delay in the program was the need to resolve spectrum use. In 1983 the DOC contracted with Telesat for a commercial viability study that Telesat delivered in 1984. The MSAT system would serve vehicles on land, on the seas, and in the air. It would also complement the fixed satellite system in rural and remote areas. Telesat recommended a joint program with a U.S. counterpart to best offset nonrecurring costs, to provide mutual backup between a Canadian and a U.S. satellite, and to provide seamless service for north–south vehicle traffic.

Telesat proposed the addition of L-band in the 1600-MHz spectrum in addition to the use of UHF (800 MHz). In the United States the allocation of adequate bandwidth in its crowded UHF spectrum proved unrealizable. In 1987 a World Administrative Radio Conference (WARC), which meets every four years in Geneva, allocated two segments of 4 MHz, each exclusively for land mobile, and two segments of 3 MHz, each for shared land and marine use. These allocations were key for the two countries to agree on a certain portion of the L-band. The back-haul to the gateway stations was going to be in the Ku-band.

In the United States the FCC approved certain license applicants, and then they were directed to form a single consortium; as a result, the American Mobile Satellite Consortium (AMSC) emerged as the U.S. entity. The Canadian federal government designated Telesat as the Canadian organization to implement MSAT service for Canada, and Telesat then formed a subsidiary for its mobile satellite business, Telesat Mobile Inc. (TMI). Procurement contracts were signed with Hughes Aircraft Company and Spar Aerospace, Ltd., for the delivery of two satellites, one each to TMI and AMSC. The AMSC and TMI satellites were launched in 1995 and 1996, respectively.

As these are long and drawn-out processes, the allocation of, and agreement on, mobile frequencies to be the same in both countries took years. It was necessary to take whatever time was needed to arrive at a satisfactory user-friendly system. Thus, North American mobile traffic is now very well served on land, on the seas, and in the air.

In the years following the start of service, the U.S. and Canadian systems were merged technically under a new operator, Mobile Satellite Ventures (MSV). MSV has been very innovative in the development of new types of service and is moving forward with the procurement of replacement satellites.

With regard to trends in satellite size, in fixed satellite services, and after the Anik Cs and Ds, Telesat moved towards larger and larger payloads with multifrequency band systems and with higher capacities and RF power. For example, Telesat's Anik F2 is the largest and most powerful of the Telesat fleet; it has 94 channels and provides service in three frequency bands.

The capital cost per channel-year decreases with increase in satellite size and capacity. This is so because the cost of a single satellite bus and the launch does not increase in proportion to the number of channels. Another advantage of a multiband satellite has been that a single orbital position can be used for more than one frequency band. However, with the precision of orbit control nowadays there is no real difficulty or risk in colocating two or more satellites within the longitudinal boundaries for each frequency band in a single orbital longitudinal location.

The single major disadvantage of a large multiband satellite is the impact of a catastrophic failure during launch or in orbit. The satellite owner usually insures against launch failure. But the backup system that would be required in order to provide continuity of service would be more difficult for a multiband than for a single-band satellite, and the cost would escalate with increased size of the multiband satellite. It also depends on the corporate size of an operator and its own fleet size for self-sustained back up, or temporary-leasing availability from neighbors if such is technically feasible.

Magnetic storms are hazards that vary in intensity with the 11-year solar activity cycle. A spacecraft accumulates a static charge from the electron flux from the sun, especially during intense solar storms, and delicate digital circuitry can be destroyed as a result of arcing. Spacecraft designers have developed an art of grounding the electronics within the spacecraft and covering the outside with a protective conductive material. The aim is to maintain the structure in a neutral electronic state or to at least ensure that static discharge is channeled along harmless paths. With added knowledge and experience damage from magnetic storms has declined, but the threat is always there, especially for a new, yet unproven bus design.

In planning replacement capacity and growth, it is unlikely that the number of channels in all the bands will be required at the same rate. Thus single-band satellites allow greater flexibility for future planning and replacement in the various frequency bands than do multiband spacecraft.

With the advantages and disadvantages of larger multiband vs smaller single-band satellites, it is difficult to guess what the trend will be. From about 2000 onward, there are launch vehicles large enough to boost any size that satellite communications users care to consider, so that is not a problem. There are several launch agencies throughout the world, and it is a matter of choice based on perceived reliability, type of service rendered, and cost. However, in the early days of the Alouette, ISIS, and first Aniks, NASA was the only game in town for nonmilitary launches in the western world. The work horse first-stage rocket was the Thor, designed and built by the Douglas Aircraft company, originally for intermediate-range ballistic missile requirements. Alouette 1, launched on 28 September 1962 on a Thor-Agena B, was the first nonmilitary launch from Vandenburg Air Force Base, California. Beginning in the early 1960s, NASA's main launch vehicle was the Thor-Delta, eventually designated simply as the Delta. The Delta evolved over time into a higher-lift vehicle with the addition of, for example, solid rocket strap-on boosters, a long-tank second stage, and a larger diameter fairing. Canadian launch services, for Alouette II through Anik D1, were provided by NASA on the Delta.

In terms of price setting, NASA operated pretty much on a nonprofit basis of meeting its operating costs. Additional costs as a result of launch failures were treated on a cost-reimbursable basis. The extra billing to NASA by its contractors was passed on to NASA's commercial customers by a backbilling process, spread over a number of years.

One of NASA's major decisions for its STS program was that the shuttle would be the only vehicle for commercial launches and all of its expendable vehicle programs would be phased out. The first four flights were dedicated to tests, and the fifth was the first commercial flight, initially scheduled for late 1981. With four shuttles available, flight planners expected to be able to schedule launches monthly to bimonthly. Telesat was taking delivery of three Anik Cs and two Anik Ds during the period of the shuttle commercial startup, and it was an ideal opportunity for Telesat to sign up, which it did, for all five satellites.

If the shuttle schedule was delayed, NASA agreed to provide Delta launches for customers in desperate need of maintaining continuity of their commercial service. A delay in the shuttle schedule did occur and Anik D1, due for launch on the first commercial flight of the shuttle, went up on a Delta in August 1982. On 11 November Anik C3 was launched on STS 5, a truly dramatic and highly successful milestone for both NASA and Telesat.

Not only did the timing of the shuttle schedule fit the Telesat launch requirements, but also the launch price was extraordinarily attractive. As an incentive to get customers to embrace the shuttle program, NASA offered the first number of launches at a price of US $10M each. This was for a launch to a nominal shuttle orbit, and the customer was required to pay for a perigee stage to boost the satellite into its traditional transfer orbit. For a Telesat satellite the cost of the stage was less than US $10M. Hence, at less than US $20M per launch it was a real bargain, especially with respect to Delta launches, which had been approximately twice as much.

There was much to be analyzed and a great deal of preparation for the first satellites on a shuttle launch. Satellite and shuttle avionics engineers interfaced for many months before the first commercial flight. A process for safely transferring satellites into the shuttle cargo bay was worked out. The satellites were delivered to the launch pad several days before a launch and stowed in an environmentally protected facility mounted on a rotatable arm. When required, the satellite in its facility was rotated into the cargo bay and transferred, all the while providing it with environmental protection.

There were many new engineering problems that needed evaluation and were different from those of expendable launch vehicles. There were concerns that there might be considerable pointing error when a satellite was released. There was every confidence that the shuttle navigation system was sufficiently accurate, but it was difficult to estimate the difference between the navigation system readings and the actual release coordinates. The body of the shuttle might be flexed somewhat by the extreme solar heat on the sunny side vs the extremely cold dark side, and the resultant angular distortion of the shuttle structure was difficult to estimate. Tenths of degrees were important.

There was much concern and controversy about contamination of the shuttle by the plume from the satellite perigee stage. To provide sufficient time for the shuttle and satellite to drift apart, it was decided that the satellite would be released at the equator crossing preceding the one at which the perigee stage would be fired. Immediately after release of the satellite, the shuttle would conduct a small evasive maneuver that would alter its orbit by the required amount. NASA assigned astronauts as dedicated payload specialists to work and train with the satellite customer engineers for many months to prepare all of the details of releasing the satellites. On 12 November the shuttle crew released Anik C3 precisely on target (Fig. 4.4) and watched, as it faded into the void. All four Telesat shuttle launches were very successful.

Following the Challenger disaster on 28 January 1986, NASA changed its policy to no more commercial launches on the shuttle. Telesat was fortunate that it was just at the beginning of its Anik E construction program and had plenty of time to procure launch services, but now at a much higher cost. The two Anik Es were launched in 1991 on Ariane IVs. These were followed by the Nimiq and Anik F series, launched on Ariane V and the Russian Proton, all successful (Table 4.1).

Fig. 4.4   The release of Anik C3 from STS 5 Columbia. (See also color figure section at the back of the book.)

Telesat was indeed fortunate to have had 18 successful launches and gotten by unscathed. In addition, with the launch of Radarsat 1, the federal government research satellites added up to six successes. This gave Canada a total of 24 successful launches from a number of different launch agencies and on a number of different vehicles.

Launches are the events when years of design, development, and construction converge for the launch supplier, the satellite contractor, and the customer. Launches are very high profile, unlike anything in any other technology that the public enjoys from the sidelines, especially as spectators at the events, or by just watching them on television. The press is quick to emerge if something goes amiss, but to a lesser extent if all goes well. Hence, depending on the outcome, one can be either a hero or a bum.

SURPRISES, FAILURES, AND SOLUTIONS

Before the space age mathematical physicists understood that a spinning rigid body is dynamically stable if it is spinning about an axis of either maximum or minimum moment of inertia. Thus, if a long slender satellite is spin stabilized about its longitudinal axis by its rocket stage and then released, it was understood that it would continue to spin about that axis. Moreover, with the conservation of angular momentum, the spin rate would remain constant, and the direction of the spin axis would remain fixed in inertial space.

The U.S. first satellite, Explorer 1, launched on 31 January 1958, was pencil shaped and was spin stabilized about its longitudinal axis. So it was a great surprise when, shortly after its release, the satellite was spinning about a transverse axis. What was discovered in this physical experiment was that a spinning body, left unattended, would end up spinning about its principal axis of maximum moment of inertia. Why was this fundamental property of spin dynamics not discovered earlier? In any Earth-based experiment the spinning body is always anchored by gravity in some way. Hence it is not free to reveal all of its properties of spin dynamics. It required the weightless medium of space to be able to observe the properties of spinning bodies completely. In fairly short order dynamicists worked out the details and showed that no matter what spin axis is chosen initially, a freely spinning body, when dissipating energy internally, will end up spinning about its axis of minimum (spin) energy. Inherently, this also corresponds to the axis of maximum moment of inertia. Real satellites (rather than the rigid bodies that were the subject of much theoretical study) all experience internal energy dissipation in some manner. Flexible appendages and onboard fuel slosh are but two examples of internal energy dissipation that will cause a freely spinning satellite to end up spinning about its axis of maximum moment if inertia. In a momentum-biased three-axis-stabilized satellite this is prevented by a high-spin-rate momentum wheel.

How long it takes a spinning body to change its body axis of rotation from a minimum to a maximum moment of inertia axis depends on the relative values of its moments of inertia and the relative amount of energy dissipation caused by the spin dynamics. In the case of Explorer 1, it happened in a few tens of minutes during the first orbit. The whipping of its turnstile antenna elements caused the energy dissipation. Fortunately, this discovery was made at the beginning of satellite technology with no significant adverse consequences.

It is often the case that when a new bus is launched there will be first-time unknowns that will arise. In the Anik A1 launch, the first time for an HS333 bus, the spin-axis attitude solutions would not converge during transfer orbit to establish the correct attitude for apogee motor firing. The points were scattered, and there was no trend. Time was marching on in the mission, and the apogee motor firing was being delayed. With the delay the propellant in the motor was getting too cool, inviting disaster. It is at times like this that the mission director needs his launch pills! Then it was realized that a fundamental principle of spin dynamics was being overlooked. If the despin motor was turned off and the communications antenna was allowed to spin with the body while taking attitude sensor readings, the longitudinal principal axis of inertia was not the geometric axis, and the Earth and sun sensors data were incorrect. The antenna despin motor was turned on periodically in transfer orbit, and each time the motor was turned off the antenna would stop relative to the body, in a different angular position, thus creating a new body spin axis. Once the inertial effects of the antenna were removed by despinning it and the body and geometric axes were coincident, spacecraft attitude was readily determined, and the mission was on its way. It was also apparent that the inertial effects of the antenna had to be removed during apogee motor firing, but that was going to be done in any case to avoid a possible pitting of the antenna platform bearing. The next two Anik A launches were pretty much ho-hum.

In the Anik B program, Telesat rejected an apogee motor that, according to an x-ray inspection, had a flawed liner. This motor failed catastrophically when it was used later on another satellite. During the Anik B launch mission, one of the two solar arrays would not fully extend. The satellite would not be at full power, and that would never do. After much discussion and analysis it was decided that the satellite could be safely slued 180 degrees about its north–south axis without loosing control, in order to expose the hinges to the heat of the sun and hopefully overcome any sluggish viscosity. The problem was solved early in the evening, and all the visiting firemen and armchair quarterbacks went home. However, there was a gremlin waiting to emerge at about 0300 hours, again to do with spin dynamics. During the period that the solar array was not fully extended, the solar radiation pressure was asymmetrical, and this being the daytime hours the result was an undetectable yaw rotation of the north–south axis. Six hours later at approximately 0300 hours this had transformed into a significant roll angle, and the Earth sensors were about to lose Earth lock. This kind of problem the mission team could do without. The roll was gradually corrected, and normality was restored.

The Anik C and D HS376 satellite bus (Table 4.1) was designed to rotate about its minimum moment of inertia axis. It had a nutation damper to constrain nutation to prevent the satellite from transitioning to a rotation about a transverse body axis. When an Anik C or D was released from a shuttle, the coast phase for the satellite was 45 minutes before ignition of its perigee stage. During that period, the configuration of satellite and perigee stage was spin stabilized about a minimum moment of inertia axis and an active nutation damper maintained stability.

In geostationary orbit, in its operational configuration, stability was actively maintained by coupling the spinning and nonspinning sections of the satellite via a small product of inertia component on the nonspinning antenna platform. The antenna platform despin motor provided the damping torques. If the antenna lost Earth lock and began to spin up, the spin axis would begin to nutate, and the active nutation control system would switch in to prevent a catastrophic transition to a spin about a transverse body axis.

The three-axis-stabilized satellites are generally momentum biased for stability, meaning they have a momentum wheel with its spin axis aligned north–south. After launch the satellite is normally drifted to its orbital longitude with its angular momentum either in the plane of the transfer or the geostationary orbit. On arrival at its destination, the angular momentum is reoriented to a north–south attitude. It is here that the principle of the conservation of angular momentum can be applied in an elegant way. By turning on the momentum wheel, its angular momentum then exerts torques at the bearings thus causing the body axis of rotation to change and to align itself with the momentum wheel axis. Because no external torques are applied throughout this maneuver, angular momentum is conserved in the north–south direction, and the momentum wheel axis will necessarily align itself in the north–south direction. But before the wheel is turned on, it is inertially inert, and its axis is at right angles to the spin axis of the body. To satisfy the inertial properties of wheel and satellite body, the spin axis of the body must change by 90 degrees to align itself with the wheel axis. To achieve this, no complex maneuvers using thrusters to apply external torques are necessary; it is necessary only to turn on the wheel and wait a few minutes for the transition. This became to be known as the dual-spin turn.

In the launch mission of Anik E2 in April 1991, the antennas failed to deploy on command. In the headlines the media was quick to declare Telesat's newly launched satellite as space junk. One of the antennas self-deployed in a couple of days, but the second one remained stowed. Various body spin rates were used to carefully apply inertial forces to try to free the stowed reflector without causing any damage, but to no avail. To configure the body to its operational attitude, the wheel was turned on, but with the solar panels still stowed this configuration had to be abandoned for charging the batteries. This required that the body spin axis be returned to the launch configuration. Between the launch on 4 April and when the antenna was freed on 3 July the dual-spin turn and reverse dual-spin turn were used several times to configure the satellite spin from one body spin axis to the other [3]. This versatility was required in order to be able to work on the problem, and when the satellite was saved, the launch insurance companies were ecstatic. Telesat was awarded the space mission recovery prize on 25 June 1992, by La Réunion Spatiale, which specializes in the underwriting of space risks. Telesat won the prize for the rescue of Anik E2 in competition with two other contestants, the Olympus and Hipparcos programs. This was the inaugural year for the prize. The prize, a sculpture of swans in flight, was a specially commissioned crystal sculpture by Marie Claude Lalique, a distinguished French artist of the Lalique family, known for its world-famous Lalique crystal. It had no monetary value, but it was quite prestigious.

Another significant problem in the history of Anik E2 was the failure of its autonomous control system with the loss of the primary and backup momentum wheels. But the satellite was kept in service throughout its contractual service life with the use of the magnetic torque and thruster subsystems. Telesat developed a novel ground loop attitude control system (GLACS) using an RF sensor system for determining spacecraft pointing. This was used as input data for a computerized automatic control loop with the magnetic and thruster subsystems. At first it was thought that the use of fuel would be excessive, but in fact the usage was such that the system operated successfully for over 10 years, and the satellite service life was over 14 years, well in excess of its contractual service life of 12 years.

These were some of the interesting and at times hair-raising experiences at Telesat, typical of space work. There were many more. Space is high-profile technology unlike any other. Many countries proudly display their successes, especially when they are first in whatever it may be, and so they should. Thus, Canadians are proud of their achievements at being first in the world with domestic satellite communications, especially the past and present employees of Telesat.

SATELLITE CONTROL GROUND SYSTEMS AND GEOSTATIONARY ORBIT CONTROL

In September 1969, when the federal government inaugurated Telesat, all technical and commercial program studies that had hitherto been performed by the government were transferred to Telesat for its discretionary use. The company became responsible and accountable for its own destiny.

The greatest challenge facing Telesat engineering at that time was the development of a geostationary orbit control system that would provide for a domestic satellite communications system with fixed Earth stations. Of all of the various types of engineering design that were required, this was by far the technology about which the least was known and that was the most undeveloped.

The state of the art in 1970 was that established by Intelsat, the only commercial geostationary communications satellite operator in the world. The Earth stations for the Intelsat system had autotrack, and therefore it was necessary to maintain the satellites only approximately on-station. Inclination was biased at launch and was allowed to drift naturally through zero while longitude was loosely maintained within a few degrees of its nominal value. Thus a system to control satellites to tight tolerances in geostationary positions for fixed Earth stations was not yet required and did not yet exist.

Theoretical analyses of the rate of change in inclination of a geostationary satellite due to the gravitational effects of the sun and moon and the longitudinal drift and librations caused by the ellipticity of the Earth had already been published in the early 1960s [4]. On the basis of these mathematical analyses, it was estimated that a domestic geostationary satellite could be designed to be controllable to within 0.1 degrees in both latitude and longitude. This was defined as the Telesat initial objective.

Another major challenge was the development of a data processing system for the launch phase and geostationary operations. Telesat consulted various sources and organizations about the type and size of system that would be required. Eventually it was determined that a Univac 1108 would meet the requirements. A scientific machine, it operated with a 36-bit word and had a 65 kB memory. However, Telesat did not wish to procure this large, expensive computer that it would require for launch missions and otherwise for only a few hours per month for the orbit control calculations. There would be little other use for it by the small group of engineers in the company. Thus, it was decided to develop a remote job entry (RJE) system [5] for flight dynamics applications.

Data for the RJE system would be provided by an in-house system of minicomputers, designed to function in real time for encoding the uplink command generators and to handle tracking and telemetry data on the downlink. The minicomputers in the 1970s had memory capacities from 4 to 16 k. A data storage computer (DSC) transferred tracking data to the orbit determination programs that were resident in a Univac 1108, owned and operated by Computel Corporation, an Ottawa-based computer utility company. Results were transferred back and printed locally at the Computer Center at Telesat. This RJE system was the first of its kind in the data processing industry whereby large quantities of scientific data were transferred directly from a minicomputer to a remotely located large data processor, with a high volume of jobs and with rapid job turnaround.

State-of-the-art computer input/output was based on punch-card technology, and jobs were executed in batch mode. An innovation to overcome the slow, cumbersome, process of punching cards was the development of a process of punch-card images for the attitude and tracking data and storing that in the DSC. With only a few, physically punched control cards the DSC transferred the punch-card-image data to the Univac. During launch missions, jobs needed to be entered in rapid succession, and punching cards would have been too laborious and slow; this would have compromised timely solutions. Development of the system progressed satisfactorily, but there was concern by flight dynamics specialists that there might be an inherent problem in using an RJE system for flight dynamics. An RJE system would function only in batch-job mode, and it was felt that it would not be feasible to filter or cull noisy data in the time-tagged tracking data. The mathematical solution was based on an iterative process, and with noisy data it might not converge. The problem was resolved with a modification to the orbit and attitude determination software on the Univac 1108 to facilitate the examination and rejection of bad input data and to resubmit a follow-up job even before the output of the original job arrived at Telesat. These kinds of quick turnaround were required particularly in launch missions, especially in preparation for upcoming maneuvers.

As the capacity and speed of computers evolved, by 1975 minicomputers had become 32-kB "powerhouses." The Univac 1108 and the RJE system were no longer required, and Telesat transferred its flight dynamics to its own in-house minicomputer system.

As various models of computers became obsolescent, Telesat systems software analysts expanded the real time and data processing systems throughout the Anik A, B, C, D, and MSAT series of satellites, while sustaining modernization. That experience in systems integration has served the company extremely well, especially for service restoration in response to in-orbit equipment failures. Nowadays (2008) Telesat uses commercially available core RTS and integrates various types of buses for the operational control of a variety of satellites; some of which are not even their own.

An outstanding feature of the Telesat-developed real time system was that, as the various satellite bus designs were procured and launched, their monitor and control functions were all integrated into a common system. A satellite controller, from a single console in the Telesat Satellite Control Centre, can access a number of satellites of different bus designs. These are all operational cost-saving features.

In early 1970, along with the need for a computer system, Telesat was facing, what seemed, a formidable task. It was going to need computerized flight dynamics programs, developed and fully tested, for a launch in about two-and-a-half years. Telesat had no significant experience and nowhere to procure flight dynamics programs. During 1970, Telesat was still working with RCA, Montreal, in the program definition phase; RCA had no experience whatsoever in this field, except that its potential bus supplier had some. A qualified prime contractor for geostationary communications satellites would deliver, along with the satellite, the data processing system and the flight dynamics programs. The contractor would also have the capability to deliver the satellite on-station and to set up the operating system for its customer.

In early 1970 Telesat approached NASA and COMSAT to procure or lease their flight dynamics programs, but neither organization was in a commercial position for that. They were all too busy developing systems for their own requirements. Thus, by the fall of 1970 Telesat found itself out on the limb with no prospects whatsoever except to develop everything from scratch. That would, indeed, have been risky.

However, when Telesat signed the procurement contract for three satellites with Hughes Aircraft, its experience and know-how in geostationary launch and operations became available to Telesat. After all, Hughes had pioneered the geostationary satellite concept. Also, it was in Hughes' interest for Telesat to succeed.

Telesat signed a consulting contract with Hughes for this technology. Although Telesat obtained some low-level software programs from Hughes that had to be modified to work on the Univac, it was the experience and know-how of its senior scientists that made the difference. That assistance was crucial in taking Telesat over a formidable hurdle in the available time frame. With Hughes experience backing the design and development of flight dynamics software, the programs required for a launch were successfully developed and tested, in a timely manner, in only 18 months.

However, the programs required for tight orbit control for a domestic system with hundreds of fixed Earth terminals were a new challenge, yet to be completed. In addition, a tighter objective was superimposed. As Earth station design studies progressed, it became evident that it would be highly desirable to control the satellites to within 0.05 degrees in latitude and longitude. If this specification could be achieved, then large-diameter network-television-receive stations would not require autotrack systems. This would result in considerable savings in capital and maintenance cost. One of the criteria of flight dynamics was that it would be necessary to determine the satellite position by an order of magnitude better than the control limits, that is, to with 0.005 degrees. This was necessary in order for station keeping to become a routine operation, guaranteed to consistently meet the requirements.

Immediately following the launch of Anik A1, analyses were begun to examine the concept of orbit determination from range data only, with measurements from two stations separated by several thousand kilometers. To provide accurate range data, the two stations were required to be accurately surveyed. The original flight dynamics software was based on an analytic method and the Earth's geopotential mathematical model was limited to the J_4 term, not accurate enough for these requirements. It was subject to failure when the orbit had either zero eccentricity or zero inclination, precisely the targets of geostationary station keeping. It was decided to replace this analytic method with a numerical integrator and to add the mathematical terms required to represent the perturbing gravitational forces that affect the geostationary satellites for these tight orbit positions. These forces included the geopotential up to order eight, solar and lunar gravity, and solar radiation pressure.

The orbit control objectives could not have been met with state-of-the-art tracking antennas providing azimuth, elevation, and range from each antenna. Variations in bias and noise in the measurements caused by wind and thermal variations would have been too high. However, range measurement biases are small, and more importantly they remain constant. They are inherently introduced by computers, and electronics and adjustments can be made to compensate for them. Range measurement is several orders of magnitude more precise than angle measurement. The cost of a range measurement system is very small in comparison to precision tracking antennas, even if antennas could serve the purpose. Dedicated Earth stations are not required as the ranging system can be added to a communications Earth terminal. Also, the maintenance and operation of a range-only system is trivial in comparison to a tracking antenna system.

The end result was that, even at the beginning with the first Anik on orbit, uncertainties in latitude and longitude were reduced to the order of 0.0002 degrees, thus exceeding the design objective of 0.005 by a considerable margin.

With that kind of accuracy achieved in orbit determinations and the ability to control the satellite motions with high precision, two or more of them can be safely collocated in a single orbital location. For instance, in order to switch traffic, without interruption, from one satellite to another they need to be collocated within the beam widths of the Earth terminals. Or, it may be desirable to collocate two partially functioning satellites as a complementary pair so that the orbital location is more fully utilized and service can be provided efficiently. Another benefit from tight orbital control is that satellite spacing along the orbital arc is much smaller than was first conceived in the early days.

Another outstanding development in flight dynamics was the introduction of the Kalman filter, a very powerful algorithm. Following some Kalman filter experiments on the Univac 1108 in 1974, it was decided to attempt to develop a real-time Kalman filter on an available in-house minicomputer. Success would indicate that it would be feasible to port the entire flight dynamics system onto minicomputers. This had not yet been done by anyone, anywhere. Not only was feasibility proven, but use of the Kalman filter demonstrated that orbit and attitude determinations could be achieved without manual intervention, and that solutions were much more precise than with the traditional weighted-least-squares method. The full system was developed and became operational by mid-1976. Some of the most useful features of the Kalman filter are that orbital elements and attitude solutions are always available to maneuver planners and that it can take account of planned maneuvers. A priori solutions of satellite attitude and orbital elements are continuously updated with incoming data. In 2007, with 14 satellites being controlled by Telesat, the Kalman filter saves about seven to nine person-years of effort. The superiority of the Telesat flight dynamics system is attributed in no small measure to Telesat's unique application of the Kalman filter. This feature alone has made the system a highly marketable product, and the flight dynamics system is a lead item in Telesat's list of goods and services in its consulting services around the world.

EARTH STATIONS AND SATELLITE SERVICES

Another major engineering challenge at Telesat was the design and construction of its Earth station system. The initial baseline system of Earth stations was installed throughout Canada (Fig. 4.5) to meet the requirements of the federal government broadcast policy. The objective was to provide telecommunications services to all parts of Canada in both English and French.

Two 30-meter heavy route (HR) antennas were installed, one on Vancouver Island and the other at Allan Park, Ontario, for east–west heavy-route traffic, which was growing at about 20% per year. These also served as the initial C-band Earth terminals for Vancouver and Toronto, respectively. Six 10-meter network television (NTV) transmit-receive stations were distributed across Canada in other large cities. Twenty-four remote television (RTV) receive-only

Fig. 4.5   The Telesat baseline Earth station system.

stations were distributed throughout the Yukon, North West Territories, Nunavut, and the northern parts of some provinces. Two northern telecommunications (NTC) thick route stations, for telephone and two-way message, were built in Resolute and Iqaluit, station numbers 50 and 51 respectively (Fig. 4.5). Two NTC thin route stations, with the added capacity to receive radio programming, were installed in Pangnirtung and Igloolik, station numbers 100 and 101, respectively. The thick route services had between 12 and 36 full multiplex circuits, and the thin route stations began with two circuits. This constituted 37 Earth stations in the baseline C-band system. A tracking, telemetry, and command station, also built at Allan Park, was required for the launch missions and for initial in-orbit operations.

There was no shortage of prime contractors for the Earth stations to satisfy the federal government policies on Canadian content. Considerable C-band experience had been developed on the Trans-Canada microwave relay system and on Earth stations built for the Canadian Overseas Telecommunications Corporation (COTC). In any case Earth stations technology was not as high profile and prestigious as the emerging space technology, and hence it was less politically sensitive. Major suppliers of the baseline system were RCA Limited for the HR, NTV, and NTC stations; Raytheon Canada Limited for the 24 RTV stations; and Philco-Ford of Canada Limited for an 11-meter tracking, telemetry, and command station.

Installation of Earth stations in the Canadian High Arctic was and still is quite different from the standard process with concrete foundations in the southern latitudes. Although some materials for the Earth stations were flown in, most sites in the High Arctic could be reached by ship during the summer, and there was always a rush to get the material to port for the first ship out in the spring. In most instances concrete was not really required, and there was a much easier way. Significantly, it would have been most inappropriate to pollute the environment and at the same time find the civil works gradually sinking into the tundra permafrost. A typical installation is the Baker Lake station (Fig. 4.6) at 64.3° N. The station is elevated on boulders so that the underlying tundra remains frozen. It is also well anchored to withstand the Arctic high winds. With real estate not a problem, installers typically selected high ground, a hill, or a ridge to establish a good line of sight to the Canadian orbital arc. In this part of the world, maintenance personnel need to maintain a wary eye for surprise visitors (Fig. 4.7).

The historic telephone call by the Minister of Communications on 11 January 1973 marked the inauguration of satellite service by Telesat Canada. Telephone traffic for Bell Canada, Canadian National/Canadian Pacific (CN/CP) Telecommunications, and the Trans Canada Telephone System (TCTS) was the first commercial service carried on the satellite system. On 5 February 1973 the Canadian Broadcasting Corporation (CBC) began television service to remote and northern communities for those with RTV stations ready for

Fig. 4.6   The Earth station on the Arctic tundra at Baker Lake, Nunavut.

service. Other locations were added throughout 1973 as their RTV installations were completed. In parallel in 1973, CBC national network television programming also began.

The federal government ensured that Telesat had the capital required for the construction of the baseline space and Earth systems, and it also arranged for the initial customers just mentioned. They would be ready to accept satellite services in early 1973 should Telesat be on schedule and fortunate enough to have a successful satellite launch. Because the CBC and the CN part of CN/CP Telecommunications were crown corporations, the government as owner had little difficulty in volunteering them as customers. The government also managed to persuade Bell Canada and TCTS to use this new technology.

But why should a crown or business corporation, well established and comfortable with its terrestrial services, which it understands well, take on a system, unknown and yet unproven and one with its main component up in space where it cannot be repaired if it fails? Thus, for some, the venture did not make much business sense, but they didn't need to outlay any substantial capital to try it out either while they continued with their existing business as usual. Little did they realize at the time, especially those in the television distribution business, that this new telecommunications technology would serve them well beyond their expectations.

With a marketing department established in 1973 and with the use of various technologies to develop a variety of cost-effective services with flexibility to meet customer needs, Telesat forged ahead. Beginning in June 1974, one of its first new customers was the COTC for back-haul of its Trans-Atlantic cable traffic from Halifax to Toronto. This was a world-first application via satellite for pulse code modulation and time division multiple access (PCM/TDMA).

There was a steady growth in Earth stations year by year. In particular, there was a requirement for transportable Earth stations that went into service, beginning in 1974. In the southern regions Telesat had a 10-meter road transportable station for transmit and receive TV at special events in the same way as did other satellite operators. In addition, Telesat built a number of 3- and 3.6-meter air transportable terminals, deliverable by Twin Otter, short-takeoff-and-landing aircraft, anywhere in Canada (Fig. 4.8). An inaugural feasibility test was the delivery, setup, and service for an oil exploration base camp on Richards Island in the McKenzie Delta at the Arctic Ocean. This was carried out in the 24-hour darkness of the Arctic winter in 1974. Personnel enthusiastically responded to the challenge of delivering terminals anywhere in Canada at any time, weather permitting, and to be on the satellite within hours of arrival. Although these thin route services were relatively low revenue producing, nevertheless they provided valuable experience, and most importantly

Fig. 4.7 A surprise visitor at an Arctic station.

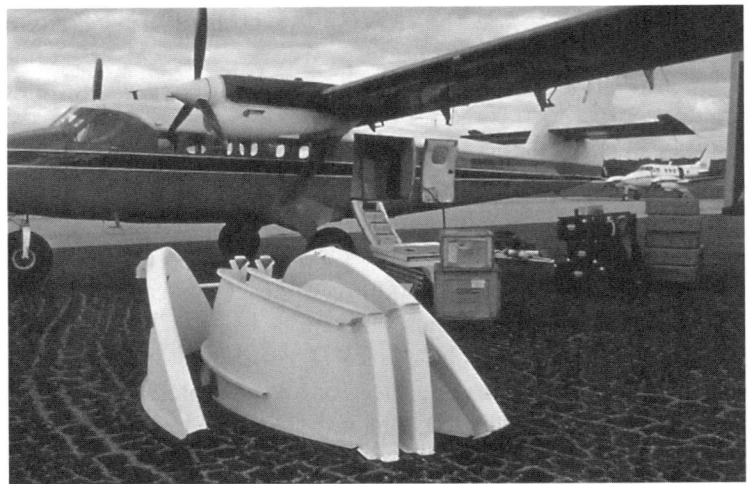

Fig. 4.8   A transportable Earth station delivered by Twin Otter aircraft.

they serviced customers in need. By 1978 the use of transportables had developed extensively. Named the Small Antenna Transportable Earth Stations (SATES), they provided users with an exceptional degree of flexibility, for emergency communications, for long- or short-term use, or for television broadcast of special community events, news gathering, or whatever. There were no interference problems with C-band terrestrial microwave for the SATES, as they were generally used in remote areas. All these various types of services were on the Anik As in C-band.

In 1976, the federal DOC initiated experimental Ku-band services on its high-powered Hermes satellite. Various experiments were conducted in the provision of social services, such as health care and education, to remote and isolated communities. To continue with these experiments beyond the life of Hermes, the DOC arranged for the addition of six channels of Ku-band on Anik B and then, beginning in early 1979, leased them for two years. In 1979 the DOC also experimented with a direct-to-home reception of video programming much like the direct broadcast satellite (DBS) service that began 20 years later with dedicated DBS satellites.

In September 1980, the world's first commercial television broadcasting service in Ku-band was inaugurated on Anik B. La SETTE, a broadcaster in Montreal, Quebec, relayed programming to cable companies in about 25 rural and urban communities. In October 1980, a major national Canadian newspaper, *The Globe and Mail*, began relaying, via Anik B Ku-band, laser-scanned facsimiles of its national edition from Toronto to printing plants in Calgary and Montreal. The benefits were the elimination of shipping costs and timely deliveries in the other cities. Satellite relay of newspapers from their production plants to other cities for printing and local delivery has become a standard in the newspaper industry.

The introduction of full-scale commercial Ku-band services in Canada really began in 1983 with the launches of the relatively high-powered Anik C satellites. With the Ku-band allocated to only satellite services and with no mutual interference with terrestrial microwave links, there was complete freedom to locate Earth terminals anywhere. When C-band satellite services were introduced in 1973, there was an enormous positive impact on the lifestyle of residents in northern Canada, but there was little affect on the public in the south because it was already well connected via microwave links. However, the introduction of Ku-band was an immediate enhancement to business and consumer services in southern Canada. With time, the northern and remote communities were also integrated into Ku-band traffic. While the Anik C transmit patterns extended northward to only the 60th parallel of latitude, the Anik Es went further, and the Anik Fs covered the north to the satellite horizon.

To be better integrated with its large, urban, business customers and to be able to provide the best of service, Telesat built teleports. The Telesat teleports, located centrally in the major cities, are sophisticated, multidish centers, capable of accessing all of Telesat's satellites and some U.S. and international satellites as well. They are updated with the latest and best equipment, with adequate spares, and they have technical personnel on the premises. This kind of high-caliber customer support would not be economically feasible if Earth terminals with the associated electronic equipment were located at customer premises.

In 1987 Telesat introduced low-cost two-way data transmission service for business communications networks, usually from a headquarters hub to branch offices. These very small aperture terminals (VSATs), as they are called, began to deploy slowly, but as the business community caught on to the flexibility and low cost of these types of systems, VSAT antennas mushroomed everywhere. By 2004 Telesat itself owned and/or maintained over 20,000 VSATs in North America. The range of customers includes motor vehicle manufacturers that want to connect directly with their dealerships, banks with their branches, hotel chains, oil and gas companies, the pharmaceutical industry, and others.

Another major development in satellite services was the provision of high-speed data services over satellite. Telesat began system design in its Research and Development Laboratory in about 2000 to evaluate Internet protocol (IP)-based broadband multimedia applications over satellite in the Ka-band. The objective was to be able to provide two-way interactive high-speed broadband services directly to the home, to the small offices in homes, to small- and medium-sized businesses, and to large users.

Broadband service began in Canada in 2004 via Anik F2 in Ka-band, the first satellite in the world on which Ka-band was fully commercialized. Fifteen

Fig. 4.9   The Anik F2 in Ka-band. (See also color figure section at the back of the book.)

of its 45 spot beams cover Canada, including some of the Arctic regions, and 30 are focused over the contiguous United States (Fig. 4.9). Telesat markets its Anik F2 Ka-band capacity in the United States through a partnership agreement with a U.S. service provider. About 25–30% of consumers and businesses in North America did not have access to high-speed services, and this is the segment of the market for which the satellite services are intended. In Canada this service became available to about 200,000 households, previously on only low-speed dial-up. Also in 2004, Telesat began upgrading its VSAT customers in Canada and the United States to broadband, and this is making an enormous difference especially to those with high-volume downloads. In Canada's remote communities, schools, hospitals, and community centers are being brought up to urban standards. Marine systems are also being outfitted with broadband. Those already upgraded are the Maritime ferry service (in 2004) and the Canadian Coast Guard (in 2006). Satellite technology closes the digital gap between city standards and those of remote communities and services throughout the country.

In broadcast services, the introduction of the Ku-band Anik Cs resulted in rapid expansions by established customers and in the startup of new services such as educational TV in Alberta and Quebec. Broadcasters could locate Earth stations right on their premises without needing to backhaul the signals.

Conversion from analog to digital television (DTV) in the early 1990s was closely followed by the introduction of digital video compression (DVC) technology. This caused significant change in satellite transmission service in the TV industry. Telesat began DVC trials in 1990 and offered commercial service in 1993. Initially it was a DVC of two TV signals per satellite RF channel, but by 2001 the CBC, working with Telesat and Tandberg Television, a Norwegian company, had achieved six TV signals on one satellite RF channel with the prospect of adding more. By 2002 it was possible to fit eight regular digital TV channels into a 36-MHz satellite channel. Besides the obvious savings in transmission costs, the broadcaster was able to centralize network control and reduce operating costs.

Telesat offered some direct broadcast services (DBS) in the late 1990s on Anik E2 with Bell ExpressVu and Star Choice as distributors. However, with Telesat's launch of its first high-powered DBS satellite, Nimiq 1, in May 1999, DBS came into its own in Canada. With this high-powered satellite, a 45-cm receive antenna usually provides excellent reception. Between July and November 1999, Bell ExpressVu switched to Nimiq 1 and increased its services from about 130 video channels on Anik E2 to about 250 national video and audio channels on Nimiq 1. ExpressVu added a host of specialty and pay-per-view TV and audio channels. By mid-2004 Telesat was operating three Nimiqs, and its DBS business had grown substantially. The Nimiq antenna pattern covers both Canada and the United States to facilitate Canada/U.S. service.

By the beginning of the 21st century, conversion to high-definition television (HDTV) by the broadcast industry was also underway. Conversion was gradual at first, but as programming became sufficient to cause the sales of HDTV sets to increase, which in turn brought the costs of HDTV sets done, the whole process began to accelerate. By 2007 HDTV programming and viewing audiences were growing rapidly. However, the impetus for the TV industry to totally convert to HDTV will come as the U.S. major networks complete their changes.

HDTV will increase satellite channel requirements very significantly. With current compression technology, HDTV requires about four times the bandwidth of standard DTV. Hence, to meet the anticipated growth in demand, Telesat has plans for additional DBS capacity. Nimiq 4 was launched in September 2008. Nimiq 5 was ordered in mid-2008 from Loral Space Systems, and is planned for launch in 2009. Nimiq 5 will be stationed in a third Canadian orbital location for DBS services.

The story continues, but perhaps one way to conclude this chapter is to leave the reader with a short description of an innovative Earth station installation in Eureka, Ellesmere Island, at 80° N. Here the elevation angle to the satellite is only about one degree, and hills can completely obstruct the path between Earth station and satellite. At elevation angles less than five degrees, serious degradation in signal can also be caused by atmospheric conditions. For this reason, Resolute (74.4° N), at an elevation angle just over five degrees, was the most northerly location in the Telesat baseline system design. However, the satellite C-band footprint actually extends to the satellite horizon.

Between 1974 and 1982 a series of experiments were carried out at Eureka. It was found that a signal, received at two locations with considerable vertical spacing, does not generally fade simultaneously. Thus, satisfactory data can be received by site diversity from one location or the other for most of the time. Because the path lengths from the satellite to the two ground antennas are unequal, the data streams are not aligned. A novel diversity switch was designed by Diversitel (information on Eureka provided courtesy of Diversitel Communications, Inc., Ottawa, Canada) in which the data streams are aligned so that either data stream can be selected at any time without introducing error. To maintain a nearly constant signal at the satellite during fades and enhancements, Diversitel designed an uplink power control system that controls the power transmitted to the satellite from each of the two antennas.

The Department of National Defence (DND) and a scientific laboratory that studies various properties of the atmosphere are linked between Eureka and the south by site-diversity configurations. Because the data, including VoIP, are handled as packets, the packets with the fewer errors are selected. The laboratory and its Earth station are located on a high ridge west of Eureka and are sometimes above the clouds in the nearby valleys (Fig. 4.10).

In 2002 Diversitel installed a Ku-band television receive station for the Environment Canada Weather Station. Although there was considerable skepticism that sufficient signal strength from Nimiq is available at Eureka, Diversitel installed a 2.4-meter antenna, and Eureka now has the world's most northern DBS service. The personnel at the weather station and others

Fig. 4.10   The Earth station at Eureka. (See also color figure section at the back of the book.)

that visit, or make Eureka their home from time to time, are enjoying the benefits of DBS with very good reception most of the time. Fortunately, low-angle fading is largely confined to the sunny season. During the cold sunless period when low-angle fading rarely occurs, the weather station receives the same quality DBS television as the south.

In addition, by employing various schemes, the Nimiq television signal is being relayed terrestrially from Eureka to Alert at 82.5° N, the northernmost permanently inhabited outpost in the world.

ACKNOWLEDGMENTS

I wish to thank all of my colleagues who assisted with the development of this chapter. In particular, acknowledgments and thanks for technical details and accuracies are due to Len Stass, Roger Tinley, Ivan Flockton, Frans Kes, Bruce Burlton, Nancy Stass, Brian Olsen, and John Strickland. In addition, special thanks go to Len Stass for his dedicated and extraordinary assistance with every aspect of the manuscript.

I wish to acknowledge the assistance provided by Telesat personnel and Telesat for generously providing photographs for Figs. 4.2 through 4.10 and for information in its Satellite Communications Newsletter.

REFERENCES

[1] Clarke, A. C., "Extra-Terrestrial Relays," *Wireless World*, Oct. 1945, pp. 305–308.
[2] Bertenyi, E., and Tinley, R. J., "Telesat Canada's New Sixth Generation Spacecraft," 55th International Astronautical Congress, Paper IAC-04-M.1.03, Vancouver, Canada, Oct. 2004.
[3] Burlton, B. V., "The Rescue of Anik E2," *CASI Journal*, Vol. 41, No. 2. June 1995, pp. 57–61.
[4] Allan, R. R., "Perturbations of a Geostationary Satellite 2. Lunisolar Effects," Royal Aircraft Establishment, Ministry of Aviation, W.C.2, Technical Note No. Space 47, London, Sept. 1963.
[5] Kowalik, H., "Telesat Satellite Control System," AIAA Paper 74–451, April 1974.

Chapter 5

INMARSAT: A SUCCESS STORY

AHMAD F. GHAIS*

INTRODUCTION

The INMARSAT story rightly deserves a prominent place among the success stories of satellite applications. Established in 1979 as an intergovernmental organization originally named "The International Maritime Satellite Organization" and later renamed "The International Mobile Satellite Organization," it morphed in 1999 into a private company: INMARSAT PLC. It has innovated in the provision of mobile satellite services (MSS) and enjoyed remarkable commercial success during the almost three decades since its establishment. Perhaps more importantly, it has contributed to the safety of life and property at sea and in the air and aided in the provision of governmental, security, peacekeeping, and disaster relief services worldwide.

This factual narrative of INMARSAT's success is derived in part from public records and in major part from INMARSAT's own archives. I am indebted to INMARSAT for granting access to its archives for purposes of historical research. Although I witnessed many of the events reported here, the story should not be read as a personal memoir or as a treatise on the technology underlying the INMARSAT system.

This story, like any story of human endeavor, is not one of unblemished success. I attempted to relate it accurately if briefly, but also allowed myself the occasional opinion. Space limitations prevented me from fully substantiating conclusions, views, and opinions. Accordingly, the story should not be taken as a critical history of a turbulent era in the evolution of satellite telecommunications in general and of mobile satellite telecommunications in particular. I leave for another forum the critical analysis of actions of governments, organizations, and individuals in these wider contexts.

---

*Former Director, Engineering and Operations, INMARSAT, London, U.K.

Finally, the narrative might appear to place undue emphasis on the roles played by U.S. entities. I have tried to tell the story from an international perspective rather than from a parochial American perspective; however, U.S. entities, governmental and private, play prominent roles.

NEED FOR MOBILE SATELLITE TELECOMMUNICATIONS

The exciting potential for mobile satellite telecommunications became apparent as satellite telecommunications technology emerged in the 1960s. Satellite communications could enhance safety and efficiency of transportation at sea, on land, and in the air. It could improve the provision of governmental services particularly in remote areas and could have a profound effect on war-making and peacekeeping capabilities.

Soon after space technology organizations demonstrated the feasibility of establishing telecommunications links between fixed points on the Earth's surface (what regulators now call Fixed Satellite Services, FSS), they began to experiment with links to moving vehicles (now called Mobile Satellite Services, MSS). NASA embarked in the 1960s on an experimental program utilizing a series of applications technology satellites numbered ATS-1, ATS-3, ATS-5, and ATS-6. NASA's Goddard Space Flight Center established links to airliners using ATS-1 and ATS-3 and later to ships using ATS-5 and ATS-6. A major international program of aeronautical and maritime experimentation was conducted in 1974 using ATS-6.

Also in the 1960s, the Communications Satellite Corporation (Comsat) demonstrated the feasibility of maritime satellite telecommunications by communicating with Cunard's M/S Queen Elizabeth II via an early Intelsat satellite. (See Chapter 3 on the Intelsat success story for background information on Intelsat and Comsat.) A little later, the European Space Research Organization (ESRO) showed interest in MSS, first by flying a balloon to simulate a satellite and linking to a research ship moored in Plymouth Harbor, England. Then ESRO participated with NASA in the1974 joint aeronautical and maritime experiments utilizing ATS-6. ESRO (now merged with the European Launcher Development Agency, ELDO, to form the European Space Agency, ESA) developed its own Maritime Orbital Test Satellite (MAROTS) intended for launch in 1977.

Concern for the safety of transoceanic airliners intensified in the 1960s. They tended to fly out of radio contact with ground controllers for hours at a time. In particular, it was feared that the existing trans-Atlantic air traffic control (ATC) system would be congested by the rapid increase in jetliner traffic coupled with the imminent arrival of supersonic transport (SST) fleets.

As experimentation continued, attention turned to the need to establish an operational aeronautical satellite system, primarily for the North Atlantic air routes. First, NASA, ESRO, and the Canadian Communications Research Centre (CRC) collaborated to propose such a system to be called NETCOS. This was eventually rejected for not conforming to long-standing U.S. government policy requiring private ownership and operation of telecommunications facilities. Instead, ESRO (now ESA) was invited to select a private U.S. party to implement the renamed Aerosat project. ESA selected Comsat as its U.S. partner. Unfortunately, the Aerosat project failed in the early 1970s because of opposition from the airlines who feared diversion of U.S. Government ATC funds and possible imposition of Aerosat user fees. Arcane funding regulations also prevented Comsat from securing multiyear government funding.

NASA embarked in the early 1970s on a major study of a satellite-based ATC system for U.S. airspace. Aircraft positions would be measured by triangulation from multiple satellites and relayed to ATC centers. (It should be remembered that this preceded the availability of the now-ubiquitous GPS navigation system.) This study was anchored at NASA's Electronics Research Center, which was soon transferred to the U.S. Department of Transportation and renamed the Transportation Systems Center (TSC). The TSC study concluded that such a system was both feasible and needed to prevent congestion of the U.S. airspace. It is instructive that nothing came of this study, and the issue is still being debated today.

ORIGINS OF *INMARSAT*

Since Marconi's time and before the S/S Titanic disaster, maritime distress and safety communications had relied on ship-to-ship radio telegraphy using Morse code. Just when attempts at establishing an aeronautical satellite system were floundering, interest arose in the maritime community in using satellite technology to improve ship-to-shore distress and safety communications. In 1973, the International Maritime Consultative Organization (IMCO) established a panel of experts on maritime satellites to explore the feasibility of establishing such an operational satellite system.

The IMCO panel of experts concluded that such a system was technically and operationally feasible, but that it would need to be established and operated by a financially viable international organization. The panel formed a working group to examine its financial feasibility. The working group reported that such an international organization would be financially feasible if it provided commercial (i.e., paid) telecommunications services, even though it would need more than 14 years to recover the required investment. In contrast, the new organization achieved positive cash flow as early as two years after initial service roll out, for reasons explained at the end of the following section.

The panel of experts debated the nature of a suitable international organization. The U.S. delegation asserted that such an organization already existed, namely, Intelsat, particularly as it already had the provision of mobile telecommunications services within its scope of activities. The Soviet delegation took the opposite view: a separate organization, which they named "MARSAT" initially and "INMARSAT" eventually, should be established to undertake the project. The Soviets were concerned to serve the interests of their considerable merchant fleet, but had declined to participate in Intelsat, which they considered to be dominated by the Americans. The East Germans and Poles sided with the Soviets for similar reasons. Other eastern bloc countries naturally sided with the Soviet Union.

Other European delegations tended to side with the Soviet Union. For instance, the Norwegians and Dutch also needed to serve the interests of their considerable merchant fleets, but their shares in Intelsat were minuscule. France and West Germany wanted to promote a European counterweight to Intelsat. While holding a substantial share in Intelsat, the British nevertheless saw an opportunity to land the prize of hosting an international organization within their shores. European governments that espoused industrial policies also saw an opportunity to promote their own aerospace enterprises in order to compete more effectively with the then-dominant U.S. aerospace industry.

In the end, the panel of experts opted for a new organization separate from, but patterned after Intelsat except in a few key respects. It recommended that IMCO (now renamed the International Maritime Organization, IMO) convene a Conference of Government to establish such an organization.

In the meantime, the U.S. Navy had encountered delays in developing its own military satellite system. A small group of American telecommunications companies led by Comsat grasped the opportunity to provide the Navy with a limited-capacity "gap-filler" satellite. To make this "MARISAT" system more commercially attractive to the Navy, and to tap the emerging market for civilian maritime satellite telecommunications, Comsat incorporated a civilian capability with the naval capability in the same satellite. Comsat also took the daring step of seeding the market for its new commercial MARISAT service. It procured 200 ship terminals from the Scientific Atlanta Company that were rapidly snapped up, chiefly by U.S. government users. Three MARISAT satellites were launched beginning in 1976, and commercial maritime service was rolled out soon thereafter. (Remarkably, one MARISAT satellite continues to operate to this day, providing service principally to Antarctica. However, service is limited to a few hours per day due to the inclined orbit of the satellite.)

The MARISAT development gave rise to considerable consternation among delegates to the IMCO panel of experts and to the subsequent conference of governments. The United States was suspected of negotiating in bad faith and of attempting to preempt the proposed new organization. To the contrary, subsequent developments showed that MARISAT served to prepare the technology and markets for the eventual inauguration of the international system.

Comsat later attempted to exploit its initial success by proposing to provide second-generation MARISAT-II satellites to serve as INMARSAT's initial space segment. This proposal did not fare as well, not only because of European opposition, but also because it did not enjoy the support of the U.S. Navy.

INMARSAT IS BORN

IMO convened the International Conference on the Establishment of an International Maritime Satellite System during 1975–1976. It concluded by adopting 1) the INMARSAT Convention, in the form of a treaty to which member governments were invited to adhere (i.e., become parties to it), and 2) the INMARSAT Operating Agreement, signatories to which would be operating entities designated by their respective governments.

Article 3 of the Convention defined

> the purpose of the Organization is to make provision for the space seg-
> ment necessary for improving maritime communications, thereby assist-
> ing in improving communications for distress and safety of life, the
> efficiency and management of ships, maritime public correspondence
> services and radio determination capabilities.

Article 5 prescribed that "the Organization shall be financed by the contributions of Signatories" (rather than of governments) in proportion to their respective investment shares, which the Operating Agreement set proportional to space-segment utilization. Article 13 granted signatories with the largest investment shares (with some minor exceptions) seats in the council, the governing body of the new organization.

It took three more years for sufficient parties to ratify the Convention. It entered into force, along with the Operating Agreement, on 16 July 1979. The INMARSAT Council then immediately established the executive body, the Directorate, to be headquartered in London, England. Soon thereafter, it appointed Olof Lundberg as the director general of INMARSAT.

The immediate task of the directorate was to acquire a space segment (i.e., satellites) and to set the standards for the new system that would utilize it. The directorate was responsible for setting standards for Coast Earth Stations (CES) and for Ship Earth Stations (SES), but their construction was left to signatories and manufacturers. Setting the standards proved fairly straightforward, as they could be based largely on those set for MARISAT. Thus, the CES standard to provide telephony and telex service to ships mirrored MARISAT Earth stations, with minor modifications to facilitate operation with multiple CESs.

The SES standard, called INMARSAT-A, also mirrored the MARISAT ship terminals, but with the added capability to operate to multiple CESs. Nevertheless, the existing MARISAT ship terminals were "grandfathered" to maintain service continuity. Signatories then proceeded to implement CESs to join three already operated by Comsat. In the end, they established more than 20 INMARSAT-A standard CESs, far more than was required for worldwide connectivity. The international organization held scant control over the proliferation of CESs, which probably contributed to the rather high end-user charges that prevailed in the early years. Some of these "national vanity earth stations" could hardly be justified on commercial grounds and have since ceased to operate.

Acquiring the first-generation space segment proved more complicated. Three competing sources of possible procurement were available. Comsat had three MARISAT satellites already in operation and was eager to see the orderly transition of its entire system, satellites, CESs, SESs, and all to the new organization. Anticipating a major role in the emerging maritime satellite business, Intelsat had equipped several satellites in its fifth-generation series that were in the final stages of production with Maritime Communications Subsystems (MCS) and was also eager to lease them to INMARSAT. And ESA developed three Maritime European Communications Satellites (MARECS) based on the MAROTS platform and intended for INMARSAT's use.

By soliciting competitive bids, the Directorate was able to negotiate leases with all three organizations on favorable terms. Within a couple of years, INMARSAT's first-generation space segment comprising three MARISATs, three Intelsat-V MCSs, and two MARECS satellites was assembled in orbit. One would serve as the operational satellite in each of three ocean regions [the Atlantic, Indian, and Pacific Ocean Regions (see Fig. 5.1)], and the other satellites would serve as in-orbit spares.

The satellite telecommunications industry is very capital intensive and financially risky. Satellite telecommunications organizations normally invest

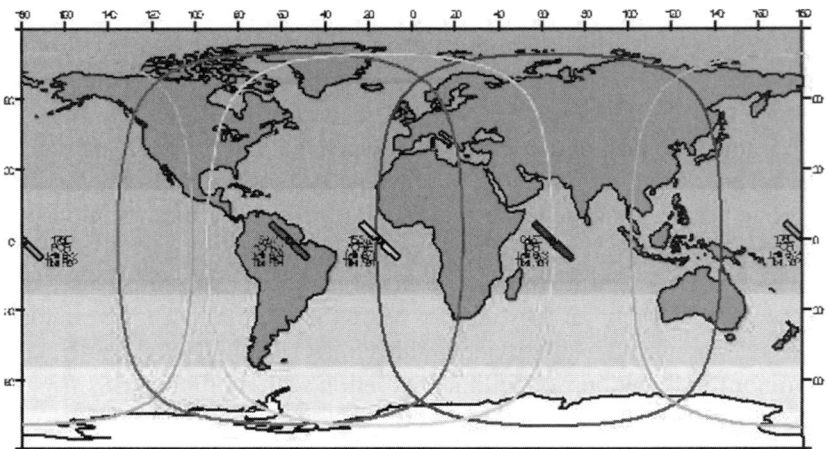

Fig. 5.1   Three or four ocean regions provide global coverage. (See also color figure section at the back of the book.)

substantial capital over several years while their satellites are designed, constructed, and launched, before earning any revenue from them. Unlike other organizations, INMARSAT enjoyed considerable business advantages and rapid startup from leasing its first-generation space segment. Not only was it able to secure favorable lease prices, but it was also able to roll out commercial service in 1982 on the MARISAT satellites and gradually add capacity as the more capable MCS and MARECS satellites became available.

AERONAUTICAL AND LAND-MOBILE SERVICES

With deployment of its first-generation space segment, INMARSAT began to enjoy remarkable commercial success in providing telephony and telex services to ships. Paid traffic and commissioning of new SESs typically increased by 30% per year. Demand for service to oil-production platforms, such as in the North Sea, was being satisfied. Governments, international, nongovernmental, and news-gathering organizations began to deploy INMARSAT terminals to remote land areas. For example, terminals were flown in to support disaster relief operations after the Armenian earthquake in 1988.

Aware of the potential value of satellite telecommunications to ATC and aviation safety, the Directorate embarked in the early 1980s on a program to develop a suitable aeronautical service. Soon the technology was being tested on airliners using the first-generation space segment, and successful test results were used to set standards for Aeronautical Ground Earth Stations (Aero GES) and Aeronautical Earth Stations (AES) for use on aircraft (see Fig. 5.2).

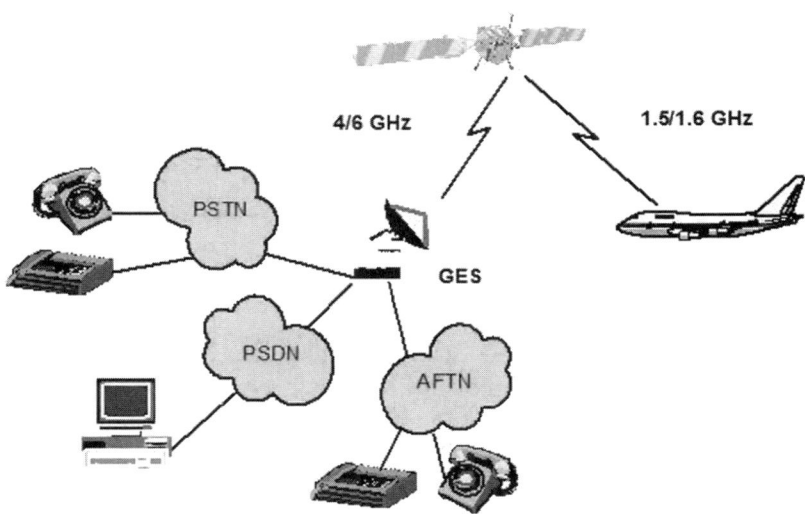

Fig. 5.2   The INMARSAT aeronautical system: GES, AES, and terrestrial interconnections.

But were these emerging aeronautical and land mobile services covered by the Convention and Operating Agreement? The organization dealt with the issues by changing the two governing documents. In 1985, they were amended to extend INMARSAT's competence so that it could provide aeronautical satellite communications. The aeronautical amendments did not come easily, however. Although the opportunity to provide such an innovative service was generally appreciated, concern was expressed about diluting the core maritime mission. In particular, the United States was starting to respond to the perception that the Convention granted INMARSAT an anticompetitive franchise to provide maritime services and took the view that it should not be granted a further exclusive franchise to provide aeronautical services.

A compromise that relegated aeronautical services to a lower priority below maritime services was reached. Whereas the original "purpose of the Organization is to make provision for the space segment necessary for improving maritime communications," it was appended with "... and, *as practicable,* aeronautical communications" (emphasis added). The aeronautical amendments were adopted in 1985 and entered into force in 1989. Aerosat, like the proverbial Phoenix, rose from the ashes to fly again.

The organization proceeded to implement the new aeronautical services. An ambitious business plan was drawn up, and signatories enthusiastically began to augment 12 CESs with Aero LES capabilities to provide voice and low-speed data services to aircraft. However, experience in the marketplace to date has not lived up to those ambitious plans. As pointed out in the second section, aviation safety authorities have been slow to adopt aeronautical-satellite technology to improve ATC. Furthermore, the cost of procuring and retrofitting AESs on airliners proved prohibitive to airlines in financial straits. Although several airlines, including Lufthansa, Singapore Airlines, and Emirates, have successfully introduced telephony and Internet connectivity services to their passengers, other initiatives called Connexion and Tenzing have faltered. Efforts are underway to interconnect passengers' cell phones via onboard minicells and INMARSAT links to their terrestrial networks. Others would connect passengers' laptop computers using Wi-Fi technology and INMARSAT links to the Internet. It remains to be seen whether extending such services to passengers might finally render aeronautical satellite services financially viable.

In 1989, the Convention and Operating Agreement were further amended to extend INMARSAT's competence so that it could provide land-mobile satellite communications. Land-mobile services were relegated to the same lower priority tier as aeronautical and for the same reasons. The further amendments were adopted in 1989 and entered into force several years later. Market experience with land-mobile services has been more positive, however. In fact, INMARSAT today derives roughly half as much revenue from land-mobile services as from maritime services. A wag once observed that "INMARSAT prospers during wars, oil crises and natural disasters," which tend to increase demand for land-mobile services.

EVOLUTION OF DIGITAL SERVICES

The INMARSAT-A terminal served the maritime market well and even crossed over into the land-mobile market. It also served as the workhorse in establishing the organization's financial viability. It was later adapted to provide what was then called "high-speed data" service (HSD, up to 64 kbps), although it would at best qualify as medium-speed service today. Nevertheless, it suffered serious limitations. It relied on dated mid-1970s technology to provide analog telephony and telex low-speed digital services. It was rather bulky and expensive and consumed inordinate amounts of scarce radio-frequency (RF) spectrum.

Soon after its establishment, the Directorate embarked on an intensive service development program. The purpose was to develop a range of terminals more suited to smaller ships and to the land-mobile market and offer a wider variety of digital services. They would be more compact and less expensive than INMARSAT-A and make more efficient use of the RF spectrum (see Figs. 5.3 and 5.4).

First came INMARSAT-C, a small telex-only terminal intended primarily for use on smaller ships to satisfy the requirements of the Global Maritime Distress and Safety System (GMDSS) being developed by IMO. Later, it was adapted to provide other slow-speed data services. It proved attractive to the maritime community for maritime safety and fleet-management applications, but did not yield substantial revenue to the organization.

Then came INMARSAT-B, intended as a direct replacement for INMARSAT-A. Although its above-decks equipment resembled that of its predecessor, it was dramatically better below decks. Beyond providing digital telephony and telex, the "B" easily supported HSD and later VHSD (very high-speed data service, up to 128 kbps). Best of all from the organization's viewpoint, it consumed much less of the satellite's capacity and occupied less than half the RF spectrum of its predecessor. When operated with the second-generation space segment (see next section), the "B" achieved substantial efficiencies that enabled dramatic reductions in user charges. It is even more efficient now when operated in conjunction with the spot beams of the third-generation space segment (see the Third-Generation Space Segment section).

The "B" was well received in the maritime market, but it replaced the "A" only gradually. The "A" continued to serve on many ships, but INMARSAT has announced that it will discontinue its "A" service at the end of 2007.

More compact than the "B" and stripped of maritime-safety features, the third new terminal, called INMARSAT-M (for mobile) was designed to

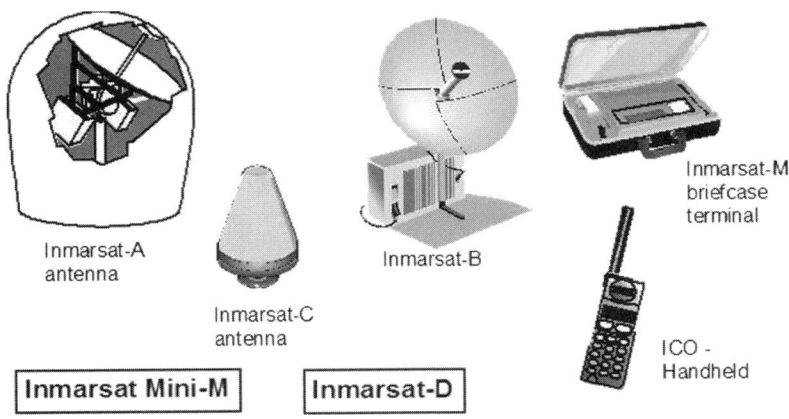

Fig. 5.3   Examples of mobile Earth stations: analog to digital.

provide digital-telephony and medium-speed data service only. Taking full advantage of the superior capabilities of the second-generation space segment, it was intended primarily for the land-mobile service, although it found use in some maritime applications as well. It made even more efficient use of space-segment resources when operated in conjunction with the spot beams of the third-generation space segment (see the Third-Generation Space Segment section). This led to the development of the Mini-M, a more compact and less expensive version of the "M" designed to take full advantage of the spot beams.

Finally, INMARSAT-D was developed initially as a paging service, primarily for use in the land-mobile market. As paging services waned in popularity, it was later adapted into a short-messaging service (SMS) and renamed INMARSAT-D + . It enjoys widespread use in asset tracking and in the Supervisory Control and Data Acquisition (SCADA) markets.

- Inmarsat-A: analogue voice, data, fax, telex
- Inmarsat-C: telex, data, fax deliver (mobile-to-PSTN)
- Inmarsat-B: digital voice, data, fax, telex
- Inmarsat-M: digital voice, data, fax
- Inmarsat-E: emergency position indicating radio beacons (Epirbs)
- Aero-H: voice, data, fax
- Aero-L: data
- Inmarsat-D messaging from 1996
- Mini-M from 1997
- ICO handheld from 1999-2000

Fig. 5.4   Evolving capabilities of MES.

SECOND-GENERATION SPACE SEGMENT

First-generation space-segment capacity became strained by rapidly rising traffic demand during the 1980s. One of the two MARECS satellites deteriorated in orbit and was retired from INMARSAT service, while one of the three MCSs exhibited sporadic increases in internal noise. The operational satellites became congested with traffic. Unlike in the first generation, the Directorate had been planning the purchase of a new fleet of four dedicated mobile satellites, one for each of the three operational ocean regions and a spare. They needed to provide an order-of-magnitude greater traffic capacity. The organization's revenue growth and financial performance would justify the purchase and could support it. But the program proved fraught with difficulty.

Signatories held divergent views on procurement strategy. Those European signatories whose countries promoted their aerospace industries endorsed the directorate's strategy. Other signatories, led by Comsat, favored replicating the first-generation experience by seeking to lease capacity rather than buy it, possibly on multi-purpose satellites as had been successfully accomplished in leasing the MARISATs and MCSs.

A compromise was reached after much debate. The organization would issue a competitive request for proposals (RFP) to provide space-segment capacity on dedicated or multipurpose satellites, by purchase or by "operational lease." In the purchase case, the organization would need to procure launch services separately and would assume the risk of successful launch and operation of the satellites in orbit. Conversely, this risk would be borne by the lessor in the operational lease case.

No proposals for multipurpose satellites were received. Intelsat might have sensed sentiments in favor INMARSAT's independence and declined to bid. Comsat also might have taken those sentiments into account and declined to bid MARISAT-II. Two proposals for dedicated satellites were received, each offering a choice between purchase and operational lease. They required a rather convoluted financial evaluation. Proposals for operational leasing were not found financially attractive, as the lease charges had to cover the risk of launch or in-orbit failure In the end, a contract for the purchase of dedicated satellites was awarded in 1985 to British Aerospace (BAe), with a major subcontract for the mobile communications payload to Hughes Aircraft Company of California. Three satellites were ordered initially and a fourth later.

The organization then sought competitive bids for launch services. Once again, frictions arose in the council as proponents of U.S. and European launchers vigorously contended their relative merits. A compromise was reached to award two launch-services contracts: two satellites were to be launched on the NASA Space Shuttle beginning in 1988 and then two on Arianespace launch vehicles.

Difficulties were encountered both at BAe and Hughes during design and development of the satellites. Not surprisingly, management difficulties arose, caused not only by distance between England and California, but also by the fixed-price nature of the contract. Being the more experienced of the two aerospace contractors, Hughes had trouble playing a supporting role to the less experienced BAe. Technical problems also surfaced, the most severe being an arcane problem called "high-order passive inter-modulation products." Hughes had previously overcome such a problem in other satellite programs but found it difficult to solve this time around. As a result, the delivery schedule kept slipping. The organization was compelled to cope with increasing traffic congestion of the first-generation space segment.

Suspecting that neither the directorate nor BAe was competent enough to conduct a successful satellite program, Comsat took a hard line in the council on how to deal with the contractual difficulties. Given its long association with Hughes, Comsat tended to blame BAe's management and demanded to "hold BAe's feet to the fire." It opposed increasing the fixed price of the contract so as to assist Hughes in solving its problems. As problems intensified, Comsat demanded and got the council to call for an independent audit of the satellite program. A group packed with Comsat alumni conducted a thorough technical, management, and funding audit of the program. It concluded that all parties shared the blame: BAe, Hughes, the council, and the directorate. It recommended that the adversarial relationship with the contractors be replaced by accommodation and that the fixed-price constraints of the contract be relaxed. The council agreed and authorized a substantial increase in funding, most of it flowing through BAe to Hughes. It also recommended replacing the directorate's senior program management. This the director general promptly did.

Other problems surfaced elsewhere in the overall second-generation program. The organization had awarded a contract to SatControl, a French consortium, to develop the computer system for ground control of the new satellite fleet. The project was an ambitious one that relied on the graphical user interface (GUI) technology now commonly known as "Windows." It turned out that SatControl was not up to the task. After further acrimony within the directorate and in the council, and after another "independent audit" of the project, the SatControl contract was terminated for non-performance, and the directorate undertook the project internally. It remains to the eternal credit of the staff involved that they successfully fulfilled the task, although they could not have done so had the launch dates not slipped by about two years.

Then the Shuttle Challenger disaster struck in 1986. NASA instituted a two-year hiatus of shuttle flights while it fixed the design problem and was obliged to review its policy for launching commercial payloads.

Delay in delivery of the satellites permitted the organization to arrange for the deployment of the first two on Delta launch vehicles beginning in October

1990, some two years later than originally planned. Against all odds, all four satellites were eventually launched successfully to the four ocean regions (see the following section) and continue to perform without a glitch well beyond their expected lifetimes.

The original operational lease proposals had been found unattractive because they would have diminished the organization's control over the operation of the second-generation space segment. However, after the project got well underway, it proved possible to secure financial backing by "finance leasing" three of the four satellites from a consortium of banks led by the European Investment Bank, which had no interest in operating the satellites themselves. Unlike "operational leasing" that had been rejected, "finance leasing" permitted the organization to retain full operational control of the satellites, while releasing capital under favorable financial terms. The financial-lease arrangements were to be repeated for the third-generation space segment (see the Third-Generation Space Segment section).

FOURTH OCEAN REGION

A satellite in Geostationary Earth Orbit (GEO) can be seen from roughly a third of the Earth's surface. So a constellation of three satellites should, in principle, suffice to cover the entire Earth's surface except for the polar regions. INMARSAT deployed its three operational first-generation satellites more or less uniformly around the equator to cover the three main Ocean Regions: Atlantic (AOR), Indian (IOR), and Pacific (POR). However, INMARSAT found it difficult to cover the entire Earth's surface, particularly its oceans, with just three satellites. For one thing, the need to share the GEO with other satellites made it impossible to space the three satellites uniformly around the equator so as to minimize overlap in their coverages and to maximize their joint coverage. For another, visibility requirements from existing CESs (now called land Earth stations, LESs) forced further deviations in the orbital locations of the three satellites. Finally, SESs needed to point their antennas to the satellite at a sufficiently high elevation angle above the horizon to avoid interference from sea-surface reflections. This had the effect of narrowing the effective ocean coverage of the satellite and opened a coverage gap between the AOR and POR to the west of the Americas and roughly a third as wide as the Pacific Ocean itself.

This infamous "coverage gap" vexed maritime safety authorities. In particular, the International Maritime Organization (IMO), which has been credited with spawning INMARSAT and was in the process of configuring GMDSS around the doctrine of ship-to-shore calling via satellite, demanded that the coverage gap be closed. In practice, this meant that INMARSAT would have to rearrange its orbital constellation and to deploy a fourth operational satellite by splitting the AOR into AOR-East and AOR-West (see Fig. 5.1).

As AOR traffic accounted for about twice the traffic flowing in each of the other two regions, splitting it between AOR-East and AOR-West would have the added advantage of leveling the traffic load over the four operational satellites. It would also solve a nagging operational problem with the Kuwait CES. To share in the lucrative AOR traffic market, the Kuwait signatory insisted on accessing the AOR satellite rather than the nearer IOR satellite. However, the Kuwait CES often suffered poor signal reception because of its low elevation angle to the AOR satellite, so transferring it to the AOR-East region would improve its reception. In the event, Saddam Hussein solved the problem permanently. When the Iraqi Army invaded Kuwait in 1990, it dismantled the CES and carted it away, never to be heard from again.

The fourth ocean region issue became yet another in a series of contentious issues in the INMARSAT Council. It was not just a question of economics. Obviously, maintaining a constellation of four operational satellites would be more expensive than one of three satellites. It had more to do with internal rivalries between signatories and, to some extent, between governments (parties to the convention).

Comsat, the U.S. signatory, strongly favored creating a fourth ocean region, not only to satisfy the IMO's GMDSS requirements, but also because it could exploit its LESs' access to both the AOR and POR to commercial advantage. Ever resentful of Comsat domination they had experienced in the Intelsat arena, European signatories, led by British Telecom, were adamantly opposed. They also feared losing their shares of the AOR traffic to the "Yankees." The old U.S./European rivalries that had simmered since the bruising debates in the panel of experts over whether to grant the maritime franchise to Intelsat, and later in the council over the procurement of the first-generation space segment, had resurfaced.

After considerable debate and several detailed operational and economic studies, the council decided, "for operational reasons," to establish a fourth ocean region. The directorate immediately began to implement the transition to four-region operation, and the coverage gap was soon closed thus assuring INMARSAT a central role in GMDSS. The decision also highlighted the need to expedite the acquisition of the second-generation space segment, but did not result in increasing the order of four satellites from British Aerospace. As part of this grudging compromise, the organization was consciously betting on successful deployment of all four new satellites. The gamble eventually paid off, as recounted in the preceding section.

THIRD-GENERATION SPACE SEGMENT

An important technical and marketing issue had to be settled when the second-generation space segment was still being defined: should the satellite cover the entire Earth's disk with one beam (called a global beam), or should it be equipped with narrower beams (called spot beams) focused on surface areas where traffic is likely to be concentrated. Spot beams offer inherent advantages in the use of scarce space-segment resources. They focus the power transmitted from the satellite to user terminals thus making more efficient use of it. They exhibit higher sensitivity to user terminal transmissions thus reducing their required power. And they enable reuse of RF spectrum by nonadjacent spot beams, thus improving the spectrum efficiency of the system. These advantages translate into smaller, simpler, and cheaper user terminals.

GMDSS requirements implied that the entire Earth's disk must be covered, either by means of one global beam or by a sufficient number of overlapping beams. Some technical experts believed that spot-beam technology was sufficiently advanced to warrant its incorporation in the second generation, but the organization considered it too risky and settled on a global-beam-only design.

Even as the second-generation program got underway, the organization embarked on defining its third generation. Continued traffic growth necessitated yet another order-of-magnitude greater traffic capacity. Because of the rising importance of land-mobile services (see the Aeronautical and Land-Mobile Services section) and because technology had advanced in the meantime, the organization decided this time in favor of spot beams. The third-generation design called for a global beam to ensure coverage of all of the oceans, plus five spot beams arrayed in a sort of crescent open to the South and aimed at the land masses to the North. The spot beams would be crucial to the development of the land-mobile market in general and of INMARSAT-M and Mini-M in particular.

The organization issued a RFP for the purchase of four satellites, this time avoiding the procurement complexities inherent in operational leases and multipurpose satellites. A contract for the purchase of four dedicated satellites was awarded in 1991 to the GE Astro Company of New Jersey (later acquired by the Lockheed Martin Company), with a major subcontract for the mobile communications payload to Marconi Space of England. Once again, various difficulties delayed delivery by about two years.

The first satellite was intended to be launched on the Russian Proton launch vehicle, the first such launch of a satellite manufactured in the West. However, delivery delays prevented INMARSAT from blazing the trail of newfound cooperation between the West and newly liberated Russia, but the Proton launch did eventually take place successfully in September 1996, and so did the other three launches on Atlas and Ariane launch vehicles beginning in April 1996.

As mentioned in the Second-Generation Space Segment section, the organization was again able to secure a "financial lease" of three of the four satellites on favorable terms. The desire to repeat the "operational lease" experience gained in the first generation was finally fulfilled in the two succeeding generations, albeit in the somewhat different form of financial leases.

Once again, the organization consciously gambled and won on successful deployment of all four new satellites. All four satellites continue to perform successfully to this day with only minor glitches. The score of eight successful launches out of eight attempts was almost unprecedented in the communications satellite industry. By way of comparison, Intelsat had never enjoyed such good fortune over a more extensive launch experience. Divine intervention might have favored the organization for its good works in protecting life and property at sea, in the air, or on land; or perhaps INMARSAT was just lucky.

PERSONAL COMMUNICATIONS

With the expansion of land-mobile services, the organization saw an opportunity to make it truly ubiquitous by providing service to pocket-sized mobile phones. The directorate had been conducting preliminary R&D to that end, but struggled to design satellites with the necessary multiple narrow spot beams. A more immediate challenge came in 1991, when the Motorola Company announced its plan to build the Iridium system. It would consist of a constellation of 77 satellites (later reduced to 66 satellites, but named after the element Iridium whose atomic number is 77) that would provide telephone service worldwide to personal phones the size of a pocketbook. They would be deployed in low Earth orbit (LEO) so as to ease the burden of establishing radio links to the personal phones below.

The story of Iridium and other so-called "Big LEO" ventures falls outside the scope of this chapter and must be relegated to another forum, but the emergence of Iridium certainly raised major concerns at IMARSAT. Some in the directorate, including the director general, saw it as a mortal threat, whereas others were more skeptical about the chances of its success. Not only did it face major technological hurdles, but major marketing challenges as well.

Signatory attitudes also varied. Some whose interest in the organization was principally maritime, notably the signatories of Norway, Russia, Greece, and Singapore, were ambivalent. Others, like the U.S. and U.K. signatories, became concerned lest their governments view the organization's efforts in personal communications as anti-competitive. Nevertheless, the council authorized the directorate to redouble its R&D efforts with a view to proposing a new system to provide personal communications service (PCS).

Under the heading of INMARSAT-P (for personal), later renamed Project-21 (for the 21st century), the directorate had been considering technical alternatives revolving around the types of orbits the proposed satellites would populate. Here again, orbital mechanics falls outside the scope of this chapter, but the essential point is that the organization's structure, experience, and service distribution networks were entirely based on GEO. In contrast, the perceived threat from Iridium and other contenders was based on LEO. Choice of orbit thus became not just a question of technology, but also one of commercial interests and internal politics. The director general was quoted as instructing his staff at one point to "give me any orbit, as long as it is LEO."

The organization eventually settled on splitting the difference by choosing intermediate circular orbit (ICO), but external events would soon overtake the directorate's efforts. Some parties to the convention, led by the U.S. government, expressed mounting concerns about the anticompetitive effects of the organization's entry into the PCS market. It is also fair to say that the U.S. government, despite its procompetitive rhetoric, was inclined to protect its native PCS entrants (including Iridium) from this privileged, dominant, intergovernmental competitor.

Internal political strains among signatories also surfaced. Although some signatories saw value in providing PCS to remote and less-developed areas, others with maritime interests saw no reason to risk their capital on a service their subscribers scarcely needed. Besides, the ICO orbital selection threatened the commercial interests of those signatories that were dominant land Earth station operators (LESOs). Their LESs maintained constant contact with the existing INMARSAT GEO space segment and hence with their markets, whereas with an ICO constellation they would have to track and hand over contact to competing LESs as the satellites rose and set relative to the horizon.

The signatories of the United States, the United Kingdom, Norway, Russia, Singapore, and some others forced in 1994 the council's fateful decision to spin off Project-21 into a separate private company to be financed by contributions of interested signatories and to be called ICO Global Communications PLC. As part of the decision, INMARSAT was prohibited from instituting a PCS service of its own and was compelled to obtain any PCS service its customers needed directly from ICO Global Communications, PLC. In his turn, the Director General, Olof Lundberg, unwisely decided to resign in order to lead the new company. Eventually, ICO would fail in the wake of the Iridium debacle, which also dragged down the other "Big LEO," Globalstar, into bankruptcy. That could hardly be blamed on unfair competition from INMARSAT.

PRIVATIZATION

More fateful decisions beyond the control of the directorate were to come. The winds of liberalization and privatization were sweeping the telecommunications industry worldwide, partly in response to U.S. government policy. In particular, several governments, led by the United States and including the United Kingdom, were actively promoting the privatization of INMARSAT's

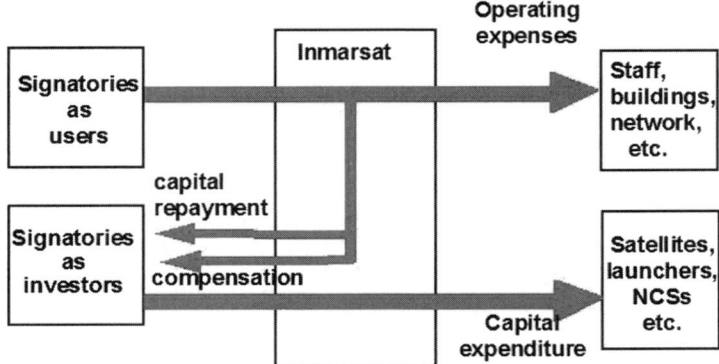

Fig. 5.5   INMARSAT and signatory cash flows.

big sister: Intelsat. The German government obliged Deutsche Telecom to go along with the privatization process.

As mentioned in the second section, in connection with the Aerosat project, U.S. government (USG) policy always espoused private ownership and operation of telecommunications facilities. Despite the fact that USG led in the creation of the unique form of intergovernmental organizations exhibited in both Intelsat and INMARSAT, it seems that by the 1980s it had a change of heart. The charitable interpretation is that it perceived that the two organizations had come of age and no longer needed nurturing government supervision. Certainly, USG representatives were known to find their supervisory efforts onerous and thought they could more profitably direct them elsewhere. The uncharitable interpretation is that the USG simply wanted to protect its native satellite ventures.

The story of Intelsat's privatization is discussed in Chapter 3. There is no doubt that it had major repercussions on INMARSAT. Our wag once observed that "when Intelsat sneezes, INMARSAT catches pneumonia."

Perceptions of INMARSAT's privileged position in the marketplace as an intergovernmental organization and its alleged anticompetitive behavior largely reflected perceptions of Intelsat. Partly because of the efforts of the founder of PanAmSat, a U.S. domestic satellite service provider, Intelsat stood accused of stifling competition in the U.S. domestic satellite market. The USG was urged to exclude Intelsat from providing U.S. domestic services until it "leveled the playing field" by eliminating Intelsat's privileges and immunities. Similar arguments were advocated in INMARSAT's case by Iridium and the American Mobile Satellite Corporation (AMSC), a company to which the U.S. Federal Communications Commission (FCC) had granted a franchise to provide domestic mobile-satellite service.

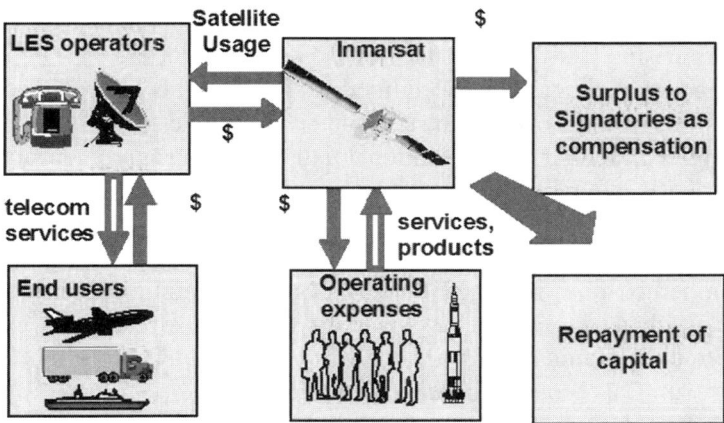

Fig. 5.6 INMARSAT cash flows from end users.

Certainly, both Intelsat and INMARSAT enjoyed treaty-based privileges and immunities. They were exempt from taxes in their host countries (the United States hosted Intelsat, and the United Kingdom hosted INMARSAT) and were granted immunity from prosecution (within certain bounds). It is notable that, unlike Intelsat, INMARSAT was not granted immunity from antitrust lawsuits in the United Kingdom.

However, the market effects of these privileges were minimal. In particular, cursory examination of the flow of funds (see Figs. 5.5 and 5.6) reveals that the market effects of the tax exemption were illusory. INMARSAT collected capital contributions from its signatories to invest in its facilities, primarily the space segment. Signatories and other LESOs paid INMARSAT space-segment utilization charges. In turn, they sold services to their subscribers (or to value-added service providers further down the food chain) at prices they set independently. Each year, INMARSAT paid back any surplus to its signatories as returns on their capital contributions. Fresh capital could be raised by calling on signatories for additional capital contributions, so surpluses rarely needed to be retained as reinvested capital. In effect, any surplus or net income flowed back to the signatories who were then subject to whatever tax regime prevailed in their home country. Essentially, INMARSAT made no net income or profit that might have been taxed by the U.K. host government. This is rather akin to the operation of a mutual fund in the U.S. tax regime: the fund invests its subscribers' capital and is not taxed on the yield or on any capital gains provided it distributes them back to its subscribers.

Nevertheless, there was some validity to critics' complaints of unfair advantage. INMARSAT staff were exempt from U.K. income taxes, although the United States (and Singapore) would not exempt its citizens employed by

INMARSAT. Staff were also exempt from prosecution in the United Kingdom on matters arising from performance of their official duties.

Some signatories egregiously claimed exclusive rights to provide services to subscribers in their own countries, and so were able to set uncompetitive end-user prices. Neither the convention nor the operating agreement addressed the exclusivity issue and left it to the jurisdiction of the parties, so that it could be granted only by national regulatory authorities. For instance in the United States, Comsat claimed exclusivity until its stranglehold was broken by a competitor now renamed Stratos Global Communications. In other instances such as in Europe, a signatory that was already the exclusive national telecommunications operator was able to hold onto a modicum of exclusivity until its broader markets were deregulated.

It was particularly difficult to enforce exclusivity in providing INMARSAT services because of the nature of mobile satellite telecommunications. A LESO could readily connect an incoming call by terrestrial means to the destination in the mobile user's home country (see Fig. 5.7). LESOs did in fact compete for the custom of large fleets, but they could not easily compete for outgoing calls from a country whose dominant telecommunication carrier was also the LESO (e.g., in Europe). In practice, this permitted competition among LESOs, more for incoming than for outgoing traffic.

It can fairly be concluded that privatization of Intelsat and INMARSAT distracted and detracted from the search for a "level playing field." This was doubly true in the case of INMARSAT as of Intelsat, as Intelsat customers did not generally benefit from competition among multiple service providers as INMARSAT customers do from multiple LESOs. Further, mobile satellite services remain a niche in the larger niche market of satellite telecommunications

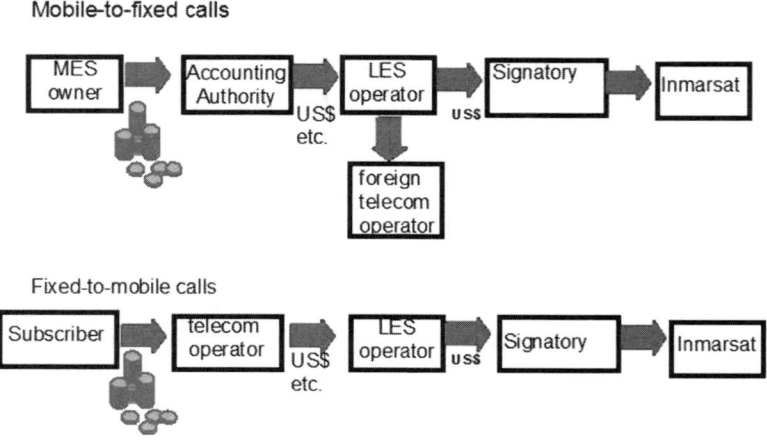

Fig. 5.7   **Call directions and LESO cash flows.**

services, which, in turn, is a niche in the much broader telecommunications markets. The benefits of competition would have to await the deregulation of national telecommunications markets.

In any event, privatization of INMARSAT gathered momentum. Seeing it as inevitable, the directorate acquiesced to privatization. Major signatories saw it as an opportunity to cash in their investments in order to bolster their balance sheets and to cope with the exigencies of deregulation or privatization in their home markets. However, some signatories, notably the Norwegian and signatories from some developing countries, lamented the potential loss of influence to private investors. Our wag remarked at the time that the organization was reverting from a unique 20th century organizational structure to a corporate structure "invented by the Dutch in the 17th century to facilitate trading in the East Indies."

In the apparently final event, INMARSAT was privatized in April 1999, and INMARSAT PLC was created to replace it. To guarantee continued support of GMDSS, the parties agreed to retain a vestige of the original intergovernmental organization under the full name "The International Mobile Satellite Organization" (IMSO), for the main purpose of ensuring the continued support of GMDSS. But IMSO was not tasked to ensure continue support of other public-service functions, such as services INMARSAT already provided in support of safety of transportation in the air and on land and in disaster relief. These fell victims to the second priority implied by the words "*as practicable*" that had been inserted in the aeronautical and land-mobile amendments (see Aeronautical and Land-Mobile Services section).

But wait! The INMARSAT privatization story is not over yet. In an apparent afterthought, the USG-proposed legislation intended to further level the playing field for both Intelsat and INMARSAT by, among other things, breaking up the signatories' surviving control of already privatized INMARSAT. (Intelsat was eventually privatized in July 2001.) Congress passed in March 2000 the Open-Market Reorganization for the Betterment of International Telecommunications (ORBIT) Act requiring privatized INMARSAT to issue an initial public offering (IPO) by a certain date. The ORBIT Act ignored the fact that the financial markets no longer favored investing in satellite systems after the Iridium debacle and in the wake of the Internet and Telecom meltdown in year 2000.

The negative reaction abroad was to be expected. The United States was suspected of bad faith, having negotiated hard for privatization terms to its liking, only to turn around later and unilaterally impose additional conditions. The USG extended the deadline while INMARSAT PLC sought to launch an IPO that would have a decent chance of success. Eventually, it was able to negotiate a leveraged buyout (LBO) by Apax and Permira, two venture capital companies that our wag promptly dubbed "Vulture Capitalists." They evidently got a bargain, as evidenced by the markups they received when they eventually floated a partial IPO in compliance with the ORBIT Act—so much for government micromanagement of private enterprise.

FOURTH-GENERATION SPACE SEGMENT

The wild proliferation of the Internet had created demand for broadband connectivity, not only for mobile terminals, but also for terminals in remote areas not easily reached by terrestrial means. Having been excluded from the PCS market, the directorate started before privatization to plan its next-generation space segment to address the broadband-via-satellite market. This was a radical departure for the organization, as it no longer looked toward the mobile markets exclusively.

Utilizing dramatic advances in satellite technology, the fourth-generation space segment was designed to provide high-speed data services (up to 400 kbps) suitable for Internet connectivity to laptop-computer-size terminals similar to the Mini-M terminals that were developed for voice services. In addition to providing a global beam and 19 spot beams to ensure downward compatibility to legacy terminals, it provides hundreds of reconfigurable "pencil-thin" spot beams aimed at traffic hot spots on the Earth's surface.

A contract was awarded to Astrium, a British company formed by merging the satellite resources of British Aerospace and Marconi Space, to provide three satellites patterned on their Eurostar Geomobile model. The satellites were delivered after privatization, and two were launched successfully in 2006 to cover the heavy traffic areas over North America, Europe, Africa, and Western Asia.

While waiting for the fourth-generation satellites, the organization primed the market for its broadband-connectivity service by launching its Regional Broadband Area Network (RGAN) using spot-beam capacity leased from the Thuraya organization based in the United Arab Emirates (UAE) and covering Eastern Africa, Western Asia, and Southern Europe. INMARSAT PLC rolled out the new service dubbed Broadband Global Area Network (BGAN) in 2007. The third satellite planned to launch in 2008. It is too early to judge the success of this new commercial venture.

With the demise of ICO and its eventual resurrection as a broadband-via-satellite venture, INMARSAT was no longer prohibited from instituting a PCS service of its own nor compelled to obtain any PCS service its customers needed directly from ICO Global Communications, PLC. Although BGAN and the fourth generation were not specifically designed to support PCS, INMARSAT PLC announced that it is adapting them to do so. Meanwhile, it acquired the troubled Asia Cellular Satellite Company (ACeS) based in Indonesia that had initiated PCS service in Eastern and southern Asia. It will be interesting to observe how PCS service piggybacked on BGAN will do when introduced (2007).

QUO VADIS, INMARSAT?

The INMARSAT story told so far is one of great success and occasional setback (see Fig. 5.8). The organization successfully developed the commercial maritime satellite telecommunications market. Increasingly, it also served the land-mobile market, particularly in remote and underdeveloped areas, but it has not opened the wider aeronautical market yet. Through a decentralized distribution network, it serves hundreds of thousands of end users that generate increasing traffic volumes and revenues. In keeping with wider telecommunications trends, it derives more revenue from data traffic than from voice traffic. It has successfully managed four generations of satellites and supported several types of terminals tailored to the needs of a variety of end users. It is also in the process of consolidating LES operations in the interest of efficiency and economy. Throughout, it has dutifully supported public-service functions worldwide: in security, peacekeeping, disaster relief, and transportation safety, most notably through the GMDSS.

The organization's achievements occurred in a complex political and economic environment. Remarkably, it managed to get the Western and Eastern blocks to work together on joint governmental and commercial operations while the Cold War raged during the 1970s and 1980s. Later, it managed to survive a turbulent privatization process as the winds of liberalization and privatization swept the worldwide telecommunications industry. At the turn of the century, it again managed to survive the financial debacle in the telecommunications and Internet industries in general and in the mobile satellite industry in particular.

The big challenge for INMARSAT now is how it will fare in private ownership. Competitors abound, and regulatory distinctions between services, as defined by the ITU, are fading. Its prospects of continued commercial success

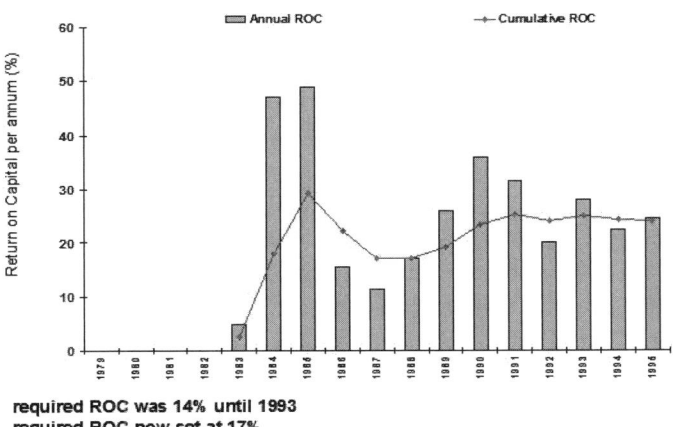

Fig. 5.8   One measure of success: return on signatory capital.

appear good, particularly if its innovative BGAN service gains traction in the marketplace and if it manages to extend its aeronautical services to passengers in the airliner's cabin. Whether it can break out of the niche markets serving hundreds of thousands of end users and into the mass market (millions of end users) depends, in part, on the degree of integration of its services within the wider telecommunications infrastructure.

To end users, INMARSAT's communications links still appear like "stove pipes," separated from and awkwardly interconnected with the terrestrial infrastructure. They crave seamless interoperability, one where a call or an Internet connection is automatically established via the handiest transmission medium at the most economical price. They would also like to deal with one telecommunications service provider, one customer service center, and one billing system. Seamless interoperability can be achieved only once commercial, financial, and organizational obstacles are overcome.

The question arises as to what will become of the public-service functions. It is unreasonable to expect private enterprise to provide public services gratis. In general, private enterprise will provide such services only if it sees a marketing advantage or some other return on investment. Imbuing a corporation with social responsibility is a delusion. Only humans have consciences; a corporation is not a human and hence is devoid of conscience. If public services are required from a corporation, then they must be mandated, preferably through democratic processes.

As a private company, INMARSAT PLC can be expected to serve the interests of its owners by satisfying the needs of its customers. Having largely escaped the direct intergovernmental supervision prescribed in the convention, it is not bound to also serve the public interest, although it has continued to do so. After privatization, IMSO was tasked mainly to ensure the continued support of GMDSS to the exclusion of certain other public service functions already being provided by the organization. As it derives its meager resources directly from INMARSAT PLC, IMSO can hardly be expected to be a tough taskmaster. The interaction between the two deserves careful monitoring.

However, one long-standing challenge remains: access to RF spectrum. Spectrum is a natural resource that has always been in short supply and high demand. Our wag explains that "spectrum is like ocean-front property. They don't make it anymore." The International Telecommunications Union (ITU) allocates some of it to the Mobile Satellite Service (MSS). National administrations (in the case of the United States it is the FCC) award licenses for use of assignments within this spectrum to MSS operators that fall under their jurisdiction. INMARSAT has been obliged to negotiate sharing of the MSS allocations with several other MSS operators, a process that bears the bureaucratic name "Intersystem Coordination."

INMARSAT has often encountered difficulty in negotiating intersystem coordination, particularly for its North American coverage where it must share use of the allocations with the U.S. company AMSC (now renamed Motient) and the Canadian company TMI Mobile that operate national mobile satellite systems. More recently, the two companies have pooled their resources, including their spectrum licenses, in a joint venture called Mobile Satellite Ventures (MSV). MSV applied to the FCC for permission to use its combined assigned spectrum for an ancillary terrestrial component (ATC).

INMARSAT opposed MSV's application for fear that ATC transmissions might interfere with INMARSAT's MSS transmissions. MSV retorted that its ATC transmissions could not cause any interference as they would be confined to its previously assigned spectrum. This subtle technical issue cannot be resolved here, but MSV went on to accuse INMARSAT of using it as a subterfuge to cover its anticompetitive intent. INMARSAT denies that.

The FCC eventually granted MSV's ATC application even though it violated the spirit, if not the ITU's intent in allocating the relevant spectrum exclusively to MSS as opposed to the terrestrial mobile service (MS). The license does not clearly spell out what constitutes "ancillary terrestrial" use, but requires that it must be accompanied by genuine mobile-satellite service. The FCC reserves the right to revisit the issue should ATC actually cause interference. Now, MSV is soliciting "strategic partners" to help it deploy the terrestrial infrastructure needed to extend ATC, particularly to areas where satellite signals are obstructed such as mountainous or heavily forested terrain and urban canyons.

Some industry experts believe that MSV's initiative is a subterfuge to monetize its ATC spectrum license, despite the FCC rule against trading in spectrum. It is suspected to be seeking partners who would pay handsomely for using the ATC license for terrestrial purposes, with little regard for the accompanying MSS. For instance, a terrestrial wireless carrier could use the license to establish a broadband service offering Internet connectivity without having to bid for the spectrum in FCC-organized auctions.

MSV's success was soon followed by a bevy of similar applications for ATC licenses in another spectrum band allocated by the ITU jointly to MSS and MS. The FCC granted several such licenses, including to Terrestar (another offspring of Motient and TMI), Globalstar, ICO, and Iridium Satellite (successor to the defunct Iridium Company). A similar ATC process is now making its way through the European regulatory process.

Having lost the ATC argument, INMARSAT PLC needs to decide whether to seek an ATC license and whether it will also seek to monetize it. A truly commercial organization might feel obliged to do so in order to serve the best interests of its investors. Alternatively, it might genuinely try to enter the terrestrial broadband Internet connectivity market in North America. On the other hand, it has always found its lucrative markets in the more remote, less developed regions of the world, and so it might forgo the opportunity to enter a North American broadband market crowded with potential competitors.

Only time will tell.

Chapter 6

THURAYA MOBILE SATELLITE SYSTEM

ALI SAEED AL MAZROOEI*

AND

ADRIAN J. MORRIS†

INTRODUCTION

In 1997, Thuraya and Hughes (now Hughes Communications and Boeing Satellite Systems) embarked on delivering high-quality mobile voice and data services to a wide geographical region covering most of Europe, Middle East, North Africa, Central Africa, and Western Asia. The widespread growth of Global System for Mobile Communications (GSM) and associated services in the late 1990s provided both opportunities and challenges for hand-held voice and data communications services to areas unserved by terrestrial systems. Thuraya required coverage for a large geographical region, primarily for voice and data communications in the initial phases, with flexibility to expand to higher-speed data services as these services become available in traditional wireless terrestrial networks, which would emerge in the next 15 years. To reach a mass market, common wisdom was that the subscriber terminals had to move beyond the traditional portable and transportable satellite terminal technology and offer a consumer-style mobile satellite handheld terminal (HHT). The HHT had to be affordable and deliver cellular-like service features and quality. The Thuraya system had to provide the flexibility to accommodate changes in Thuraya's traffic types and evolving capacity needs across more than 100 countries. Service requirements demanded postlaunch optimization of performance over geographic areas where growth of user population traffic might occur.

*Chief Technology Officer, Thuraya Telecommunications, Dubai, United Arab Emirates.
†Executive Vice President, Hughes Network Systems, Germantown, Maryland.

Another daunting challenge was to develop and deploy a total system within a three- to four-year period. The Thuraya coverage area is extensive, and although a single geostationary satellite deployed at an altitude of approximately 36,000 km above the Earth's surface can meet the needs, it required high spacecraft power and high-performance spot beam technology with extensive frequency reuse to optimize use of the limited spectrum. Single satellite hop direct HHT-to-HHT communications was a prerequisite because there was a projected large demand for mobile user-to-user communications in the absence of either landline or wireless communications in many of the regions. Hence, there were significant satellite-to-ground segment design issues to ensure clear sky link margins. Thuraya requested that the HHT could select either a cellular network or the satellite network to allow for a single homogeneous network across many countries, many languages, and time zones. Low-delay signal transmission was required to ensure good quality of service. The HHT antenna and electronic components should not require the use of expensive components. Hughes understood that the development of digital signal processing algorithms and techniques that could be realized in hardware and real-time software was critical to system realization. The major program plan with milestones was as follows: 1) preliminary system design, 2) preliminary segment design, 3) critical segment design, 4) critical system design, 5) factory integration, 6) field deployment and preliminary testing, 7) system and service test and launch, and 8) additional production of gateways and HHTs.

THURAYA BUSINESS AND OPERATIONS CONCEPT

To comprehend the business and operations concept of Thuraya, it is vital to return to their business plan to explain what Thuraya was thinking in 1997, and why it chose a certain direction.

At the outset Thuraya decided to create a mobile satellite system that was complementary to, not competitive with GSM. The key concept behind the visionary project was to enable any GSM user who cannot access a terrestrial network to switch to the satellite mode seamlessly. Thuraya decided to begin its commercial activities on a regional rather than a global scale to ensure operational and business success; precedent examples of networks launched globally had ended up with many difficulties. Furthermore, Thuraya selected a simple technical system of one gateway and one satellite, whereas other satellite operators have used several low orbiting systems (LEOs), which meant that implementation could be quicker and licensing much easier.

Hence, Thuraya chose the most flexible of mobile satellite services (MSS) systems with the help of Hughes and ensured that it had a competitive advantage when it came to technology. This technology was innovative and pioneering compared to that of other operators. Technically, this simple and flexible system was based on the digital beam forming concept, which allows the operator to change its coverage system. This has allowed Thuraya to gradually expand from 246 beams to 296 beams, and finally to 351 beams.

Thuraya's concept also relied on starting as a regional system that would gradually evolve and expand. With its third satellite (Thuraya-3), Thuraya will cover the Asia-Pacific region to double the population currently in its footprint. It also relied and focused on acquiring GSM users from the terrestrial community resulting in providing a GSM capability over satellite telephone. It selected a satellite-air interface, which was closely aligned to the upper layers of GSM, which meant that it was easy to have applicable and seamless systems. The services were familiar to users, and it eventually allowed Thuraya to offer a GPRS packet service, an additional factor that gave Thuraya headway in the market.

Furthermore, Thuraya had a contract that was very clear and focused on functional requirements. The system capacity and requirements were well described. This meant that the scope of work in terms of Thuraya and its contractors was explicit. It involved three exhibits, whereby exhibit A was the statement of work, exhibit B was the technical description of the system, and exhibit C was the verification method. To further ensure having a smooth project implementation, the project was contracted on a turn-key product basis, whereby the contractors had to deliver everything in one go. This included the space segment, the ground segment, the user segment, and the billing segments.

The space segment involved two spacecraft (one of them was a backup), launch, and a satellite operation center. The ground segment consisted of a network switching subsystem and a gateway station subsystem. The user segment included the production of 235,000 user terminals by two separate companies, Hughes and Ascom. The billing system was designed by LHS, a German company.

In terms of operations, it is important to note the Thuraya capacity requirements span from 13,000 up to 25,000 simultaneous calls. Furthermore, Thuraya ensured that the design can support multiple gateways to ensure redundancy and additional ground capacity if required in the future.

As for services and features provided by the Thuraya system, the use of global positioning system (GPS) gave Thuraya a competitive edge, and it aided in accurate billing because Thuraya could always locate the subscriber. Additionally, prepaid cards, which allow subscribers to be charged real time, were much easier to use than a post-paid billing system and provided another great unique advantage to Thuraya. Similarly, the integration of GSM and all of its features, such as call waiting and call forwarding, was another differentiation factor. Finally, Thuraya acquired an air interface that was open to

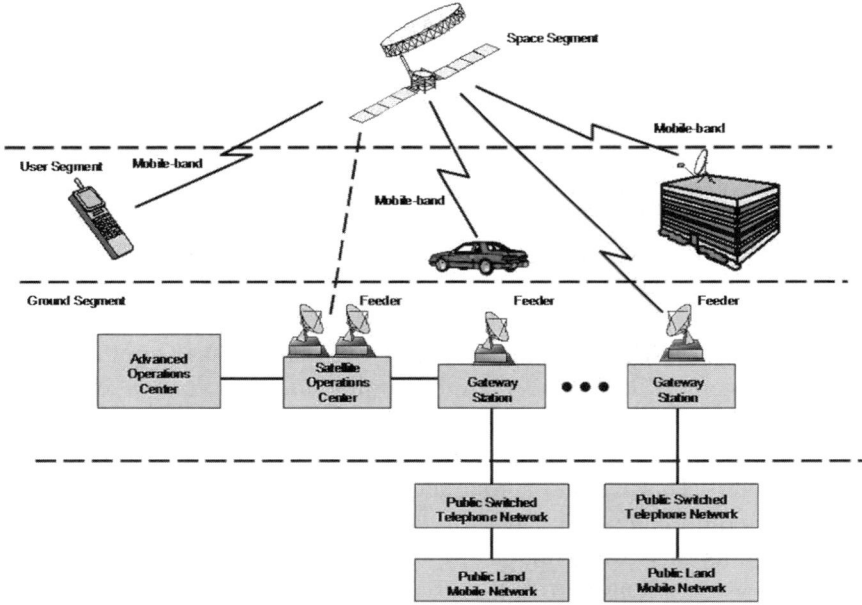

Fig. 6.1   The major elements of the satellite communications system.

anyone where there were no technical restrictions, which later proved useful with the introduction of the second-generation handheld terminals.

All of the facets just mentioned were the deciding factors that affected the business and commercial concepts of Thuraya at the time.

SYSTEM ARCHITECTURE

*The satellite communications system shall deliver mobile telephone, data, facsimile, and messaging services to the Thuraya coverage region using digital communications techniques. The architecture shall support subscriber roaming between gateways, between satellite and GSM, and potentially multiple satellite systems using cellular like roaming procedures.**

The system comprised the following major elements (see Fig. 6.1):

1) **Space segment:** A state-of-the-art spot beam space system design supporting gateway-mobile, mobile-gateway, and mobile-mobile communications. Hughes Electronics were pioneers in advanced satellite communications and had established a clear leadership position in geosynchronous satellite design and manufacture.

2) **Ground segment network:** This segment comprised one or more gateway stations (GS) and an advanced operations center (AOC). A GS provides

---

*Throughout this chapter, italized text denote quotations from Thuraya and Hughes documentation.

access to the public switched telephone network (PSTN) for originating and terminating calls with mobile users and provides access to the public land mobile network (PLMN) and public switched data network (PSDN) via the PSTN. The AOC controlled and allocated all spectrum resources of the system and allocated such resources to gateways.

3) **Operations support systems (OSS):** The OSS provided the functionality for customer care and billing and the SIM card personalization and system network management. System performance monitoring requirements were clearly defined by Thuraya. The architecture supported a billing system for all services. Operation in each country could be different, with factors like number of subscribers, associated GSM operators, national switching systems, etc.

4) **User segment:** The user terminal (UT) provided the subscriber interface with the system to obtain network services. Multiple types of UT were accommodated and planned, including handheld terminals, vehicular terminals, and fixed terminals. The handheld and vehicular terminals were dual mode, providing access to both Thuraya and terrestrial GSM networks. HHT size and weight were mutually agreed by Hughes and Thuraya, consistent with battery capacity, and talk times. Target mean time between failures (MTBF) were defined, consistent with cost targets, battery lifetimes, etc.

All mobile terminal equipments support dual-mode operation, both Thuraya and global system for mobile communications (GSM) modes. *The Handheld Terminal (HHT) shall be the baseline configuration for all terminals.* The UT design presented a common interface to the user, independent of the satellite or GSM air interface. The services supported by the terminal include voice, data, fax, SMS, supplementary services, high penetration alerting, and voice mail. The HHT can be used in a vehicular application by mating it with a vehicular docking adapter and an external vehicular antenna. The adapter allowed the terminal to be used in hands-free mode, with an external speaker and power adapter. The HHT can be used in a fixed application by mating it with a fixed docking adapter.

SYSTEM PERFORMANCE

The system shall be able to cope with changing traffic demands over the coverage area and over time by flexible capacity allocations to individual spots and regions.

The system shall provide up to 256 equivalent voice circuits per MHz of allocated spectrum. In addition, system modeling of equivalent voice circuits per MHz per 1000 square kilometers of traffic ensured optimization of capacity density parameters.

The system design accommodated users anywhere in the allocated mobile link L-band spectrum: 1626.5 to 1660.5 MHz uplink and 1525 to 1559 MHz downlink. Allocation of small increments of channel bandwidth was desired because of the unpredictable and often costly acquisition of spectrum. Spectrum assignment across the coverage area allowed for time variances and could be reconfigured.

The system shall connect UT-UT voice calls to satellite subscribers using a single L-band to L-band satellite hop. The small HHT profile shall have sufficient clear sky link margin to provide for cellular like quality of service, utilizing a lot less bandwidth.

Low-latency call setup and call-duration times were specified. This set of collective parameters and performance requirements, plus many others, kept a lot of engineers awake at night. A fundamental prerequisite that current and future system competitiveness would be significantly improved if optimization of the layer 1 impacting spectrum, bandwidth consumption, and HHT RF power were a prime focus.

End-to-end performance definitions were required to tie in all aspects of system design, and end user quality of service was a continual dominant factor in trading off specifications. With this focus in mind very early in the program, it allowed detailed derived requirements that are sometimes missed in first-pass system designs to be determined and allow the engineering teams to get a total realizable design picture in mind. The optimum system designs achieve a balance between meeting tough specifications and practical, commercial viability that are schedule compliant and can be integrated and verified in a reasonable manner. Often, overly complex specifications drive solutions that are unreliable.

Both customer and supplier worked well together towards a common end goal to achieve commercial success.

PRELIMINARY PHASE

In typical Hughes tradition, technical challenges were a daily part of life, and usually a tough set of problems brought out the best and the brightest. It quickly became widespread at the Hughes engineering and operations groups that this was a project not to be missed, and there was no shortage of volunteers to bridge the continental divide between cellular and satellite communications. Although Hughes had been in the satellite communications business for over 30 years, and the engineers believed that all the technical system and service challenges could be solved, it was still a massive challenge to roll out the service in the beginning of the 21st century. It would require combining the traditional satellite communications technologies with new digital signal processing techniques and forming integrated satellite systems, ground systems, and customer teams that could continually optimize and refine both the system and service concepts. Several hundred engineers and several million man-hours of effort were required to design, integrate, deliver, and get the product into service in a little more than three years.

From a Hughes perspective, the Thuraya team paid attention to the details and was very precise in defining their requirements. Thuraya team members focused on "what they wanted" and did not try to dictate implementation. It allowed Hughes the flexibility to invent and make smart trades on technical feasibility, cost, and schedule. Thuraya had done its homework and continued throughout the program to finetune its needs with focus groups, provided specific market data of various markets, and stayed abreast of the evolution of the trends in the GSM cellular markets. To remove any ambiguity of requirements, Thuraya clearly specified the service quality and features at the beginning of the program.

Some questions started to unfold. The satellite design is state of the art, and both weight and power requirements are demanding. How does a handheld phone track a geostationary satellite located 36,000 km above the Earth? How do two handheld phones that are communicating with each other achieve the same objectives? How does one ensure that the extensive hardware designs and use of very large-scale integration (VLSI) can be implemented with a "get it right first time" methodology? There is no time to do it over. Can the capacity requirements of Thuraya be achieved and allow many thousands of simultaneous phone calls across the regions? Clearly, early risk retirement of key technologies was a must, and in addition, it quickly became apparent to all of the team at Hughes that the Thuraya team was going to make valuable contributions to the overall program success, and polite gentle nudges were generally well received.

The program management challenges were considerable. To deliver a total system and provide true end-to-end communications demanded very close coordination on detailed schedules, milestones, a variety of frequent unit test and subsystem compatibility tests, and sometimes specification trades. In the early days of a program, a critical success factor is being able to identify and retire through analysis means as many of the critical design performance parameters. Mistakes in this phase tend to become much larger problems when the hardware fabrication and build phase starts later on. Hughes and Thuraya spent a lot of time and effort on this phase, with a constant stream of checks and cross checks, culminating with a preliminary design review.

Third-party suppliers of subsystems and components were identified early and tracked on a regular basis for both schedule and technical compliance. It is often the small stuff that ends up driving the schedule and performance and not what is anticipated.

PHYSICAL DESIGN AND PHYSICAL LAYER

Satellite system choices ranged from low Earth orbiting (LEO) satellites to medium Earth orbiting (MEO) satellites and geostationary (GEO) satellites. The higher-altitude GEO satellites have the advantages of covering a larger geographical area with a single satellite, assuming the added path length and associated added latency can be accommodated as part of the system and design to provide the required service quality. A single GEO satellite can therefore cover a larger regional footprint, compared to using multiple MEO or LEO satellites, and also meet the required traffic capacity needs. In addition, the latitude requirements of the Thuraya coverage area did not require either MEOs or LEOs. The key remaining technical system challenges were link design and system capacity. Hughes took maximum advantage of digital signal processing, low-rate high-quality voice coding and forward error correction and the latest techniques in solid-state RF and advanced antenna designs. The spectrum available to Thuraya for the system consisted of portions of the WARC allocated L-band. This is very precious spectrum. The Thuraya system design utilized spot-beam technologies to provide for a frequency reuse capability, similar to frequency reuse deployed in cellular networks, but over a wider geographical footprint.

Hughes shall specify the common air interface (CAI) defining all protocol layers and performance of the satellite interface to be implemented by the user terminal. The lower layer physical interface shall be optimized for satellite resource efficiency and user performance over the satellite link.

However, with a weight limit of approximately 200 gm for the HHT and a size profile comparable to cellular phones at the time, a new CAI would be required to achieve the desired quality of service. Hughes and Thuraya embarked on the Global Mobile Radio-1 (GMR-1) standard for mobile satellite communications. The GMR-1 MSS standards process was convened at the European Telecommunications Standards Institute (ETSI), and later at the Telecommunications Industry Association (TIA), with the intent to additionally gain the checks and balances provided by the industry technical forums of the standards bodies for the products and technology and provide both service providers and UT manufacturers with the same confidence.

COMMON AIR INTERFACE

The design approach utilizes frequency division multiple access (FDMA)/ time division multiple access (TDMA) mode of operation. In the late 1970s and early 1980s, Hughes had pioneered both TDMA and FDMA techniques in satellite communications for many years and solved many of the unique technical problems associated with timing, synchronization, Doppler

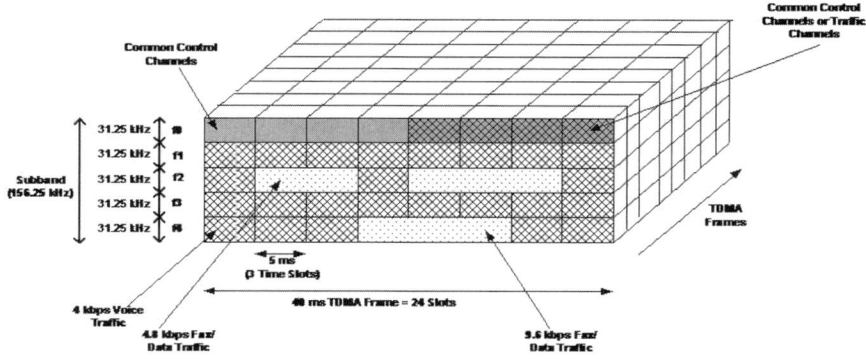

Fig. 6.2   The TDMA frame format.

correction, delay, latency, etc. In the early part of the 1990s, these techniques were applied to develop the latest generation in low-cost very small
aperture terminals (VSATs). With TDMA, each user terminal can transmit
a set of data for a precise period of time on a single frequency, and multiple
terminals have access to the same frequency. System synchronization and
individual precise timing by each terminal ensures that no two transmitted
bursts overlap. With the potential fractal nature of available spectrum across
the area, a minimum assignment of as little as 31.25 kHz was defined for
each single frequency. The TDMA frame format is shown in Fig. 6.2,
allowing up to eight voice users to transmit on a single frequency. With
careful coordination between satellite and gateway stations (GS), up to 32
such frequency carriers can be accommodated in 1 MHz of spectrum, across
the coverage area. GEO satellites have the longest delay path compared to
other lower Earth orbits. However, in GMR-1, unique single satellite hop
UT-UT calling was possible, and this saving of a second satellite hop was
a primary capability in ensuring cellular phone-like quality of service to
areas with no adequate landline coverage or wireless coverage.

GPS capability shall be part of system and user terminals capability.
Although GPS systems and stand-alone GPS devices were available at the
time, the deployment of dual-mode HHTs with integrated GPS receivers
was state of the art. Satellite communications networks typically operate in
a line-of-sight coverage, unlike cellular communications with its large
population of base stations. For the Thuraya system, the GMR-1 design
provided an additional capability such that *the system shall provide the
capability to increase coverage to users in disadvantaged positions such as
when the user is in a building.* By reducing the actual information data rate
and increasing the amount of forward error correction and clear sky link
margin, a user could be notified in building on his user terminal display of
a pending phone call.

The design demands of such a complex system forced physical hardware designs to be largely determined and concluded very early in the program. Although digital signal processing permits flexibility in requirements and reconfigurability, the Thuraya system capacity and scale meant that extensive use of very large-scale integration (VLSI) was required. This placed a burden on designer and customer alike to ensure that a number of requirements and derived requirements were determined, analyzed, and validated by the completion of the preliminary design phase in early 1998. VLSI designs can take from two to three years to complete from concept through production. First-pass success is a must, and mistakes can add many months to a schedule. The test and validation phases are as important as the design phase to ensure requirement completeness.

One of the more stressful situations for an engineer is the design areas that have to be done flawlessly, where there are no reprieves and no software design bailouts. This is particularly true with the complex digital communications designs for the physical layer of layer 1 of the Thuraya program. Many design implementation decisions could be delayed or made programmable; however, this is not the case with some of the fundamental physical layer definitions: the team got one chance to get it right to avoid significant delays and expensive repetitions.

The fundamental system performance specifications and hence derived specifications in the areas of system capacity, link performance, and design margin, and elements dominating end-to-end delay and quality of service to users had to be largely retired as risk areas by this time. The physical layer design driving the users/MHz and the UT received signal-to-noise (E_b/N_o) performance were largely determined at this time. *The subjective quality of voice in terms of mean opinion score (MOS) shall be statistically equivalent to that obtained by GSM full-rate vocoder in clear channel.* Subjective quality was also characterized in impaired channel conditions. *Subjective tests shall be administered in an independent laboratory using relevant ITU-T P series recommendations.*

Although it was still about two years away, significant time was now applied to planning both the factory and field integration and test programs. Subcontract management of all physical components is at a high priority in this phase. Operational concepts, although still not in full swing, start to take on more significance. Emphasis will soon shift to software and services from a design perspective. Unlike the urgency to tie down hardware requirements and design, software engineers in many areas prefer to wait until the last possible moment before starting to code, to ensure coverage of all system requirements, including any recently identified items not covered by hardware.

CRITICAL DESIGN PHASE

The hardware designs were mostly completed by the end of 1998. Test programs were in full swing for satellite and ground hardware elements. Now the focus turned to systems, software, services, and network management and control. Whereas at the preliminary design phase, the system architecture and design aspects of the program were conducted ahead of the various segment designs, now at CDR timeframe, the system critical design phase should wrap up and complete all activities prior to full-scale integration. On the satellite front, the Hughes engineering, program, and operations teams were focused on completing any outstanding final designs, with unit and subsystem testing in full swing. The mix of engineering, program, and operations people was in itself a new experience for many.

A full suite of customer care support systems was envisaged and planned as follows: real-time rating, flexible tariffs, GSM compliance, account management, equipment provisioning, order entry and processing, purchase orders, translator support for a variety of languages, mediation and interfaces to system HLR, bill generation and handling, user access, and security management.

System performance monitoring and summary status and detailed status of system components was a requirement as follows: statistical data on calls, resource usage, and user location information; statistical data on calls on a per beam basis trending traffic capacities and densities; and radio interference management for both external and internal interference.

The CAI shall preserve the upper layers of GSM consistent with satellite functionality. Although the CAI upper layers shall be consistent with GSM at the time of contract award and system design, the design shall facilitate the future introduction of new services in the future.

HANDHELD TERMINAL

DUAL-MODE OPERATION. *Handheld (HTT) and vehicular terminals shall operate as dual-mode devices for public land mobile network (PLMN) and geomobile system access. The cellular mode shall be implemented according to the GSM specification and the satellite mode according to the GMR-1 CAI.* Given that the GMR-1 and GSM CAI provided equivalent network service functions, they have identical layered protocol architectures. The differences between the GMR-1 and GSM CAI were determined by the design optimizations required for satellite resource efficiency and user performance over the higher latency satellite link. *The dual-mode GMR-1 user terminal equipment shall utilize a unique SIM for both satellite and GSM mode.*

HHT SIZE. Everyone wanted a small phone, with lots of features. The design requirements meant that the Thuraya phone had to be considerably smaller than equivalent satellite phones used on the LEO satellite systems. Conventional wisdom says that the closer the satellite, the smaller the phone. Thuraya business and marketing wisdom said something different: in order to sell and market the service, it had to be comparable to cellular service. Traditional GSM cellular capability was also a requirement. Additionally, it would be great to add a GPS receiver into the phone. In 1997, GPS receivers were traditionally external stand-alone devices, not embedded in a trimode satellite phone. The first-generation Thuraya dual-mode HHT with GPS receiver is shown in Fig. 6.3.

SERVICES MAN–MACHINE INTERFACE. The UT architecture maximizes reuse between the two modes, GSM and GMR-1, which results in ease of use to the subscriber. To the man–machine interface (MMI) display, the Thuraya network is an alternative to the GSM PLMN, or alternatively, the GSM PLMN network is an alternative to the Thuraya network. Terminal registration can be done either manually or automatically.

Thuraya conducted extensive focus groups on the user segment, including HHT aesthetics, the MMI features, and the variety of languages required. By product launch timeframe, Thuraya engineers were very familiar with the HHT operational concepts and contributed to many of the sequences of operation.

Fig. 6.3   The first-generation Thuraya dual-mode HHT with GPS receiver.

SERVICES

Service features, network interfaces, authentication and privacy shall follow the cellular GSM standard. Technically, this meant that the physical and data link layers would be optimized for the satellite link, but the higher layers should follow PLMN. *The offered services shall follow GSM and ISDN standards, with adaptation for satellite use, and both mobile-to-fixed and mobile-to-mobile services shall be offered single hop delay through the satellite.*

Asynchronous data and facsimile calls shall be supported, in addition to a range of supplementary services. This includes user latitude and longitude location, to be displayed to the user upon request on the UT display. Additional features for the system as it unfolds include the following: variety of call modification services such as call waiting, call holding, and multiparty capability; and variety of call charging services, including advice of charge.

INTELLIGENT NETWORK SERVICES. *The gateway station and associated mobile switching centers shall support interface capability to equip future intelligent services as they unfold.*

SHORT MESSAGE SERVICES. *The system shall support mobile-originated short message service and point to point to subscribers.*

MODE SELECTION. *Service providers shall be able to program user terminals with network selection rules on the SIM card. Automatic network selection shall allow satellite mode priority, GSM mode priority, and the enabling or disabling of network selection.*

ROAMING AND REGISTRATION. *Subscribers shall be able to roam in accordance with GSM standards and agreements to be established between Thuraya and national service providers. Exchange procedures between GEM gateways and between Thuraya and GSM service providers shall be accommodated in the central OSS.*

The system shall support single number roaming of UTs according to GSM procedures. Location registration procedures were defined.

THURAYA PROJECT PERSPECTIVE IN 1999

As stated earlier, Thuraya's initial contract was very clear. However, to complement the contractual part, several Thuraya staff were sent to the United States to work on the program and to perform both periodic reviews and quality assurance throughout the various stages. So, when the system was delivered in year 2000, the process was very smooth. A proof of this is that Thuraya was able to make its first call six days after spacecraft deployment.

This was a result of proper program planning, which was implemented with entry and exit criteria to each stage and milestone. At the onset, the contract was signed with milestones, and one of the major ones was the preliminary design review (PDR), which was completed in February 1998. At PDR, Hughes described the preliminary system design to Thuraya. This was followed by the critical design review (CDR), which was completed in October 1998, and this was for the closure of the design phase. There were regular quarterly review meetings that enabled Thuraya to review the progress.

SYSTEM INTEGRATION AND VERIFICATION

Hughes and Thuraya had been planning the integration campaign since the beginning of the program. In late 1999, however, the pace quickened. A four-stage system integration and test campaign was conceived, namely, SIV1 (system integration and verification), SIV2, SIV3, and SIV4, with increasing scope as the program progressed through the final integration stages. The campaigns began in the factory. Where possible, verification was completed by either inspection or analysis. As a follow on, either a demonstration or test was performed for particular requirements verification, with the latter requiring performance data. As is typical with complex programs, performance data validation was more easily completed in the factory prior to commencement of the field test program.

A full suite of system operational concepts was developed and, where possible, partially verified in the factory by engineering staff. A number of tests were performed with multiple segments, using simulators and in some cases real hardware, and the end-to-end methodology helped retire a lot of system risk. With a well-thought-out multilayered architecture, with clear interfaces between layers, thread testing was accomplished with some success.

Even with the solid preparation and detailed design reviews, etc., the factory programs uncovered problems, not surprising given the complexity and the parallel nature of so many designs to meet the schedule constraints. For the most part design changes were not associated with the critical physical layer, which in no small way is a tribute to everyone involved.

In 2000, the system staging complexity in the factory was increasing as more pieces of the system came together, but at the same time the scale of things still to be accomplished also became more apparent. Striking the right balance between shipping equipment to the field for final operational tests versus continuing performance testing in the factory is often driven by commercial launch needs. Even in these scenarios, ongoing factory feature testing continued at the same time as full-scale system field deployment. Once the field program started, different software releases were carried between factory and field, and system stability and test coverage becomes the driver, before upgrading to new features.

Hughes established a system problem reporting (SPR) system, categorizing bugs into various categories, such as defects–revenue affecting and deficiencies–deviation from specifications.

DEPLOYMENT AND SERVICE LAUNCH

The Thuraya team believed that it was in the best long-term interests of the service to have a preliminary field deployment, integration, and test of several months before going live with revenue-bearing services. In the end, it proved to be the right strategy. Over several months, engineering improvements to the system and both operations and marketing processes allowed fine tuning of the system.

Satellite launch was successful on 19 October 2000, and the collective Thuraya and Hughes team celebrated in good fashion from Sea Launch headquarters in Southern California. For the Hughes engineering teams, there was still a lot of work to be done, as the field program moved into high gear.

Dave Roos is a senior Hughes transmission engineer and recalls this observation:

> A field complaint observed that the control channel signal levels were incorrectly changing. The SIV team proved there was a problem. The Gateway Base Station engineers believed that the GS signal quality was stable and non-varying. The Radio Frequency Terminal (RFT) engineers believed that the RFT signals were stable. Everyone knew for certain that it was someone else's problem yet each could not prove it.

All agreed there was a problem, and after a lot of creativity and soul searching on the most advanced state-of-the-art mobile satellite communications system to be deployed worldwide, the problem was isolated to a pinched cable between two ground subsystems. The lesson learned is "sweat the small stuff."

Channasandra Ravishankar (Ravi) is a senior transmission engineer at Hughes and recalls the following:

"I got an email from one of the engineers in the field, noting voice quality problems. This was a recorded voice test call between the Chairman and the Chief Technical Officer at Thuraya noting a problem in the compressed voice quality." Ravi (and his colleagues) spent the entire day and night trying to unravel the problem. He went home late that night and the next few nights. His wife Jayshri made note of the fact that Ravi was working on a different time clock than their family. Suddenly Ravi had a clue. He again skipped the late-night dinner, went back to the office, and suddenly realized that he had not checked one of the clock timing signals between the mobile switching center (MSC) and the gateway base station controller. Ravi simulated the potential impact of a mismatch and concluded that a voice corruption could occur with this scenario.

Ravi called the network control center and within minutes figured out that there was an incorrect configuration parameter set between the MSC and the PSTN. Problem solved, and normal dinner hours resumed for Ravi and family.

On 28 October 2000, the first over-the-air call was successfully transmitted, kicking off the start of SIV3. Marshall Fisher, a senior Hughes networking engineer recalls the following: "Placing the first call over the satellite, we had an external antenna on a preproduction HHT because we expected the signal quality to be insufficient without it, but after the first few successful calls, it was disconnected and it worked. This was cool!" For so many people who had worked with such dedication and focus, this one moment was an enormous achievement and a relief. Some time later, single-hop UT–UT communication was achieved interconnecting, basically, all of the system elements. Although much work was still to be done, the millions of man-hours of dedication were validated.

SIV4 began in December 2000. This was the full end-to-end system integration. On 1 May 2001, Thuraya announced that they were open for business. On 23 June 2001, the sample traffic data analysis for the previous seven days showed a total of 193,000 calls, with a 98% call success rate. Most of the requirements specifications were achieved, with some margin, and even though some things took longer than planned to complete, and usually in areas that could not be concluded in the factory, overall system performance was outstanding.

THURAYA ACCEPTANCE TESTING

A preliminary acceptance test was conducted to verify the requirements. After Thuraya accepted the system, it gave out the preliminary test certificate. The commercial trial operation took place in 2001. It was largely similar to a soft launch phase. This involved Boeing and Hughes operating the system for two months while Thuraya mostly took a back seat. Two months later, Thuraya operated the system while Boeing and Hughes took a back seat. Six months afterwards, there was the final acceptance of the system by Thuraya, where the system was contractually handed over to Thuraya with a five-year warranty.

THURAYA SYSTEM AND SERVICE LAUNCH

In preparation for the launch, Thuraya had to complete several technical, commercial, and regulatory milestones. It had the obligation to prepare the primary gateway with all of its key subsystems. It had to undertake a strong marketing roadshow to introduce its project in many potential markets and make the business case for it. This was essential for creating market awareness and establishing the necessary commercial and service distribution infrastructure, including appointing service providers (SPs). Additionally, the company had to get commercial licenses to operate in each country under its footprint, with importance given to those countries identified as priority markets according to marketing research. Securing orbital slots from the ITU for its satellites was also part of the regulatory challenges because the spectrum, which is increasingly a scarce resource globally, was coordinated by Thuraya itself.

It also had to prepare the billing system with certain service providers as well as the point-of-sale connections. It also had the responsibility to train its SPs and partners for customer provisioning and customer care. That massive activity was well managed and was highly momentous for Thuraya's success with commercial launch.

Thuraya's commercial service was launched very smoothly, and it was a well-organized process aided by the clear contract that explained all of the necessary milestones. There were few delays, mainly because of the good support from key suppliers, a strong team spirit, and a qualified team of consultants. In addition to the technical side, there was a strong marketing push whereby so many potential markets and regional GSM operators have already been acquired as Thuraya's partners.

Thuraya was initially launched with a few service providers among its key founders. However, after the final acceptance, the system coverage was made available to 99 countries with the use of 246 beams. Figure 6.4 shows an example of Thuraya users.

Fig. 6.4   Thuraya phone users on Mount Everest. (See also color figure section at the back of the book.)

USER TERMINAL PRODUCTION 2002–2007

A total of approximately 400,000 first-generation terminals were manufactured and delivered to Thuraya. The baseline terminal was the HHT, but, in addition, a variety of other terminal types, including both fixed and mobile, were produced. Although these volumes were lower than the several million of GSM terminals produced annually, they were significant enough to allow the manufacturer to accomplish the necessary significant advances in VLSI techniques and miniaturizations required to realize the desired size and weight constraints. In addition, the cost and reliability goals could be achieved by completing detailed automated manufacturing processes, production test fixtures, and warranty support.

As a follow on to the voice terminals, Thuraya procured several thousand data terminals for both fixed and transportable uses. These terminals deliver IP-like services across the coverage region at data rates in the range of 150 kbps.

The CAI, discussed earlier, meant that Thuraya built an air interface that was open to any technology developer without any technical restrictions. This has provided Thuraya with so much flexibility in its business operations by capitalizing on having many options when it comes to the development of its satellite products and applications. This was evident in the production of its second-generation handheld terminals launched in 2006. In essence, the CAI was another technical feature in Thuraya's technical design, which later proved to have a huge business advantage. Figure 6.5 shows a subset of the latest Thuraya product offerings.

CAI STANDARDIZATION

In June 1998, ETSI created a new GMR working group to establish a common air interface for geostationary satellite handheld terminals. In March 2001, a full set of GMR-1 Release 1 specifications were published by ETSI. In April 2002, the first updates to Release 1 were published by ETSI. These specifications contained all of the radio interface specifications for circuit mode mobile satellite services. In March 2003, the initial set of Release 2 specifications was published by ETSI. In March 2005, the first update of Release 2 specifications was published by ETSI. Release 2 contained extended specifications and new specifications for packet mode services (GMPRS). The specifications adopted a similar structure to that used by GSM, reflecting the correspondence between some of the GMR-1 specifications and the GSM specifications. However, because of the differences between the terrestrial and satellite channels, particularly in the lower layers of the radio interface, the standards diverged, and so, although some GSM standards applied, some GMR specifications had no corresponding GSM specification. The current

Fig. 6.5 (a) The second-generation handheld Thuraya SO phone; (b) the second-generation handheld Thuraya SG phone; and (c) the Thuraya IP data terminal.

status of the specifications is covered under ETSI TS 101 376. As a follow on, the GMR-1 specifications were approved by the TIA in November 2001, and the current status of these specifications is covered under TIA/EIA/IS-J-STD-782.

THE FUTURE

The Thuraya network today has several hundred thousand subscribers, with a variety of user terminal types for both fixed and mobile usage. Quality of service is second to none, and the system has been gracefully expanded to meet the growing subscriber demand. Wider bandwidth data services will follow in the 2008–2009 time frame, compatible with the evolution of equivalent services in cellular and fixed wireline systems.

Thuraya high-speed packet services provide high-speed Internet protocol (IP) connectivity between transportable/fixed terminals and Internet service providers. With the advent of IP-based services and the evolution of both 3G and 4G services and PLMN systems in the future, the Thuraya system is well positioned to provide an expanded range of features and capabilities to its subscribers.

Applications already been envisaged and deployed, including the following: interactive Internet access from a device attached to the Thuraya terminal; FTP in either direction between the Internet and a device attached to the UT; e-mail; streaming applications and lower latency streaming applications; video applications; VPN support for the widely used VPNs in the marketplace; TCP/IP applications; and higher speeds, higher QoS, consistent with efficient satellite system resource uses.

CONCLUSION

Ten years on from concept, Thuraya is the most widely used, and most resource-efficient, geomobile satellite communications system worldwide. The Thuraya HHT delivers best in class mobile voice and data services, and the system achieved both technical and commercial success. Such an achievement is of course worth celebrating in any field; however, in the mobile satellite field during the early 2000s it was especially noteworthy because it vindicated the flexibility and reach of the satellite medium once again after some unfortunate commercial failures only a few years earlier. Success of this magnitude always owes a lot to the hard work and perseverance of many professionals working tirelessly both in the customers' and manufacturers' organizations. A strong additional enabling factor was that Thuraya had the foresight to contract the complete project to one single entity and worked side by side throughout the project. This chapter is a tribute to all of the contributors to the success of the Thuraya program and to their talent and dedication.

Chapter 7

Eutelsat: An Impressive Success

GIULIANO BERRETTA*

INTRODUCTION

"A true broadcast service, giving constant field strength, at all times over the whole globe would be invaluable, not to say indispensable in a world society." Many readers instantly recognize this visionary and simple declaration by Arthur C. Clarke in his groundbreaking article in *Wireless World* in 1945. It took another 20 years, with the launch of Early Bird, for the world's leading rocket scientists to master the propulsion that was needed to put an artificial satellite into the orbit identified by Clarke. His genius was to realize the amazing potential of the virgin real estate in geostationary orbit and the cost efficiency of using space-based rather than terrestrial relays to build a global network. His reward is to see the scale of the impact of using the geostationary orbit on our everyday lives.

To a journalist who questioned him in 1982 on his role in the worldwide history of communications, Sir Clarke replied: "If I had not myself formed the idea of geostationary repeaters, at least half a dozen other people would certainly have taken my place at almost the same time. My only contribution, I think, is perhaps to have accelerated the progress of telecommunications by about fifteen minutes, let's say twenty."

For all of us who since then have made some contribution to commercial satellite communications, we would agree that the dimension of our sector has surpassed all expectations. Clarke thought three satellites with 120 degrees spacing would be enough to satisfy the requirements of a global communications network. More than 30 years later, in 1977, when Eutelsat was created, the general belief was that two in-orbit satellites (one operational and the second a spare) would fulfil Europe's satellite requirements over the next

*Chairman and Chief Executive Officer, Eutelsat Communications, Paris, France.

10 years. The reality is that satellite communications have grown in pace with what today is a vast and ever-growing TMT (technology, media, and telecommunications) sector. The trick behind the "success stories," which is the title of this book, is to understand that commercial satellite communications have functioned best when they have adapted within the overall sector of which they form a modest but essential and highly profitable subsegment.

From their initial key activity of transatlantic transmissions, followed by intracontinental telephony and data and analog television broadcasting, satellite services have gradually moved towards digital video distribution for national and regional markets in developed and emerging regions and towards VSAT corporate data services for multinationals. More recently, we have seen the emergence of Internet protocol (IP) broadband services mainly in regions not served well by terrestrial broadband infrastructure.

This commercial development underscores the particularly high levels of adaptability of satellites that stems from their ability to provide ubiquitous coverage, from their high levels of flexibility and their capacity to transport all service protocols. Equipped with these fundamental assets, satellites are the most powerful tool for content distribution to terrestrial networks and direct to users in areas not served by terrestrial infrastructure. As such, they are guaranteed a privileged place in telecommunications and broadcasting. Moreover, the satellite operators who have played to these assets are the ones who have most thrived, whereas those who stuck to the initial set of services have found themselves overtaken by more competitive technology solutions.

Today, telecommunications and media landscapes are diversifying and fragmenting way beyond the three basic elements of 25 years ago of PSTN, long-distance network and broadcast network, to a myriad of solutions, which include GSM, GPRS, UMTS, ADSL, digital cable, WiFi, WiMax, DTT (digital terrestrial television), and FTTH. Each can support a range of applications that are increasingly diversified, both in terms of utilization and underlying technology (compression, encoding, software, middleware, and terminals). On top of this fragmentation, the emergence of substantial new geographic regions with important potential for growth is also broadening the landscape.

In this context, the ability for players in the sector to be able to adapt to a frequently, if not constantly changing environment, is even more critical. For Eutelsat, the capability to adapt our satellites to fit the requirements of an evolving market was a key factor driving our development and underlying our success. Initially designed to support a pan-European long-distance telephony network, our satellites were key to the launch of cable and satellite television in western Europe as well as eastern Europe and North Africa, and they also facilitated the introduction of VSAT corporate data networks.

This was followed by the launch and rapid takeoff of digital television, starting in the mid-1990s. The number of digital video channels on our satellites increased over 10 years from 50 to over 3000, of which more than 1100 today broadcast from our HOT BIRD video neighborhood at 13° E. Since 2007, the HOT BIRD satellites are also broadcasting Europe's first high-definition television channels, which represent the next step change in the broadcasting environment. More of that will be discussed later.

START OF EUTELSAT

Set up by Postal Telegraph and Telephone (PTT) administrations and telcos from 26 countries in western Europe and growing into an intergovernmental organization with 48 members in a landmass stretching from Portugal to Kazakhstan, Eutelsat is now a private company listed on the Paris stock exchange, with a particularly international "DNA" via 540 employees of 27 different nationalities. For many, this level of diversity might represent a handicap in view of the challenges of forming a homogeneous, coherent whole. But at Eutelsat, we believe cultural diversity is a source of innovation as well as a competitive advantage in addressing local markets.

My own inside experience of Eutelsat dates from 1990. I arrived from ESA, the European Space Agency, where I was a part of the first group of research engineers who created the European Space Applications program. At this point, Eutelsat had completed the EUTELSAT I (Fig. 7.1) program of satellites (also known as ECS, European communications satellite) supplied by ESA and had embarked on its own EUTELSAT II program of more powerful satellites. My first task was to set up and drive a commercial department.

Fig. 7.1 The EUTELSAT 1 satellite. (See also color figure section at the back of the book.)

This was a bold undertaking for an organization whose income was derived mainly from leasing capacity to its own shareholders who were telecommunications administrations and incumbent telcos.

Modest though they were, the EUTELSAT I series ignited a spark in Europe by proving that satellites could provide a new network resource to television channels with no access to terrestrial frequencies and also enable public broadcasters to extend their reach to international and expatriate audiences. The company that was emerging as our principal competitor in Europe had also identified this potential, and by the late 1980s SES Astra was offering capacity for DTH broadcasting in Germany and the United Kingdom.

Some thought that Eutelsat had no future in the television sector. Their belief was that the best markets had already been taken, and that in other countries, such as Italy, the abundance and dominance of terrestrial channels closed the door to new electronic media and space-based communications. As for developing countries in central and eastern Europe, North Africa, and the Middle East, they were not thought to be ready to afford the "luxury" of satellite television.

We were not deterred by these assumptions. On the contrary, we believed that southern, central, and eastern Europe as well as Africa and Turkey, where there was little cable and large appetites for television viewing, held the right set of ingredients to begin to move into the satellite television age.

The emergence of this new electronic media played out against the backdrop of the fall of the Berlin Wall. This had two consequences: the rapid arrival, which we strongly supported, of central and eastern European countries as members of Eutelsat and the emergence in these countries of entrepreneurs who used electronic media to complete an historic transformation to market-led economies with media plurality. Our immediate contribution was to go back to the drawing board and modify the core coverage of our future satellites, as well as redesign one already in construction, in order to stretch it across eastern Europe and the Middle East to accompany public and private broadcasters in expanding their audience and launching new channels.

MOVE TOWARDS TELEVISION

By moving our center of gravity towards television, we radically changed the mission of Eutelsat. This transition did not happen without long and intense debate with those who insisted that it was better to leave television to others and for Eutelsat to concentrate on telephony, which was considered our natural vocation. In the end, our convictions won the day, and video was progressively solidly established as our primary source of income.

The next big challenge on our agenda was digitization, which some believed would see the business of commercial satellite companies shrink drastically. By switching to digital, up to 10 channels could be accommodated in the bandwidth occupied by a single analog channel. If the market had remained unchanged, it would only have required one satellite or two at the most to transmit over Europe all of the analog channels then transmitted in orbit in digital format.

We wagered that compression would reduce transmission costs, making capacity more commercially attractive by lowering the barrier to entry into broadcasting. We backed up our conviction by ordering a string of HOT BIRD broadcast satellites that would enable us to exploit the full potential of Ku-band frequencies at our key neighborhood of 13° E. From a maximum 16 analog channels in 1990, we banked on the potential of over 1000 digital channels once the switchover from analog to digital was complete. By mid-2008, this switchover was almost behind us. Our HOT BIRD neighborhood broadcasts over 1100 channels, and only three channels continue in analog.

This strategy of building a multisatellite video neighborhood has enabled us to offer capacity to a wide range of television channels, which can be received with small antennas, thereby attracting viewers and broadcasters in a virtuous circle that has brought excellent results. Today, HOT BIRD represents not only a constellation of high-power satellites with high intersatellite redundancy, but also a well-known international brand that reaches into over 120 million homes throughout Europe and the Mediterranean Basin. Anchor pay-TV platforms using the HOT BIRD neighborhood include SKY Italia, Cyfra Plus, and MultiChoice Hellas while Al Jazeera, the BBC, Bloomberg, CNN, France 24, Discovery, National Geographic, Polsat, RAISat, and Viacom are just some of the many broadcasters also using the satellites to build their own global networks.

These key clients are also among the first to pioneer high-definition television (HDTV), which marks the newest change in broadcasting. High definition is a logical development in sophisticated image delivery, following in the wake of the path carved by DVD and consumer appetite for plasma and LCD screens. Assuming a replacement cycle of TV sets over six years on average in most countries in Europe, penetration of HD-ready sets expected to approach 10% in 2008. HDTV should be further supported by the arrival of HD DVD players and growing interest in home cinema.

In this new HDTV world, satellite is the only "pipe" capable of broadcasting high definition in MPEG2 and MPEG4 with no constraints on bandwidth or geographic location, unlike cable, terrestrial spectrum, and ADSL networks. In the emerging European market, Eutelsat, by mid-2008, was already broadcasting over 60 HD channels. Key sporting events such as the 2008 Olympic Games triggered strong demand with consumer take-up only conditioned by the availability of MPEG4 decoders.

HDTV is also a unique product in that it assembles players along the complete broadcasting chain in a common aim to create value creation from filming, to distribution, to end-user viewing experience. For satellite operators, it indisputably constitutes an important engine for growth, which we believe will occur for some time in parallel to continuing growth of standard digital (SD) channels. For pay-TV operators it presents an unquestionable opportunity to sell new premium product in addition to a choice of hundreds of SD channels. However, we do not believe that HDTV will be limited to pay television. Public service broadcasters around Europe, including the BBC, RAI, and France Télévision, are generating and broadcasting flagship content, and it is reasonable to assume that the overall level of mobilization and motivation will in time establish HDTV as the standard for television. We also believe that central and eastern Europe will advance at a similar pace to western Europe into HDTV.

The key assets of bandwidth availability and geographic coverage that have already secured the role of satellites in an environment populated by SD and HDTV also apply for other fixed and mobile high-bandwidth video networks where satellites have a card to play and where Eutelsat in particular focuses strong attention. The ability of our satellites to deliver content to terrestrial headends, which began with cable, has expanded into feeding DTT transmitters, DSLAMs for IPTV, and DVB-H transmitters for video reception by handheld mobiles.

For Eutelsat, feeding DTT networks is a logical evolution of a longstanding business of analog terrestrial channels to transmitters and direct to homes located beyond quality reception (or even any reception) through the regular over-the-air network. In France, our 5° W location, occupied by ATLANTIC BIRD 3, has been fulfilling this task for many years in the analog world. Since 2005, ATLANTIC BIRD 3 has also been delivering digital multiplexes to the network of DTT transmitters, which will ultimately number 115 across France, serving about 95% of the population. This still leaves several million television homes in mountainous or border regions beyond DTT reception, an issue that will be resolved through a satellite solution in order to complete digital switchover in France in 2011.

The solution emerging in countries such as Switzerland is to go direct to home via satellite in the areas not served by direct to home DTT. This hybrid solution has the dual benefit of considerable cost savings (serving 10% of homes in rural or frontier regions through miniterrestrial transmitters is almost as costly as serving the rest of a population) and facilitating analog switch-off to release spectrum for additional digital services.

IP-BASED SERVICES, BROADBAND, AND OTHER INITIATIVES

The same business logic applies in the IP world. In the overall TMT sector, IP is the area most radically and fundamentally changing business and consumer habits, not to mention the "A list" of global communications empires.

Here, as I indicated at the beginning of this chapter, satellite is a niche product that can be profitable if it identifies the right markets. Our core strength lies in delivering IP connectivity to areas that for economic and geographic reasons are likely to remain permanently, or for the foreseeable future, beyond terrestrial broadband. It also applies in locations where networks have been disabled by a disaster and who need emergency response applications. Experience in North America with Hurricane Katrina and following the tsunami in Southeast Asia were dramatic reminders of the critical nature of communications for disaster relief.

Standardization, shared bandwidth solutions, and automatic antenna pointing are some of the technology developments that have enabled satellite-based solutions to enter the catalog of emergency products used by public service agencies and aid organizations around the world. For example, firefighters in the south of France are using the D-STAR turnkey broadband service, commercialized by our Skylogic affiliate, for establishing VOIP, fax, and IP connectivity to central offices in Marseilles. Within 20 minutes, equipment delivered by helicopter can be set up in a region with no GSM and no radio in order to download key information including maps and the location of firefighting units.

The benefits of serving regions beyond terrestrial broadband networks also apply in the enterprise market. VSATs are a familiar and long-standing solution for multinationals in retail, banking, and the oil and gas sectors for applications such as Internet access, data transfer, and videoconferencing. The value of broadband in today's highly competitive markets is such that we will soon get to the point where it is virtually unthinkable for even small enterprises to operate efficiently without broadband access. However, we might never get to the point where enterprises in all locations have a local ISP offering "always on" of 50 Mbps.

This need for speed to maintain a competitive edge is driving the industry expectation that satellite broadband access for small businesses will be the fastest-growing segment of the broadband satellite market. At Eutelsat, through Skylogic, and service providers using our capacity, we are seeing demand from users in western European countries where terrestrial broadband infrastructure is well developed but still incomplete. However, the big dynamic is playing out in regions in eastern Europe, the Middle East, and Africa, where we have strong satellite coverage and where we continue to build up our broadband distribution networks.

Another developing frontier for broadband is the maritime market. In the same way that high-speed video networks can combine satellite with DTT and DVB-H, we see the logic in combining technologies that enable the end user to have a seamless and single interface. With this in mind, Skylogic has teamed up with Maltsat International in order to bring GSM and broadband connectivity to passengers and crew on ferries in the Mediterranean. A single

D-STAR antenna connects with GSM base stations located on ferries and also provides VOIP and IP access. Passengers pay for Internet access through fixed PCs or a Wi-Fi network using prepaid calling cards. For the crew, the service also enables real-time connectivity with headquarters for supply-management logistics and on land control of safety systems.

I believe our imagination has no limits on how satellites can fit into an expanding communications universe. Ironically, however, space does have its limits. The Clarke Belt is by definition finite, giving us room for 120 to 180 satellites at two- to three-degree spacing. Eutelsat is among a few satellite operators that have optimized frequencies at an orbital location by stacking a number of satellites into a virtual box where they orbit at distances of 70 km from one another. To receive antennas on the ground, they appear to be a single transmitting device beaming many hundreds to channels and services.

This means operators have had to find smarter ways to exploit available slots and frequencies in order to extract maximum value out of our available real estate. At Eutelsat, some of the initiatives we have pursued to do this include the following:

- We have built new neighborhoods for targeted regions through other multisatellite positions, notably 36° E, where we have collocated two satellites, W4 and SESAT 1, which serve Russia and the Ukraine through one beam and sub-Saharan Africa through a second beam. This has enabled NTV Plus to build a subscriber base of 560,000 pay-TV homes in European Russia. The second arm is used by MultiChoice Africa for its DSTV platform, which serves regions in sub-Saharan Africa stretching from Gambia to Mozambique.
- We offer high-power steerable spotbeams that can operate like a virtual national satellite on an international satellite platform or be pointed in-orbit to a region experiencing sudden demand.
- We have partnered with other satellite operators at shared orbital locations in order to optimize frequencies, including with Nilesat with whom we have collocated at 7° W the ATLANTIC BIRD 4 satellite with Nilesat 102 and 103.
- We have fostered reception from more than one orbital position with double-feed antennas. Our most recent move towards this direction came in 2007 with the EUROBIRD 9 satellite whose three-degree proximity from the HOT BIRD neighborhood makes it possible for homes to receive channels from both positions with a single antenna.
- We are entering new frequency bands. For example, our KA-SAT satellite, to be launched in 2010, will enable homes equipped for DTH reception of TV channels to use the same dish for a host of IP-based services, including Internet access, VOIP, and IPTV. Until then, we are already using Ka-band capacity on our HOT BIRD 6 satellite to support a

broadband service called Tooway, which is delivering a broadband service comparable to ADSL in terms of speed and cost.

• We are exploiting the potential of hybrid networks for new-generation mobile video services. With this in mind, Eutelsat is collaborating with SES in a joint investment in Europe's first satellite infrastructure for broadcasting video, radio, and data to mobile devices and vehicle receivers. A 50/50 joint venture company called Solaris Mobile has been set up by both companies to operate and commercialize an S-band payload on Eutelsat's W2A satellite, which will be launched early 2009. The S-band (2.0 and 2.2 GHz), which is located next to the frequency band used in Europe for 3G mobile services, provides a set of frequencies optimized for wireless distribution networks for delivering video and other services to mobile devices, including phones, PDAs, laptops, and vehicle receivers. The S-band payload on W2A will be an essential building block for a hybrid infrastructure over Europe, combining satellite and terrestrial networks, to provide universal coverage and indoor penetration for mobile video services that are expected to be used by 25 million people in western Europe by 2010. The payload has also been optimized for a broad range of business applications such as security surveillance and two-way commercial data services and also ancillary services linked to the Galileo positioning system.

And what of the satellites themselves? Today's commercial communications satellites have an average life in orbit of more than 15 years, compared with seven years in the 1980s. If you factor into this the two to three years needed for the design and delivery process, satellite operators need to have a pretty clear view on market developments over two decades to protect themselves from any market misassumptions. With this in mind, our satellites are designed to meet the today's demand, but also, and above all, they are as transparent and as flexible as possible in order to adapt to technologies and services to ensure they can be a profitable investment right to the end of their life cycle.

Eutelsat has championed new satellite technologies, including the following: lithium-ion batteries that can store the same amount of energy as nickel-cadmium or nickel-hydrogen batteries in a much smaller and lighter package for longer satellite life; greater onboard power for smaller on-ground receive and transmit antennas; steerable spot beams to modify coverage in orbit; onboard processing (Skyplex) to integrate single uplinked carries into a digital multiplex; and multiple spot beams associated with high frequency reuse to deliver radically enhanced economic performance for consumer interactivity.

Our efforts in space have been matched by initiatives on the ground that include our work on standardization within the DVB (Digital Video Broadcasting Group), particularly the working groups developing DVB-S, DVB-RCS, DVB-IP, and DVB-SH, as well as development of double-feed antennas and add-on LNBs to receive channels from more than one orbital location with a single dish. These technological innovations are based on a policy of openness aimed at the standardization of systems and products and the creation of systems facilitating satellite access under the best financial conditions.

CONCLUSION

I believe that the fundamental principles required for continuing to rise to new challenges in the TMT sector and extracting maximum value from new opportunities are in fact quite simple. They are driven by the need to over-come technical and technological issues associated with introducing new networks and services and by the need to limit the costs of network deploy-ment while enhancing performance and capacity. Options in our sector for achieving this second objective include the following:

1) Reduce launcher costs and raising launcher reliability. The consequence of the high cost and risk related to access to space has a direct impact on satellites that need to be built to last a number of years, resulting in costly programs. Research and design (R&D) initiatives are needed to achieve new breakthroughs in launcher technology, which could relieve pressure on satel-lite programs.

2) Reduce costs of terminals, hubs, and application software, and enhance transmission techniques. This could include using DVB S2 or other advanced systems for more efficient use of available spectrum or optimizing frequency bands, notably the Ka-band in association with ACM (adaptive coding and modulation), which makes very good use of onboard power through real-time correction of transmission parameters in response to weather conditions.

3) Develop solutions to address the gradual filling up of the geostationary orbit. These could include initiating use of new frequency bands such as the S-band for video to mobiles or new orbits such as high elliptical orbits, which are well suited for mobile services.

At Eutelsat, our "success story" continues to be driven by the need to maintain technological excellence and our determination to identify and anticipate markets and to work with other technologies towards customized satellite-enabled applications. We can remain confident in the fact that satel-lites remain the most efficient, secure, and ubiquitous solution for delivering video and data any where in the world.

However, we need to recognize that the Space Age, which was the great frontier of 30 years ago, has been replaced by the Broadband Age and the rich media Mobile Age, which are the two key technology developments driving our overall sector. The burning issues that companies are now grappling with are how they fit into a converged media marketplace populated by time-shifting and place-shifting devices, social network websites, and self-generated content. Any company in our sector that considers itself immune from the fundamental consequences in consumer habits ushered in by these innovations would be wise to reconsider its opinion.

Personally, I believe that that these new technologies are much less of a threat and much more of an opportunity for our sector. The need to shift, deliver, control, and store content has never been higher, and this brings us closer to content groups, telecommunications providers, Internet service providers, broadcasters, and terminal suppliers in delivering programming, Internet, and interactive services through multiple transmission-agnostic platforms.

We have gone way beyond the three-satellite global network identified by Arthur C. Clarke as an alternative to terrestrial relays, and we are still scratching the surface of the potential of our real estate. This first chapter in our success has been exciting enough, and we are fired up for writing the next in which our role will continue to grow and improve with new products, new services, and fresh innovation.

Chapter 8

THE MOUSE THAT ROARED: VIA LUXSAT AND GDL, HOW SES BUILT ASTRA INTO THE WORLD'S PREEMINENT SATELLITE OPERATOR

YVES FELTES*

SES is one of the world's most remarkable companies, having grown from a risky, and some claimed foolhardy, venture to today's market leader in satellite broadcasting and communications.

When Jacques Santer (president of the European Commission between 1995 and 1999) became head of the Luxembourg government in 1984, there were some who declared that he was risking his political career in lending support to one of the most ambitious projects of his eminent predecessor, Pierre Werner, and indeed one of the most courageous for the country of Luxembourg itself: propelling one of the smallest countries in the world to the forefront of commercial space exploitation. Most industry observers admit that they would never have dared to imagine that the then tiny Luxembourg ASTRA satellite system would one day not only be the uncontested leader of direct-to-home (DTH) reception in Europe, but also, with its giant fleet of 40 satellites in 26 orbital locations under the SES banner today, emerge as one of the world's leading satellite operators.

SES has come a long way from those early days. Local political pressure and diplomatic quarrels surrounded the initial project, in particular with the country's French and German neighbors and many so-called "experts" of the time who bluntly claimed that signals from a medium-power satellite of the ASTRA type could never be received on small-sized satellite dishes. SES has come a long way from initial cultural fears that imported American TV would swamp Europe. Indeed, ASTRA was accused of participating in an invasion of American cultural imperialism as illustrated by this quote from former French Post and Telecommunications Minister Louis Mexandeau who, in 1984,

Copyright © 2008 by Yves Feltes. Published by the American Institute of Aeronautics and Astronautics, Inc., with permission.
*Vice President Media Relations, SES and SES AMERICOM-NEW SKIES, Betzdorf, Luxembourg.

stated, "We are certainly not willing to allow Coca-Cola satellites to undermine our linguistic and cultural identity."

THE EARLY DAYS

Marcus Bicknell, SES commercial director from 1986 to 1990 and member of the board of directors of SES today, had this to say about the early days of SES:

> In 1985 SES had no money, no frequencies, no regulatory approval, no satellite, no rocket, no TV channels, no clients, no reception equipment and no viewers. We did have a small office next to the Luxembourg railway station in which the company's first engineer kept his soldering iron plugged in and smoking on his desk. Our critics said the company was fragile and the foundations hollow. Indeed the governments of Germany, France, and the United Kingdom were intent on sinking us. The French Minster of Communications in 1988 announced that he was going to put up spoiler signals to prevent the reception of ASTRA channels in France.

It was a long and winding road to reach the overwhelming success of SES today. So let us outline some of the crucial stages in the development of Luxembourg's satellite ambitions, detail the ingenuity of some of the technological choices that had to be made, and reflect on the political and economic context in which the project unfolded.

The country of Luxembourg looks back at a long history of liberal audio-visual policy dating back to 1933. Unlike its European neighbours, Luxembourg had decided against setting up a public radio system, instead granting the right to exploit the radio spectrum allocated to the Grand-Duchy to a commercial entity, CLR or Compagnie Luxembourgeoise de Radiodiffusion. In 1955, with the advent of television in Europe, CLR became CLT (Compagnie Luxembourgeoise de Télédiffusion), and under the brand name of RTL (Radio TV Luxembourg) developed into Europe's largest commercial television and radio broadcaster. It was within CLT (today RTL Group) that Luxembourg's first ambitions to become a satellite player first flourished: the LuxSat project was intended to give RTL's channels extended reach beyond terrestrial broadcasts of Luxembourg, which were by nature limited.

Jacques Santer was Luxembourg's Prime Minister in 1984, and there were many at the time who declared he was risking his career in lending support to projects of his eminent predecessor, Pierre Werner. (Luxembourg has a population size comparable to Rhode Island in the United States.) Most industry observers admit that they would never have dared to imagine that the then tiny Luxembourg satellite system would one day not only be the uncontested leader of direct-to-home (DTH) reception in Europe, but also, with its giant fleet of 38 satellites, emerge as one of the world's leading satellite operators.

However, in the same way as the other European projects at the time for "heavy" satellites of the French TDF type or the German TV-Sat, the initial Luxembourg project had some serious drawbacks. First, there was the very limited number of five transponders per satellite because of the very high power levels mandated for these satellites. Then there was the problem of restricted areas of coverage, the stipulation being that a satellite's transmission area should correspond more or less to the national borders of the country of origin. Each European nation had been allocated five dedicated frequencies in the WRC 1977 agreements. This was much less the result of any technical limitation and much more to do with political considerations. These discussions were well ahead of the eventual European "Television without Frontiers" Directive, and very few countries were inclined to accept that their citizens might access audiovisual channels broadcasting from "abroad" and, perhaps more importantly, outside the control of the respective national authorities.

It was probably the major financial commitment as much as any political pressure that finally led CLT to distance itself from LuxSat, encouraged, it must be said, by its French shareholders on CLT's board of directors, who argued in favour of the group aligning itself with the TDF project and were thus expressing the official point of view of France, which saw in LuxSat a potential and sizable competitor to be fought off using every available means.

CLT was still equivocating when, in March 1982, Pierre Werner found himself taking a tough line with CLT, setting a deadline for informing the government as to whether or not the company was intending to embrace the LuxSat project. If necessary, said Werner, the government could seek an alternative licensee. But Werner's request, just like a similar initiative from the Chamber of Deputies (National Assembly), did not even receive the courtesy of a reply.

FRESH THINKING

It was at this point that Pierre Werner turned to an American expert who proposed a new satellite design that had already proved itself in the United States: the famed "medium-power" satellites. The advantage of this type of satellite was that it had less transmission power, but the payload could carry a far greater number of transponders: 16 as compared with 5 on a "heavy" satellite! At the same time, thanks to the technical progress made in terms of reception equipment, the signals transmitted by these satellites could still be received using small-sized satellite dishes. Another considerable advantage of the fresh thinking was to use frequencies in the FSS (fixed satellite services) band, enabling circumvention on the geographical restrictions established in the BSS (broadcasting satellite services) band and allowing pan-European satellite footprints.

At that stage Luxembourg's satellite ambitions changed not only in terms of concept, but also in name: the system was called GDL, for Grand Duchy of Luxembourg. In November 1983, GDL was formally notified by the Luxembourg authorities to the International Telecommunication Union. In parallel, the research company Coronet was incorporated in Luxembourg, as was the Société Luxembourgeoise des Satellites (SLS).

However neither the change of name nor the change of concept succeeded in alienating the opposition that the project continued to generate, both outside and inside the country. In 1984, CLT announced publicly that it had reserved two transponders on the French TDF satellite and, through the mouthpiece of its then director-general, declared publicly that CLT would do everything "both legally and politically" to ensure that the GDL project failed.

PRESSURES: ECONOMIC AND POLITICAL

At the same time, new opponents entered the scene, in particular Eutelsat, which today is a privately held company, but at the time an intergovernmental organization of the various European post office and telecommunications administrations modelled on a similar operation, Intelsat, another international governmental organization (IGO). Eutelsat's director-general at the time publicly claimed that GDL was going to cause "significant economic damage" to the international organization and called upon all signatories of the Eutelsat Charter to boycott GDL. In the end it was only Luxembourg's threat of referring the matter the European Court of Justice that finally made Eutelsat abandon this course of action.

In the run-up to the legislative elections of 1984 in Luxembourg, GDL and its political as well as financial risks quickly became a topic in the electoral campaigning. The leader of the political opposition went as far as denouncing the project from the Speakers Desk of the National Assembly as a "danger for Luxembourg, and a danger for Europe."

These pressures, and the seemingly nonstop negative press that the project generated, made it increasingly difficult to complete the capital structure of the future operating company and to assemble the necessary funds for financing its first satellite. By mid-1984, the GDL project had come to a standstill.

The problems were recognized as being serious and involved the then newly elected Prime Minister, Jacques Santer, who called a crisis meeting held on a Sunday at his home. Following discussion with fellow cabinet members, he took the decision to continue the project.

But the crisis forced the government to recognize that the Luxembourg state would have to play an even more active role in order to bring together a group of investors, if it wished to carry out its satellite project. Santer therefore asked two public Luxembourg financial institutions, Société Nationale de Crédit et d'Investissement (SNCI) and Banque et Caisse d'Epargne de l'Etat (BCEE) to make the first move and take a holding in the company that was to be formed.

The Prime Minister then contacted a number of banks in Luxembourg to round off the financial "tour de table," consisting of two state-owned institutions and several industrial investors and foreign financiers. Several Luxembourg subsidiaries of leading German banks came forward, rapidly joined by two Luxembourg banks. They were thus following in the footsteps of SNCI and BCEE, which for their part had truly played the role of founding shareholders. This role was moreover honored by the allocation of category B shares when, on 1 March 1985, Société Européenne des Satellites (SES) was incorporated. As its name suggests, the company was positioning itself from the start as a European undertaking with exclusively European shareholders. The European character of the company was essential in order to alleviate the anxieties of those who continued to fear that Luxembourg's satellite ambitions were a Trojan horse for U.S. cultural imperialism.

Funding for the company in terms of equity capital was ensured, but the risks involved in the project were considerable, and SES was initially unable to find a banker prepared to lend it, without any guarantee, the additional capital required for purchasing and launching the first satellite. Once again the Prime Minister had to step in, taking the initiative to break the deadlock and enable the project to progress. He managed to convince the government and the parliamentary majority to adopt a law in January 1986 authorizing the State of Luxembourg to provide a State guarantee for a sum of 3.6 billion Luxembourg francs (more than 100 million US$). From that point on, the fate of Santer's political career was irrevocably linked to the success of the project.

However, problems remained. There were technical setbacks, notably with a failed deployment of a solar panel on the German TV-Sat, as well as commercial setbacks and uncertainties with its French TDF counterpart, which generated anxieties about the reliability as well as viability of satellite platforms in general. The discussions for and against heavy satellites and comparing them with medium-power satellites appeared to tail off, but other challenges still had to be faced, especially at the regulatory level.

It was only in May 1987, more than two full years after SES had been incorporated, that Luxembourg finally succeeded in having ASTRA registered as the first private satellite system with the IFRB (International Frequency Registration Board). After this major breakthrough, ASTRA was accepted first of all by Intelsat in October and finally by Eutelsat in November 1987. ASTRA, at least on paper, was now a recognized satellite player.

BUILDING THE SATELLITE

The ASTRA 1A satellite had, however, become seriously delayed following a launch failure that left the European satellite launch provider Ariane grounded from May 1986 onwards. It was not until January 1988 that Arianespace was able to offer a new launch date, planned for October/ November of that same year.

It was a period of intense work, late nights, and more than a few grey hairs. Everybody—staff at ASTRA, politicians in Luxembourg, and the financial community—was aware of the political and commercial stakes when, during the night of 10 December 1988, they watched a live video transmission from Kourou, French Guyana, of Ariane 4 on its equatorial launch pad with ASTRA 1A onboard. Seeing Ariane soar upwards into the skies of French Guyana was the apotheosis of a long journey! In truth, however, the journey and the hard work had only just begun.

As the first (and only) private operator of satellites in Europe, the future of SES/ASTRA was far from being ensured over the long term. Certainly, with ASTRA 1A, SES had a satellite that was optimized for offering direct-to-home services, and, certainly, the fact that high-profile names like Rupert Murdoch's group as well as broadcasters like Filmnet and Kinnevik were on the client roster augured well. But these operators were targeting specific markets such as the United Kingdom or the Scandinavian markets, whereas Société Européenne des Satellites, as its name suggested, aimed to serve the European market as a whole.

For the whole of the following year, 1989, there were still no German, French, Spanish, or Italian clients. ASTRA lacked attractive offerings for the audiences of several key European markets. Besides, the few channels already on air were embryonic, themselves struggling to find audiences.

Some help was on the way. The "Television without Frontiers" Directive, which finally came into force in the same year, successfully established the principle of the free movement of audiovisual programs in the member states and therefore fitted perfectly with the pan-European broadcasting operations of ASTRA, but SES had to admit that the vast majority of consumers in the various linguistic markets wanted most of all the availability of channels of national origin and/or in their own language before installing satellite dishes and decoders for a new mode of television reception. It was thus necessary to proceed step by step, one linguistic market after another.

By December 1989, a year after launch, the strategy of the company under its then Director-General, Pierre Meyrat, started bearing fruit. With his teams he had taken the strategic decision to focus sales and marketing efforts on the biggest linguistic market in Europe, Germany, and to reserve four of the remaining transponders on ASTRA 1A for German clients.

THE BERLIN WALL FALLS, AND GERMANY SIGNS UP

After months, if not years, of heated negotiations and much prevarication—more extreme with some than others, it has to be said—Sat 1, Pro Sieben, Teleclub, and RTL+ , Germany's only existing private private broadcasters at the time, announced that they would begin broadcasting from ASTRA in December 1989. Barely one month after the fall of the Berlin Wall, the announcement was perfectly timed: because of its ubiquity, ASTRA finally enabled the citizens of the new German states, and particularly those of the infamous "Valley of the Innocent" where terrestrial TV broadcasts from the West were unavailable, to access television programs from the "free world." The number of ASTRA dishes sold in Germany massively increased to such an extent that within two years the public channels were more or less forced to broadcast from ASTRA in order not to miss out on this growing audience.

The trend towards satellite dish take-up in the market was further bolstered by the judgment of the European Court of Human Rights, which, in May 1990, stipulated in the historic "Autronic" case that "the reception of audio-visual programmes by satellite originating from third countries [was] ... covered by Article 10 of the Convention on Human Rights." These developments led European countries such as Germany to finally abolish taxes on the installation of private satellite dishes. Although it is true that these provisions were akin to hidden regulatory assistance aimed at slowing down the advance of the potential competition represented by direct reception by satellite, and thus to support the state-backed—and financed—German cable plan, it became more and more difficult to justify to taxpayers the payment of taxes for the reception of private channels originating from a foreign private satellite system.

A LEAP INTO THE UNKNOWN

Sales of dishes were timely, as ASTRA was about to take a huge leap into the unknown. But boosted by its initial success, and in order to meet a growing demand for its services on the part of an ever-growing number of broadcasters and program makers, in March 1991 SES launched its second satellite, ASTRA 1B, under the slogan "The power of 2" and completed its first copositioning at the orbital position of 19.2° E. The company's copositioning strategy is one of the keys to ASTRA's success. In fact, it consists of positioning several satellites on the same orbital position and using adjacent frequency bands. This concept enabled SES to offer an ever-increasing number of TV programs on a single fixed satellite dish and thus become increasingly attractive in the eyes of consumers while offering unrivalled intersatellite backup capacity to customers. This concept was revolutionary at the time, and SES ASTRA today is still the only operator in the world to fly seven copositioned spacecraft simultaneously at its prime Continental European orbital slot of 19.2° E.

But new challenges were already appearing on the horizon, this time from Brussels and the European Commission. SES had quickly carved out a market of several millions of viewers watching signals based on the common European PAL transmission standard. They had witnessed the undeniable commercial failure of "heavy" satellites such as those used by French TDF or German TVSat. But the regulators seemingly wanted more, and all of a sudden the specter of D2Mac/HD-Mac, potentially including a new higher-definition standard, emerged.

In 1986, an initial draft European Directive had stipulated the use of D2Mac signals for all satellites using the BSS frequency band. Following intense lobbying on the part of leading European electronics groups, who perceived in the new standard a means of generating new sales of TV equipment, Brussels published in 1992 a second draft that provided for use of D2Mac for all new audiovisual services with effect from 1 January 1995, and this was to be by way of intermediate stage prior to the introduction, planned for 1999, of HDTV in the HD-Mac standard.

What would have been the result of implementing such constraining regulations? The answer is simple: it would have forced millions and millions of households to dispose of obsolete reception equipment and buy new in order to continue watching the existing channels! Incidentally, the result would have been to kill off, or at least seriously hamper, the development of the nascent market for direct-to-home reception via ASTRA in PAL mode.

It is therefore hardly surprising that SES/ASTRA quickly emerged as one of the most fierce and persuasive opponents to the adoption of the D2Mac/HD-Mac Directive. Continuously voicing its position, participating in professional colloquia, issuing press releases, and engaging in lobbying activities, SES gradually succeeded in rallying other players to its point of view. The company even organized a press trip to California, bringing together around 100 specialist European journalists, in order to make them aware, in the laboratories of Silicon Valley, of the technological developments on the other side of the Atlantic, where the all-digital era was already taking shape. For many impartial observers it became evident that the future of audiovisual media lay in going fully digital, a path that SES backed strenuously. This led to the company becoming a founder member of the DVB (Digital Video Broadcasting) group, which subsequently established the technical standard for digital transmissions worldwide.

Whereas ASTRA 1C and ASTRA 1D (launched in May 1993 and November 1994, respectively), with their additional capacities, served to boost the number of analog channels at 19.2° E, the launch of ASTRA 1E in November 1995 signalled the advent of digital television in Europe for the general public.

Digital transmission meant ASTRA's Betzdorf headquarters was once again subject to major changes. It was in 1987 that the first technical staff moved into Betzdorf, imbuing new life into the château, which had earlier been home to the Grand-Ducal family between 1953 and 1964. The initial control facilities for the ASTRA satellites had been put into service in 1987. As early as 1991, in view of the launch of the satellites 1C and 1D, the technical installations had been considerably enlarged. On 19 September 1995, SES, under the aegis of its new Director-General Romain Bausch, was able to officially inaugurate its new DTF (digital technical facilities), a veritable technological gem, allowing for the reception, monitoring, multiplexing, and, where needed, uplinking for hundreds of digital channels on the ASTRA system. Then, in 2001, the company opened its SOC (Satellite Operations Center) 2000, a state-of-the-art control center enabling simultaneous in-orbit supervision of 16 satellites from Betzdorf. At the same time, SES renovated the château from top to bottom and began construction of a new administrative building in order to cope with its ever-growing workforce.

With the arrival of digital television, ASTRA was also finally delivering on one of its original objectives: extending the appeal of ASTRA to new linguistic markets. The first client to launch a digital offering on ASTRA 1E, in 1996, was in fact Canal Plus of France, with its multichannel digital package Canal Satellite. The digital era also attracted Spanish and Polish clients to ASTRA, ensuring that ASTRA had a presence in the main linguistic markets—but still with the notable exception of Italy.

The success of ASTRA, and the demand on capacity, was such that it rapidly became obvious that a second orbital position, at 28.2° E would be needed. Once again, the company's expansion plans provoked a fierce battle with eternal rival Eutelsat, which was seeking to obtain the same orbital position—and the outcome of this altercation can still be observed today, with SES and Eutelsat locked up in a frequency-sharing agreement at the orbital position of 28.2/28.5° E over Europe.

DIVERSIFICATION—AND EXPANSION

There were other changes afoot, which would quickly lead to the creation of SES in its current form. Despite the undeniable success of ASTRA over Europe, the management at SES was aware of the medium- and long-term risks because SES was essentially a company with one product (the transmission of audiovisual programs by satellite) and a single market (even though this did consist of the whole of Europe).

Diversification of products and services as well as of geographical spread was needed. SES's first initiatives began in 1996 with the introduction of new services such as ASTRA-Net and, today, ASTRA2Connect. Although high-selling satellite services have still not produced the success initially expected among the general public, these technologies developed at SES ASTRA form the basis on which SES continued to diversify the portfolio of its services, in particular, in the fields of government applications, corporate networks, and future mobile applications.

In terms of geographical expansion, SES ventured outside Europe for the first time in 1999 with the acquisition of a 34.10% holding in AsiaSat, based in Hong Kong. In the following year, SES established its presence within NSAB in Scandinavia and Star One in Brazil. And in 2001 it achieved its current status as leading satellite operator worldwide by taking over the huge satellite activities of General Electric (GE) in the United States and bringing Americom into the fold. In 2006 finally SES acquired New Skies Satellite, adding extra capacity to the overall fleet and bringing new clients to the portfolio.

But there's more to SES's expansion strategy. It is developing new markets, not the least over Canada and Mexico. It has secured global coverage through SES NEW SKIES and continues to expand its overall fleet with currently 10 additional spacecraft under construction. Although it has unashamedly expanded its geographic coverage, it has not fallen prey to private equity investors, which is not the case with all of SES's main competitors. SES is publicly quoted on the Paris Euronext exchange and of course in its home country, Luxembourg.

Luxembourg is the tiny mouse that roared, and in barely 25 years SES has grown to an impressive size. In so doing SES has put Luxembourg firmly on the broadcasting map of the world. The achievement of the government, the ministries, the financiers, and even the royal family cannot be overstated. Tomorrow's technology-based world will throw up just as many challenges. Viewers, and our broadcasting and transmission clients, know we cannot stand still. The next 20 years will see HDTV become commonplace and probably also see other technological marvels for the broadcasting world to wonder at. SES looks forward to that future confidently.

Chapter 9

THE SUCCESS STORY OF JSAT

Yutaka Nagai*

WITH

Taketo Furuhata,[†] Susumu Kitazume,[‡] Shigeru Miyakawa,[§]
Eiichi Matsumoto,[¶] and Osamu Kato**

JSAT TODAY

As of the end of March 2007, JSAT Corporation owned a fleet of nine satellites in orbit (Fig. 9.1), providing a global service that included not just Japan, but also the rest of the Asia/Pacific region, Hawaii, and North America. JSAT successfully launched the JCSAT-9 satellite (operating in orbit as JCSAT-5A) in April 2006 as the successor to N-STARa and the JCSAT-10 satellite (operating in orbit as JCSAT-3A) in August to replace JCSAT-3. During 2007, JSAT planned to launch JCSAT-11 as a replacement for the backup satellite JCSAT-R, as well as Horizons-2, a satellite to be launched into orbit above North America in a joint project with Intelsat, which recently had been formed through a merger with PanAmSat. After successfully launching these spacecraft and deorbiting end-of-life satellites, JSAT at the end of 2007 owned and operated a fleet of 10 satellites. In terms of scale, this will make JSAT fifth worldwide behind Intelsat, SES Global, Eutelsat, and Telesat, and the largest satellite operator in Asia.

Japan Communications Satellite Company (JCSAT) was formed in 1985, launching its first satellites in 1989 and 1990 and commencing service with

*Director of the Board, Senior Executive Vice President, SKYPerfect JSAT Corporation, Tokoyo, Japan.

†Board Director, EMOBILE Ltd.; former President, International Digital Communications, Inc., former Vice President, ITOCHU Corporation.

‡Executive Adviser, Jepico Corporation; former Vice President, NEC Corporation.

§Former Board Director, General Manager, Sales and Marketing Department, JCSAT; former General Manager, Space and Electronics Business Development Department, ITOCHU Corporation.

¶Chairman and CEO, Hawaiiana Group Incorporated; President, Satellite Culture Japan, Co., Ltd.; former Manager, Space and Electronics Business Development Department, ITOCHU Corporation.

**Managing Executive Officer, SKYPerfect JSAT Corporation.

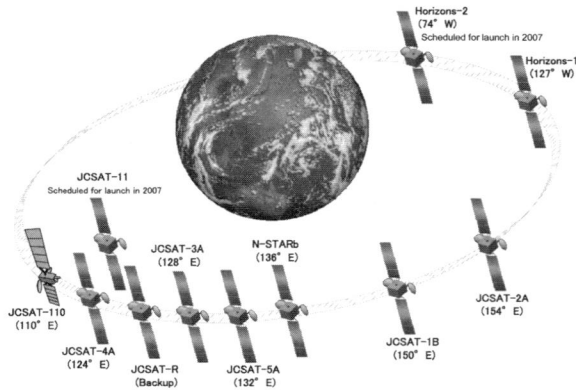

Fig. 9.1   The JSAT satellite fleet (as of March 2007).

two satellites. In 1993 the company merged with Satellite Japan (SAJAC) and changed its name to Japan Satellite Systems (JSAT). The company began providing digital satellite broadcasting service with the launch of its third spacecraft in 1996. In 2000 JSAT took possession of two satellites, N-STARa and N-STARb, owned by the NTT Group, rapidly expanding its operations. In August of that year, the company successfully listed on the First Section of the Tokyo Stock Exchange and at the same time changed its name to JSAT Corporation. JSAT had expanded its business through services related to multichannel CS digital broadcasting (SKY PerfecTV!) and revenue growth from the NTT Group (Fig. 9.2). However, the spread of fiber-optic networks and the Internet in recent years has intensified competition between satellite and terrestrial services in the fields of communication and broadcasting, underlining the need for new growth strategies. Faced with this situation, in October 2006 JSAT announced its plan to merge with SKY Perfect Communications, Inc., the provider of SKY PerfecTV!, in April 2007.

JSAT has a primary control center in Yokohama (YSCC: Yokohama Satellite Control Center) and a backup facility in Gunma (GSCS: Gunma

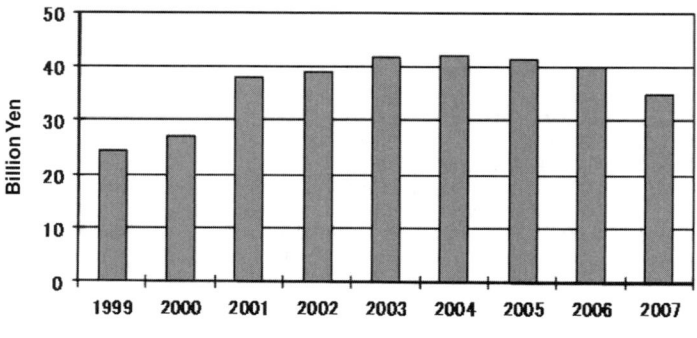

Fig. 9.2   JSAT revenues.

Satellite Control Station), from which it operates the satellites in orbit and monitors the satellite links. During the initial years of service, JSAT had equipment to handle only two satellites, but it has since upgraded and expanded its land, equipment, and facilities in step with the increase in the number of spacecraft. It also used to simply lease transponders to clients and Type II telecommunications operators who provide communications via their own ground equipment. However, satellite communications operators have been called upon to offer end-to-end services in recent years, and so it installed a teleport facility with uplink equipment for customers adjacent to the satellite control facility at YSCC, offering a variety of services.

More than two decades have passed since the company's formation in 1985 through a U.S.–Japan joint venture, and JSAT has risen to be the fifth largest satellite company in the world, though the road has not been smooth. The following sections retrace JSAT's steps and examine how it has grown to become the company that it is today.

BROADCASTING, COMMUNICATIONS, AND GEOSTATIONARY METEOROLOGICAL SATELLITE PROGRAMS

DAWN OF SATELLITE COMMUNICATIONS IN JAPAN

The history of space development in Japan began with rocket development pioneered by Hideo Itokawa of Tokyo University. The Institute of Space and Aeronautical Science was established at Tokyo University in 1964 mainly to develop satellites and rockets to launch equipment for astronautical observations into space. That same year, the then Science and Technology Agency established an Office for Space Development. In 1969, the National Space Development Agency of Japan (NASDA) was created to promote space development in Japan. NASDA took over the activities of the Office for Space Development. This was, of course, the same year that the U.S. space agency NASA launched Apollo 11 and successfully landed a man on the moon.

Space development at NASDA differed from that of Tokyo University, which was aimed at launching satellites into space for space science. The commitment of invaluable public tax revenues meant that space development must make practical use of space to improve the daily lives of the public. From this was born a program for three types of satellites for practical use: broadcasting satellites (BS), communications satellites (CS), and geostationary meteorological satellites (GMS). The Japanese-made N-I and N-II rockets, and the H-I and H-II rockets that followed, were developed out of the need to launch theses satellites. Cooperation in planning for the missions of these satellites was received from the Nippon Telegraph and Telephone Public Corporation (NTT) for communications satellites, from Japan Broadcasting Corporation (NHK) for broadcasting satellites, and the Japan Meteorological Agency (JMA) for weather satellites. These public agencies dispatched technical personnel to NASDA to support the development of these satellites. NASDA intended to develop Japanese satellite technology through these three projects and placed orders with three domestic electronics manufacturers to build the satellites. Of course, Japanese manufacturers at the time were unable to design and build satellites by themselves, so that each of them partnered with a satellite manufacturer in the United States to offer joint proposals to NASDA. The communications satellites were awarded to Mitsubishi Electric Corporation in cooperation with Philco Ford [now Space Systems Loral (SS/L)], the broadcasting satellites to Toshiba and General Electric (now Lockheed Martin), and the weather satellites to NEC Corporation and Hughes (now Boeing).

Having developed and launched Syncom, the world's first geostationary communications satellite, Hughes (HSCG: Hughes Space and Com) already had a strong track record in manufacturing many communications satellites. NEC at the time had a close relationship with Hughes by providing certain onboard communication components for Intelsat-IV series satellites, and so the NEC-Hughes group was at first expected to pursue the order for the communications satellite as a matter of course. However, Hughes at the time questioned the future prospects of satellite communications in a country as small as Japan and, determining that it would be more efficient and provide better future potential to bring to Japan the weather satellites that Hughes had been successfully awarded in the United States, suggested NEC choose the GMS. As will be shown later, because Hughes ultimately partnered with a Japanese trading company and started a joint venture for a satellite communications business in Japan, this is one of the ironies of history.

START OF SATELLITE COMMUNICATIONS IN JAPAN

Satellite communications in Japan started with the launch of CS-1 (an experimental, medium-capacity geostationary communications satellite called Sakura) in December 1977 (Fig. 9.3). CS-1 was a satellite for experimental purposes in accordance with the government's plan. The development of the onboard communications equipments and the planning of communications experiments were conducted by NTT's Yokosuka Electrical Communications Laboratory (YECL) and Ministry of Posts and Telecommunications' Radio Research Laboratory (RRL), while NASDA handled the manufacture and launch of the satellite.

The long-distance communications network in Japan at the time was dominated by microwave communications, and NTT had microwave links using C-band and Ku-band throughout Japan. A portion of the Ku-band was also used for terrestrial communication links by government-related organizations, local governments, and others. Because of this situation, the Ministry of Posts and Telecommunications and NTT decided to use the Ka-band for satellite communications, which was not yet used anywhere in the world, in order to avoid interference with terrestrial microwave communications links. Following

Fig. 9.3   The first Japanese satellite CS-1 (Sakura) [copyright held by Japan Aerospace Exploration Agency (JAXA)]. (See also color figure section at the back of the book.)

this policy, NTT began research and design (R&D) activities for Ka-band onboard transponders and ground station equipment. CS-1 was launched in order to conduct experiments with these communications equipments and systems. Satellite manufacturing and rocket technologies in Japan had yet to reach a practical level, and so the satellite was procured from Philco Ford through Mitsubishi Electric, and the rocket used was a U.S.-made Delta 2914.

Following on the success of the CS-1 experiments, the CS-2a and CS-2b satellites were launched in 1983 to provide commercial service. Despite their purpose, they carried just six Ka-band and two C-band transponders and were small-scale, spin-stabilized spacecraft with an in-orbit dry mass of 350 kg. The onboard telecommunications equipment was developed at NTT's Yokosuka Electrical Communications Laboratory (YECL), the transponders manufactured by NEC, and the antenna by Mitsubishi Electric. For the manufacture of the satellite itself, Mitsubishi Electric was the main contractor, though the actual construction was made by Ford Aerospace (now SS/L). The satellite integration and testing of CS-2b were conducted at Mitsubishi Electric's Kamakura plant following a licensing of technology. The satellite was launched using a domestically produced N-II rocket (Fig. 9.4), developed through the licensing of Delta rocket (Delta 3914)

Fig. 9.4   The N-II rocket that launched the satellite [copyright held by Japan Aerospace Exploration Agency (JAXA)].

technology from the United States. The users of CS-2a and CS-2b were NTT and certain public agencies, while the procurement cost for the satellite was covered through a scheme based on the proportion of use of the transponders. The use of the satellite at that time was limited to such purposes as emergency and disaster recovery, temporary communications links, and communications for remote islands.

FIRST STEPS FOR PRIVATE-SECTOR SATELLITE OPERATOR IN JAPAN: IT ALL BEGAN WITH A MEETING

In 1979, four years prior to the launch of CS-2a, the experimental geostationary communications satellite ECS-1 (called Ayame 1) was launched for the purpose of conducting communications experiments using millimeter wave and the ECS-2 in 1980 (Fig. 9.5). Both spacecraft were launched on N-I rockets developed by NASDA through the licensing of Delta 2914 technology from the United States. During the launch operation of ECS-1, however, the third stage of the rocket collided with the separated satellite, while a problem developed with the apogee motor on ECS-2, and the launch of both spacecrafts unfortunately ended in failure.

The meeting of two individuals during this ECS program, however, marked the starting point for the establishment of the first private-sector satellite

Fig. 9.5 The experimental geostationary communication satellite ECS-1 (Ayame 1) [copyright held by Japan Aerospace Exploration Agency (JAXA)]. (See also color figure section at the back of the book.)

communications company in Japan. These two persons were Mr. Kitazume, the representative from NEC (commissioned to develop the millimeter wave-band transponder for the ECS), and Mr. Miyakawa, the representative from ITOCHU Corporation (contracted as the agent for Hughes EDD to supply the TWTA, the heart of the transponder). Mr. Miyakawa, who at that time was new to the satellite business, was being held responsible by NEC and NASDA for a delay of more than one year in purchasing the TWTA for the ECS-1 from Hughes EDD. As a result of the sincerity he showed during this association, he and Mr. Kitazume developed a close friendship, and Mr. Miyakawa succeeded in securing the business of transporting the transponder for the ECS-2 from Japan to the satellite manufacturer Ford Aerospace.

From this experience Mr. Miyakawa developed an interest in the space business and approached Mr. Kitazume, and they worked together to sell NEC-made transponders to satellite manufacturers in Europe and the United States. He also sold Hughes-made TWTA units to the NTT Yokosuka Electrical Communications Laboratory and other places.

As his friendship with Mr. Kitazume grew, Mr. Miyakawa, who had become deeply involved in the space business, entered ITOCHU's in-house essay contest at the end of 1979, arguing that ITOCHU should look ahead to the 1980s and make a strategic move toward the future by investing in the space business. He received a prize from the contest. His essay also prompted ITOCHU to establish in 1983 a special division within the company to conduct development work related to the space business, which later played a central role in the establishment of Japan's first private-sector satellite communications company. It is said that Mr. Kitazume gave Mr. Miyakawa the idea of developing a satellite communications business in Japan and of partnering with Hughes to do so. Mr. Kitazume presented a similar idea to NEC, but because the company took no notice of it, he encouraged Mr. Miyakawa to pursue the matter at the trading company.

MAJOR TREND TRIGGERED BY U.S.–JAPAN TRADE ISSUES

Around the time of the launch of CS-2a and CS-2b in 1983, the trade issues that had flared up between the United States and Japan had reached its peak. The bias in procurement by Japanese government organizations in favor of domestic companies was seen as a problem, with the brunt of criticism directed at NTT, which made large amounts of purchases. NTT had pursued cooperation with Japanese manufacturers on the development of communications equipment and material procurement. To deflect the attacks directed at its research laboratories, NTT urged them to consider whether there was anything that could be procured from the United States. NTT finally settled on purchasing Cray supercomputers for research purposes and on making a purchase in the satellite communications field, where research and development were being conducted on large-capacity satellite communication systems by developing large satellites, expending an extensive research budget at the time.

In the summer of 1983, NTT sent a team of procurement mission to the U.S. Hughes Aircraft Company (HAC). The team was headed by Reijiro Fukutomi, director of reseach and development division of NTT, and the team members were Mr. Murotani, who was responsible for the research activities on satellite communications at YECL, and other key employees. One key employee of the YECL asked E. Matsumoto (a member of Mr. Miyakawa's staff) from ITOCHU, who handled relations with the YECL and whom he had met in the course of purchasing TWTA units from Hughes EDD, to intermediate between the mission team and HAC and to accompany them to HAC. In this way, ITOCHU Corporation became involved in the historical meeting between NTT and HAC. HAC chairman Dr. Wheelon handled NTT's procurement mission for Hughes, while Mr. Miyakawa from ITOCHU also attended the meetings. The person who arranged this meeting was W. Kamei, the Japan representative for HAC's satellite division.

The NTT side kicked off the discussions by saying that they were interested in purchasing software for satellite design, but HAC responded that they were unable to sell the design software on a stand-alone basis because the software contained critical expertise and urged NTT to purchase the satellite itself instead. The Japanese government at that time, as part of its national strategy to foster industry, had adopted a policy of developing its own space technology so as to be able to produce satellites and rockets domestically. NTT, as a public corporation, was therefore unable to purchase satellites from the United States.

The negotiations themselves broke off at that point, but ultimately this meeting helped to pry open one of the doors of history. These negotiations put stronger political pressure on Japan to purchase U.S.-made satellites, and the issue found its way into the agenda of U.S.–Japan trade issues talks between the then Japanese Prime Minister Yasuhiro Nakasone and U.S. President Ronald Reagan during a summit meeting held at the Hinode-sanso (Sunrise Lodge) in western Tokyo. Nakasone promised to pave the way to allow for the purchase of U.S.-made satellites in the future.

ITOCHU CORPORATION BECOMES THE JAPANESE DISTRIBUTOR FOR HAC SATELLITES

The U.S.–Japan summit meeting in 1983 did pave the way for Japan to purchase U.S.-made satellites, but the issue required a shift in Japan's policy, so that making this a reality was by no means easy. Perhaps recognizing this, in December 1983 HAC Japan representative Mr. Kamei contacted ITOCHU Corporation. HAC chairman Dr. Wheelon was pleased with the job that ITOCHU had done and gave his consent for the trading company to become HAC's satellite distributor in Japan. Mr. Furuhata and Mr. Miyakawa from ITOCHU immediately flew to Los Angeles to meet with Dr. Wheelon and agreed to be HAC's agent in Japan. The agreement was formally signed the following year in February 1984, becoming the lead story on the front page of the *Nihon Keizai Shimbun* (Nikkei Newspaper), Japan's foremost business newspaper.

HAC, ITOCHU, AND MITSUI ESTABLISH A THREE-WAY TASK FORCE

Once ITOCHU had become HAC's Japanese agent, Mr. Miyakawa, his subordinate Mr. Matsumoto, and others made sales calls throughout Japan, but no companies were interested in buying satellites, deterred by the extensive investment necessary. The only possible options were to sell communications satellites to NTT and broadcasting satellites to NHK.

The concept of a "Keidanren satellite" then surfaced, a proposal led by Mr. Kobayashi of Fujitsu, the chairman of the telecommunications division of the Japan Federation of Economic Organizations (Keidanren) at the time. This idea initiated a three-way struggle among the companies acting as distributor of satellites in Japan, ITOCHU for HAC, Mitsubishi Corporation for

Ford, and Sumitomo Corporation for RCA. However, the Keidanren concept was plagued by an internal struggle between those who wanted to import satellites and those who wanted to produce them domestically, and the project eventually became a dead issue.

Around this time, HAC proposed that if there were no companies willing to buy satellites in Japan, then HAC and ITOCHU should establish a joint U.S.–Japan venture company to lease transponders in Japan. HAC was already operating a transponder leasing business for communications satellites through its subsidiary HCI (Hughes Communications, Inc., later PanAmSat and purchased by Intelsat) and had a plan for the launch and leasing of satellites for the U.S. Navy called Leasat. As HAC's agent ITOCHU had been focusing on the business of selling satellites in Japan, this proposal caused a fair amount of commotion. However, it was still in line with ITOCHU's basic policy of fostering the space business, and the company ultimately began to study the issue. This was around May 1984, just one year prior to the birth of JCSAT, Japan's first private-sector satellite telecommunications company.

At this time Mitsui was also becoming interested in the satellite business and contacted Hughes through its Los Angeles branch about the possibility for a satellite communications business in the United States or Japan. Through these negotiations HAC and HCI came to feel that it would be necessary to bring Mitsui into the fold as a partner in order to ensure the success of a U.S.–Japan joint project and made such a proposal to ITOCHU.

The immediate reaction from the ITOCHU team members was that the commercial rights they had established were simply being handed over to a rival company, and they resisted the idea. However, finally deciding that there were significant merits in the idea from the standpoint of sharing the risk for an investment that at the time exceeded approximately ¥100B, and gaining the assistance of Mitsui in terms of establishing the business and marketing, a three-way joint venture that included Mitsui was established.

The three companies then set up a task force on 17 September 1984 to conduct feasibility studies, establishing an office in Mori Building No. 33 in Kamiyacho. HAC sent Mr. Judge to lead the task force as COO. He had successfully headed the Palapa satellite project in Indonesia.

MOVES TOWARD TELECOMMUNICATIONS DEREGULATION

During 1983–1984, in parallel with the moves toward importing U.S.-made satellites prompted by U.S.–Japan trade issues, liberalization of the telecommunications business was also being debated in the Ad Hoc Commission on Administrative Reform led by Keidanren Chairman Doko. Telecommunications services in Japan at this time, in accordance with the NTT law and the KDD law, were provided exclusively by NTT for domestic communications and Kokusai Denshin Denwa (KDD) for international services. The commission determined that this situation produced a lack of diversification in services and invited price rigidity, and debate focused on the privatization of NTT and promoting the entry into the market of new private-sector telecommunications carriers for both domestic and international services. Bills for telecommunications deregulation were drafted following the recommendations of the Doko commission. Three new common carrier laws were passed by the Diet on 24 December 1984 and went into effect on 1 April 1985.

The debate over importing satellites from the United States that was triggered by U.S.–Japan trade issues ultimately led to the idea of establishing a joint venture that would own satellites and lease transponders. The liberalization of telecommunications, which happened at the same time, permitted the entry into the market of new private-sector telecommunications carriers and allowed this idea to become a reality. The venture was truly blessed by perfect timing.

DEBATE OVER KU-BAND VS KA-BAND

As mentioned earlier, C-band frequency band (4–6 GHz) was widely used for domestic, long-distance wireless relay links in Japan. Ku-band frequencies in the 14-GHz band were used for short-distance relay links in urban areas, whereas local governments and other public agencies used the Ku-band at 12 GHz. As a result, the Ministry of Posts and Telecommunications adopted a policy of allocating Ka-band frequencies (20–30 GHz) for the satellite communications frequencies used in Japan. In accordance with this policy, development of onboard satellite communications equipment and ground station equipment using Ka-band had been started for the CS program. Hughes, however, proposed the Ku-band as the frequency to use for satellite communications in Japan. The background of Hughes' proposal to use Ku-band was as follows.

At the time C-band was the frequency widely used for satellite communications, mainly for international telecommunications around the globe. However, with the introduction of the Open Skies policy in the United States in the early 1970s, many companies entered the market for satellite communications, for domestic as well as international services. As a result, greater use of the C-band led to problems of interference between satellites and with terrestrial radio stations, and the trend toward use of the new Ku-band was widely spreading in the United States.

The Ku-band experiences greater attenuation due to rain compared with the C-band, but there is less of a problem with interference between satellites and with terrestrial radio systems. The fact that ground antennas could also be designed smaller and that telecommunications systems could be built economically were also advantages. Research and development on the Ka-band for satellite communications had been conducted in Japan from early on, but the world was moving in the direction of using the Ku-band. As a result, there were few manufacturers around the world capable of producing Ka-band onboard satellite communications equipment or ground station transmission equipment, and therefore the price was high. Also, the biggest problem was that rain attenuation was significantly more pronounced in the Ka-band compared to the Ku-band, which were huge disadvantages in terms of link availability and economic efficiency.

The task force from HAC, ITOCHU, and Mitsui strongly urged the Ministry of Posts and Telecommunications and NTT to use the Ku-band, but because this would reverse the previous government policy, it could not be easily accepted. HAC forcefully asserted that it would be impossible to establish a business in Japan unless use of the Ku-band was accepted, but ITOCHU and Mitsui, which had a better understanding of the situation in Japan, were doubtful that use of the Ku-band would be approved. Faced with this state of affairs, the group went ahead with satellites for its business with specifications that assumed the Ku-band. Even when it was on the verge of receiving a formal satellite order, the Ministry of Posts and Telecommunications still had not approved use of the Ku-band. In the end, formal approval for use of the Ku-band had to wait until the launch of the joint venture company JCSAT, when the company received its operating license. It is reasonable to assume that HAC, which had forged ahead with the manufacture of the satellite during this uncertain period, judged that a political solution to the situation would be reached.

ESTABLISHMENT OF JAPAN COMMUNICATIONS SATELLITE COMPANY

Under the new telecommunications business law, foreign investment in Type I telecommunications carriers would be restricted to an equity interest of one-third or less. The three-company task force, deciding on a holding ratio of 40% for ITOCHU, 30% for HAC, and 30% for Mitsui, established the Japan Communications Satellite Planning Company, Inc., on 18 February 1985. On 1 April 1985, when the new telecommunications business law went into effect, the company dropped the word "planning" from its name to become Japan Communications Satellite Company (JCSAT), and Japan's first private-sector satellite communications company was born. JCSAT submitted an application on April 9 to the Ministry of Posts and Telecommunications to be a Type I telecommunications carrier business and received an operating license in June of that year, becoming the first satellite-based Type I telecommunications carrier (Fig. 9.6).

Other Type I telecommunications carriers to emerge from the liberalization of telecommunications included the terrestrial domestic telecoms DDI

Fig. 9.6   The JCSAT opening party on 17 July 1985.

Corporation, Japan Telecom, and Teleway, as well as the international telecom providers IDC and ITJ. Satellite-based carriers, in addition to JSAT, included Space Communications Corporation (SCC) created by the Mitsubishi Group and Satellite Japan (SAJAC), in which such companies as Sumitomo Corporation, Nissho Iwai Corporation, Marubeni Corporation, Sony Corporation, and NEC Corporation were involved.

ADVANCE SALES OF 64 TRANSPONDERS COMPLETED, THE FULL COMPLEMENT OF TWO SATELLITES

The FS Task Force made up of ITOCHU, Mitsui, and HAC was established around May 1984, with the mission of clarifying whether demand for satellite transponders really exists in Japan. As a result of marketing activities conducted over roughly six months by sales staff sent from ITOCHU and Mitsui, the task force concluded at the end of 1984 that there was sufficient demand in Japan. Based on this conclusion, it was decided to officially establish a planning company. However, top management at HAC, which had continued to harbor doubts about the use of satellite communications in a country as small as Japan, was not easily convinced and demanded more substantial evidence. The decision was then reached to demonstrate the desire of users by taking tentative reservations. Once this plan was implemented, the task force accumulated tentative reservations for more than 50 transponders in two to three weeks, just as it had expected. The user companies that had exhibited a desire to purchase transponders were taken on a kick-off tour to observe the state of satellite telecommunications in the United States, which was conducted immediately following the establishment of JCSAT in April 1985.

Despite this success, HAC stressed that tentative reservations that were not backed up with money were meaningless and insisted that the companies pay a deposit. The Japan side felt that somewhere around ¥1M would be in line with accepted business practice, but HAC was adamant that to truly gauge a company's desire a commitment of at least ¥10M, an amount requiring formal internal authorization, would be required and that anything less would be meaningless. It was ultimately decided to collect a deposit of ¥10M per transponder. The Japan side was extremely anxious about how Japanese customers would react to the tough U.S.-style sales methods. Once they began to ask for an advance, however, a steady stream of companies appeared ready to pay the deposit and make a formal reservation, and all 64 transponders were quickly sold.

The company that reserved the greatest number of transponders at this time was NTT. Having been privatized as part of telecommunications deregulation, NTT, at the urging of Toshio Doko, the chairman of the Keidanren who had led the Ad Hoc Commission, accepted as its president Hisashi Shinto, who had previously served as president of Ishikawajima-Harima Heavy Industries Co., Ltd. (IHI). At the time, Mr. Shinto was urging NTT to move into satellites "even if it has to pawn its telephone offices" and to promote the use of satellite communications. He pushed for NTT to use its influence not just with the CS-3 spacecraft it owned, but with transponders on private satellites as well, and the company applied for use of 13 transponders from JCSAT. Other users included business companies supplying programs to CATV operators, TV stations, Type II telecom carriers (Uplinkers) established as subsidiaries of trading companies, preparatory schools, religious institutions, electronics manufacturers, automobile sales companies, and securities companies. Nearly all of these organizations, with the exception of NTT and certain Type II telecom carriers, intended to use the service for video transmission.

It is true that the Japanese economy was entering the bubble period, and with the appearance of the first private satellite communications company there was an atmosphere of wanting to use the new media and not to miss out on a business opportunity. The deposits received were ultimately returned to all customers a year later at the direction of the Ministry of Posts and Telecommunications, and wanting to return in some way the interest earned on these deposits, JCSAT formed study groups made up of small groups of users who had reserved transponders and conducted tours to observe the state of satellite communications in the United States. These tours deepened ties between the representatives of JCSAT and user companies, helping to foster a number of loyal JCSAT customers in the years ahead.

HIRING OF KEY ENGINEERS

Japan Communications Satellite Company (JCSAT) began as a joint venture between three companies, the U.S. firm Hughes (now Boeing) and the trading companies ITOCHU and Mitsui. At the firm's outset personnel from the trading companies handled the business plans, market surveys, and marketing activities, while engineers dispatched from Hughes took care of the technical aspects. However, it became necessary to hire engineers skilled in satellite communications and/or operations technology for the continued operation of the business after the actual launch of the satellites. These engineers were required to maintain ground equipments, operate the satellites using these equipments, and provide technical support to customers. Acquiring such engineers became a major issue.

The number of Japanese engineers skilled in satellite communications and operations was extremely limited at the time and could be only found at companies or organizations such as KDD, which used satellites for international telecommunications, the Ministry of Posts and Telecommunications' Radio Research Laboratory (RRL), NTT's Yokosuka Electrical Communications Laboratory (YECL), which were involved in the CS project, and manufacturers of satellites and ground equipment. Faced with this situation, between 1985 and 1986 the human resource directors for JCSAT at the time, Mr. Morita from ITOCHU and Mr. Morokuma from Mitsui, began using whatever connections they had to recruit these engineers.

Particularly urgent was the hiring of the core engineers who would operate satellites at the YSCC control center then under construction in Yokohama. Several engineers had been successfully recruited by the fall of 1986, drawn from such organizations as the Ministry of Posts and Telecommunications' RRL, NTT's YECL, KDD, NEC, Toshiba, JRC, and the Self-Defense Forces. That fall 10 core members were dispatched to HCI (Hughes Communications, Inc.) in Los Angeles to study the design and operational technology for JCSAT-1 and JCSAT-2, then under construction at Hughes. Among these key members was Mr. Hashimoto, who at the Ministry of Posts and Telecommunications' Radio Research Laboratory (RRL) had been involved in satellite relay experiments using Relay 1, which first transmitted news of the Kennedy assassination, and relay of Tokyo Olympics using Syncom 3, as well Mr. Nagai, who had been involved in research and development of the onboard communications equipment and operations technology for CS-1, CS-2, and CS-3 at NTT's YECL. Once these 10 core members had completed their training at HCI, they returned to Japan in 1987 and began conducting operator training and readying the ground equipment at the newly completed YSCC control center and preparing for the launch of JCSAT-1 and JCSAT-2.

LAUNCH OF JCSAT-1 AND JCSAT-2

The initial plan was to launch both JCSAT-1 and JCSAT-2 on the space shuttle, with JCSAT-1 scheduled for launch in December 1987. However, after the tragic loss of the Space Shuttle Challenger on 28 January 1986, the route of commercial use of the space shuttle was closed. There was a further launch failure with an Ariane-III rocket in May 1986 and the successive failures of the two major launch vehicles in the world at that time effectively suspended commercial launches for some time. In response, JCSAT changed its plans and in order to mitigate the risk decided to launch JCSAT-1 with an Ariane-IV from France's Arianespace and JCSAT-2 using the first Titan-III Commercial Version rocket from the U.S. firm Martin Marietta Corporation, which had just decided to enter the commercial launch market.

Incidentally, the satellite bus employed for JCSAT-1 and JCSAT-2 was a Model HS393 spin-stabilized satellite from Hughes, with a cylindrical shape. The outside diameter of this cylindrical satellite was designed so that it could be neatly loaded sideways in the space shuttle cargo bay, with the idea that using the space shuttle would allow it to be launched economically. The shuttle launch scenario collapsed with the decision to suspend commercial use of the shuttle, and in the end only three Model HS393 units were manufactured, two for the JCSAT-1 and JCSAT-2 satellites for JCSAT and one for the SBS-6 satellite for Satellite Business Systems (SBS) in the United States.

Fig. 9.7   President Kamiya celebrates the successful launch of JCSAT-1.

LAUNCH OF JCSAT-1

The launch date for JCSAT-1 was delayed from the initial plan and set for 28 February 1989, but was delayed twice more for two reasons. The first delay was caused by a strike by workers at the French company servicing the downrange station that tracked the Ariane rocket during launch operation. This was finally resolved, but then just as the launch stage was reached strong winds at the launch pad in French Guiana in South America caused a further delay. The weather calmed on 7 March 1989 (Japan time), and JCSAT-1 was finally launched. JCSAT President Hiroshi Kamiya, who was at the launch site, celebrated the success in an interview afterwards by likening the successful launch of JCSAT-1, after repeated delays, to the idea that "babies that have a difficult birth grow up to be strong" (Fig. 9.7).

COMMENCEMENT OF SERVICE VIA JCSAT-1

The first user to access JCSAT-1 (Fig. 9.8), on the day the service started on 16 April, was TV Asahi. The broadcaster was planning a new program at that time called Sunday Project, and though spring was in the air it sent its Satellite News Gathering (SNG) vehicle to Kurobe Canyon in Toyama Prefecture, where the snows were still deep for a live report. The screen for the TV Asahi broadcast included an illustration of the JCSAT satellite logo along with the live images, indicating that it was relayed via JCSAT satellite. This was the first live transmission using a private communications satellite in Japan.

Fig. 9.8   The JCSAT-1 satellite.

LAUNCH OF JCSAT-2

JCSAT-2 was launched using a Titan-IIIC from the U.S. firm Martin Marietta Corporation (now Lockheed Martin). The Titan-III rocket was originally developed for military use, designed to deliver large, heavy payloads into low-altitude orbit. Consequently, it is composed of two powerful core stages and large solid auxiliary rocket boosters. It has no third-stage rocket to place the satellite into transfer orbit, so like with the space shuttle it is necessary to attach a third-stage rocket to the satellite, which is placed in the rocket fairing and launched together. The Titan-III, reconfigured in this way for launch of commercial satellites, was called the Titan-IIIC, and was first used to launch JCSAT-2.

The launch of JCSAT-2 was initially planned for the fall of 1989, but various circumstances pushed back the launch to December 1989. The United States was experiencing abnormal weather that year, with unusually low temperatures and strong winds persisting at the launch pad for Titan-III in Florida. (This was the year that the Florida citrus industry was reported to have been dealt a devastating blow.) Another problem was the wind. Strong winds in the upper atmosphere during launch can significantly disturb rocket attitude and orbit, producing a large margin of error in the placement of the satellite in orbit. Weather balloons were released into the atmosphere before each launch to measure the wind speed aloft. The data on wind speed at each altitude recorded by the balloon were sent to a computer at Martin Marietta's headquarters in Denver, and the effects were analyzed. Launch of the JCSAT-2 had been attempted since early December, but the situation with the low temperatures and strong winds did not improve, the launch was delayed day after day, and even after endless waiting the launch could not be executed. The "go" signal for launch was finally given on 31 December 1989 local time or 1 January 1990 Japan time. The work conducted for the launch, stretching over more than a month, caused everyone to forgo their Christmas and New Year holidays.

JCSAT AND ITS RIVAL, SCC

Space Communications Corporation (SCC) is a satellite-based Type I telecommunications carrier established by the Mitsubishi Group, like JCSAT formed following the liberalization of telecommunications in the spring of 1985. The Mitsubishi Group was involved in space development in Japan from the early stages. Mitsubishi Heavy Industries was engaged from the initial stages in the development, manufacture, and launch of the domestically produced N-I, N-II, H-I, H-II, and H-IIA rockets being developed by NASDA at the time. Mitsubishi Electric was involved with the government's communications satellite program, and through a cooperative relationship with the U.S. satellite manufacturer Philco Ford (currently Space Systems Loral) developed and manufactured the CS-1, CS-2, and CS-3 satellites.

Mitsubishi Corporation, along with Mitsubishi Heavy Industries and Mitsubishi Electric, was also engaged in the import of space-related equipment. It was from this background that the Mitsubishi Group, led by Mitsubishi Corporation and Mitsubishi Electric, formed SCC after the establishment of JCSAT by the ITOCHU-Mitsui-Hughes group.

SCC's Superbird satellite was originally supposed to carry a Ka-band transponder. The company had a long history of involvement in Japan's CS program, which used the Ka-band, and with the Ku-band not approved for use in Japan, it considered using the Ka-band. However, once the news reached SCC that JCSAT was likely to receive a Ku-band license from the then Ministry of Posts and Telecommunications, it hastily made a significant reduction in the number of Ka-band transponders, altered the design to focus mainly on Ku-band transponders, while adopting a strategy of strengthening the output power of the Ku-band transponders. The satellite bus adopted for Superbird was a three-axis-stabilized Model FS1300 from Ford Space Systems (formerly Philco Ford), which had been used for the Intelsat-V satellites.

SUCCESSFUL LAUNCH OF SUPERBIRD-A

SCC's first satellite, Superbird-A, was launched on 5 June 1989—three months behind JCSAT—using an Ariane-IV rocket. JCSAT had originally envisioned the launch of its first satellite to be in December 1987 and had begun readying its satellites and ground control centers. But with the Space Shuttle *Challenger* accident in January 1986 and the failed launch of an Ariane-III rocket that followed in May that same year, JCSAT was forced to make a hurried change in launch vehicle, and with the investigation into the cause of the failure of the Ariane-III, the launch was delayed by a year to 18 months. Consequently, SCC was able to launch its first spacecraft just three months behind JCSAT despite the fact that SCC had fallen well behind in terms of satellite launch preparations. SCC also made the most of its position as a latecomer to alter the satellite specifications during the design stage and make other modifications, which together resulted in success. Space communications services commenced in July, the same year that Superbird-A was launched successfully.

TRANSPONDER COMPETITION: 27 VS 36 MHZ

As described earlier, when JCSAT was taking reservations for transponders prior to the launch of service, it received more applications than the capacity of the 64 transponders. The situation changed substantially by the time for the final formal contract, however. This was because the latecomer SCC had begun to catch up in terms of attracting customers. JCSAT was well ahead of SCC in terms of studying commercial viability and marketing activities, but with the delays caused by accidents with the delivery vehicles the launch of the satellites was delayed substantially, narrowing to just three months the advantage JCSAT enjoyed in being first to initiate service. Furthermore, SCC maximized its latecomer status to successfully change the specifications for its own satellite. The changes included a substantial reduction in the number of Ka-band transponders, which were initially supposed to be the main component, in favor of a Ku-band main system and an increase in transponder power. It also fitted the Superbird with Ku-band transponders with a wider bandwidth of 36 MHz, while all of JCSAT's Ku-band transponders had a bandwidth of 27 MHz.

The design for JCSAT's satellites, led by Hughes, was based on the so-called "cable bird," considering that video program transmission to CATV operators was the single biggest application for satellites in the United States at that time. Video transmission at the time was conducted entirely by analog FM modulation, and to accomplish this with the greatest efficiency, each video channel was transmitted at 27 MHz, the minimal bandwidth that could be used without any technological difficulties. Hughes's aim was to maximize the number of transmission channels per satellite. This is because if the entire 500-MHz Ku-band allocated for satellite communications is shared using both vertical and horizontal polarization, and 27-MHz bandwidth transponder channels are established, then the maximum number of transponders that can be placed on a satellite is 32. If the transponder channels are set at the 36-MHz bandwidth, the maximum number of transponders is only 24, and the economic efficiency of the satellite deteriorates.

JCSAT, which was first to market, had in its application to the Ministry of Posts and Telecommunications separated the service fee per 27-MHz transponder into "protected" and "preemptible," and the service area into A-area transponders that include Okinawa and B-area transponders that do not. SCC, meanwhile, had come out with a strategy whereby it offered the majority of its wider bandwidth, high-output 36-MHz transponders at a price that was in effect equal to that of JCSAT's "preemptible" transponder fee. Following the logic that one transponder is still only one transponder even if it has a wider bandwidth and slightly higher power, the company had the idea to provide a discounted price. SCC's customer base was Mitsubishi Group companies and government agencies, but this strategy proved effective, and SCC succeeded in stripping away customers who had initially made reservation applications with JCSAT. In particular, television stations that wanted two SNG video signal transmissions on a single 36-MHz transponder, government agencies that desired more VSAT station capacity from each transponder, and program suppliers looking for larger discounts on transponders crossed over to SCC.

JCSAT found itself in a difficult situation in terms of transponder functionality and price, but managed to retain a majority of its customers by convincing companies from the same industrial groups as its parent companies

ITOCHU and Mitsui to stay onboard, winning over electronics manufacturers other than Mitsubishi, including NEC, Toshiba, Hitachi, Fujitsu, and Matsushita, as well as a variety of marketing efforts. One reason in particular was the relationship of trust cultivated with customers by using the interest on deposits to conduct observation tours of the U.S. satellite communications business.

FAILED LAUNCH OF SUPERBIRD-B AND THE SECOND ATTEMPT

The launch of SCC's second satellite, Superbird-B, was attempted on 23 February 1990 using an Ariane-IV rocket, eight months after the first spacecraft (Superbird-A) and two months later than JCSAT-2. Directly after launch, however, the first-stage engine began exhibiting abnormal combustion, and 101 seconds into flight the rocket was destroyed by command from the ground. In a curious coincidence, the payload being launched by this Ariane-IV rocket included, in addition to Superbird-B, the BS-2X, the twin of Japan's broadcast satellite BS-2. Incidentally, the launch failure is believed to have been caused by a foreign object that found its way into the engine water pipe during rocket booster integration work on the launch pad. As a result of this accident, SCC was forced to provide service solely with Superbird-A until the launch of the replacement Superbird-B in February 1992.

IN-ORBIT INCIDENT FOR SUPERBIRD-A

With the failed launch of Superbird-B in February 1990, SCC had to rely on Superbird-A alone, but on December 23 that same year the Superbird-A was also lost to an incident suffered while operating in orbit. The situation originated some time around December 18 with a malfunction in the electronic circuitry that regulated the attitude control. The satellite began to deviate from its normal attitude. The cause was reported to be a single event upset (SEU) involving a digital circuit malfunction triggered by the impact of a heavy particle in a cosmic ray. In trying to restore attitude, most of the oxidizer in the attitude control propellant was consumed, and in the end continued operations became impossible.

JCSAT-1 AND JCSAT-2 SUPPORT SCC CUSTOMERS AFTER FORCED SUSPENSION OF SERVICE

SCC, which had been providing service solely with Superbird-A, had no choice but to suspend service entirely because of this in-orbit incident. It became clear on Sunday, 23 December 1990, that operations would be suspended, and though Monday was a national holiday many of the JCSAT employees who had received word of the incident voluntarily gathered at the head office in Tokyo to discuss a response strategy. To continue to provide service to customers using Superbird-A, the only option was to move them to either JCSAT-1 or JCSAT-2, already operating in orbit.

There were two problems: which transponders from JCSAT-1 and JCSAT-2 to provide to customers who had been using Superbird, and how to handle customers' radio licenses for transmission Earth stations. Positioned as a "Cable Bird," JCSAT-2 served all of the company's CATV program suppliers at the time. By transferring CATV program suppliers using Superbird to JCSAT-2, it stood to reason that these customers could make use of existing JCSAT-2 antennas already installed at CATV stations nationwide, enabling JCSAT to realize its original "Cable Bird" conception for JCSAT-2—the idea of bringing all of Japan's CATV program suppliers onboard a single satellite. However, this would require transferring other existing users of JCSAT-2 to transponders on JCSAT-1. Therein lay the problem. The transfer of existing customers contracted to use JCSAT-2 to different transponders on JCSAT-1 merely to suit JCSAT's own ends was expected to trigger a backlash from them. Discussions on response strategies continued throughout the day, after which then-President Nakayama took several employees to the bar in the nearby Hotel Okura to thank them for their effort. Over drinks one of the employees proposed to the president that despite the range of opinions from the sales department, unless JCSAT found some way to help all of the customers who were using Superbird by opening up capacity on either JCSAT-1 or JCSAT-2, the company would draw severe public criticism. In the end JCSAT, respecting the position of its existing customers, decided to distribute available transponder capacity on both satellites to all Superbird customers. Consequently, JCSAT was unable to realize the "Cable Bird" conception.

The other problem of the radio licenses for Earth stations required very intensive work from the next day on a 24-hour basis. With the New Year holidays approaching, and numerous events scheduled, such as special program transmissions to CATV stations and satellite news reports from TV stations, there was a flood of requests to begin transmissions from Superbird customer Earth stations to JCSAT-1 and JCSAT-2 as soon as possible. The JCSAT radio license team went back and forth with officials at the then Ministry of Posts and Telecommunications and representatives at SCC in an attempt to make license applications for all Earth stations in an extremely short period of time. The ministry processed the applications with unprecedented speed, and transmissions were able to begin between the end of the year and into the new year. Immediately after the New Year holidays, the network operations manager at the YSCC made a copy of the frequency spectrum being transmitted from JCSAT-1 and JCSAT-2 and sent it to the head office. This showed that there exist uplinked signals on almost all of the 64 transponders on the two spacecraft. Thus just one year after the launch of the second spacecraft on 1 January 1990, JCSAT's two satellites were operating at full loading.

SCC RESUMES SERVICE AFTER SUCCESSFUL LAUNCH
OF THE REPLACEMENT SUPERBIRD-B

Superbird-B1, the replacement satellite readied following the failed launch of Superbird-B on an Ariane rocket, was launched on 27 February 1992, again onboard an Ariane-IV rocket. It was two years to the day from the failed launch in February 1990, and approximately 14 months after Superbird-A was lost in an in-orbit incident. With the successful launch of Superbird-B1, SCC resumed its own satellite service from the spring of that year. The customers who had for nearly 14 months been saved by JCSAT-1 and JCSAT-2 all gradually returned to SCC and Superbird-B1.

The approximate ¥50B in losses accumulated by SCC during this time was covered, with the help of its shareholder the Mitsubishi Group, through adjustments to capital. Meanwhile, JCSAT, which recorded more than ¥25B in revenue during fiscal 1991 as a result of the exceptional demand from the transfer of customers from SCC, saw revenue decline from fiscal 1992 to fiscal 1994 as all of these customers gradually drifted back to SCC.

On 2 December 1992, SCC successfully launched Superbird-A1, the successor satellite to the Superbird-A lost while in orbit, and from early 1993 finally began operating with a fleet of two satellites.

START OF CS ANALOG BROADCASTING

From the outset, one of the major customer segments for the private-sector communications satellite service provided by JCSAT and SCC was program suppliers for cable TV. Transmission of programs via satellite to cable TV providers was also one of the most successful applications for satellites in the United States. At the time, cable television in Japan was dominated by small-scale providers serving areas where reception of terrestrial signals was difficult, and there were only a few cable TV operators offering multichannel services in urban areas. The Ministry of Posts and Telecommunications, which was in a position to promote cable television, was trying to encourage the spread of multichannel cable television services everywhere in Japan using satellite program distribution to cable operators, a concept it called the "space cable network." In fact, it had become possible to transmit a variety of programs to cable operators via satellite, and multichannel services began to spread.

With the launch of private-sector communications satellites, it became technologically possible for the businesses that owned the programs to use these satellites to make broadcasts that could be received directly by users. Program suppliers, which wanted to gain as many direct customers as possible without going through the CATV networks, partnered with a receiving equipment manufacturer (Sony) and in 1989 established the Skyport Center for the purpose of direct broadcasting via communications satellite. At that time, broadcasting license was only granted to the operators who owned both the program content and broadcast equipment, and so initially the program suppliers positioned themselves as a user of satellite communications service, who is delivering programs only to a limited number of specific recipients. However, that same year the Ministry of Posts and Telecommunications, in response to requests from these operators, introduced a "consignment broadcasting system" that allowed the program suppliers who owned the content to broadcast it using the equipment of satellite communications companies, which in effect allowed for the separation of content and equipment.

In response to this change, in 1992 program suppliers established platforms for fee collection, customer management, and scrambling control to allow full-fledged direct broadcasting. These were CS BAAN, established by JCSAT and the group of program suppliers using its satellites, and Skyport, launched by SCC and the program suppliers using its satellites. For the scrambling system, CS BAAN adopted the government-created COATEC system used for WOWOW on BS satellite, while Skyport embraced the Skyport system developed by Sony, so that there was no compatibility between the two systems. That year only six companies were approved as consignment broadcasters to make broadcasts using communications satellites, and so CS analog broadcasting began with six channels (Fig. 9.9). The following year, 1993, four more companies received approval for consignment broadcasting, making a total of 10 channels. At that time, not all program providers sending programs to CATV stations were immediately approved as consignment broadcasters.

Fig. 9.9   **President Nakayama receives the contracted satellite broadcasting license from the Minister of Post and Telecommunications.**

In accordance with the Ministry of Posts and Telecommunications' basic plan for promoting the broadcasting service, program suppliers were approved to become consignment broadcasters only after undergoing a rigorous screening process, with other suppliers only able to transmit programs to CATV stations or public organizations as communications services.

CS analog broadcasting did not require a different system from that used for transmitting programs to CATV operators, but simply scrambled the signal being sent for cable TV and made it available for reception by individuals. Video broadcasting at the time was transmitted in analog form, and one transponder could only transmit one channel. Receiving the broadcasting signal also required a dish antenna around 25 inches in diameter, and sales of the equipment were slower than expected.

JCSAT'S MOVE TOWARD CS DIGITAL BROADCASTING

Around 1992, just as CS analog broadcasting was beginning in Japan, the world started moving toward digital broadcasting. Theoretical research on digital compression technologies for video and voice had been widely conducted since the 1970s, but there had been no economical way to produce the hardware, that is, vast amounts of memory capacity and high-speed processing in a small device, that would make the technology a reality. The rapid development of semiconductor technology during the 1980s brought the hardware closer to reality, increasing the momentum to utilize digital compression technology for video and voice in specific products and services. The International Telegraph and Telephone Consultative Committee (CCITT, currently the ITU-T, or International Telecommunications Union-Telecommunications Standard Sector), set up by the United Nations, and ISO/IEC (International Organization for Standardization/International Electrotechnical Commission) proceeded with the examination of digitization standards, and from this effort came standardization of the MPEG-2 standard, a video and voice compression technology that took into account the TV broadcasting in the future. Standardization of MPEG-2 took place between 1992 and 1993.

Building on this momentum, in 1991 United States Satellite Broadcasting Company, Inc. (USSB), established by Stanley Hubbard in 1981, and DirecTV, a subsidiary of Hughes Electronics Corp., announced a plan to jointly launch a satellite to offer a digital satellite broadcasting service that would transmit programs up to 175 channels that could be received by individual with just 18-inch-diameter antennas. Plans were formed in Europe as well for satellite broadcasting services using the DVB standard approved for digital video broadcasting in Europe, along with the MPEG-2 standard.

In Japan, meanwhile, CS analog broadcasting was making little headway, and the CS BAAN business that had been established by JCSAT and program suppliers was incurring large losses. The Skyport service begun by the SCC group, on the other hand, had acquired a comparatively larger number of subscribers than CS BAAN. Around this time JCSAT took a hard look at the reality of the recently begun analog CS broadcasting service, and reconsidering its business decided that it should begin offering a multichannel satellite broadcasting service in digital format that would become the main pillar of its business in the future. Then in November 1994, after JCSAT's merger with Satellite Japan Corporation (SAJAC), discussed in the next section, DMC (Digital Multi-Channel) Planning, Inc., was established for the purpose of offering digital satellite broadcasting service and began full-scale feasibility studies. Mr. Nito played a central role in forming the company and conducting the studies. The fact that JCSAT was behind the SCC group in CS analog broadcasting service ultimately allowed it to make a clean break with analog broadcasting early and be one of the first companies to consider the digital broadcasting business, which offered much more potential for the future.

MERGER WITH SATELLITE JAPAN

SAJAC, another satellite operator established in April 1985 with the aim of becoming a satellite-based Type I telecommunications carrier like JCSAT and SCC, lagged behind the other two companies in business development and had experienced significant delays in its business operating application and satellite launch plans. SAJAC's shareholders included companies that were unable to participate in the establishment of either JCSAT or SCC, including the trading companies Sumitomo Corporation, Nissho Iwai Corporation, and Marubeni Corporation, as well as the electronics manufacturers Sony Corporation and NEC Corporation. Despite SAJAC's slow start, the other trading companies undoubtedly felt that they needed to be involved in the new business of satellites in order not to lose on any business opportunities. In February 1991, SAJAC finally received license as a Type I telecommunications carrier and began preparing to procure and launch its first satellite. This was just at the time when SCC had suspended its service after the loss of Superbird-A following the in-orbit incident, and all of its customers had migrated to JCSAT-1 and JCSAT-2. By 1992, SAJAC had formulated its business plan, envisioning its main application to be the digital satellite broadcasting being pursued in the United States and had begun the selection process for a satellite manufacturer.

By this time the exceptional demand JCSAT had enjoyed while aiding SCC customers was ending, and its revenue was declining. On top of this, SCC was preparing to launch Superbird-A1 at the end of the year, and a successful launch would give the company a two-satellites operating structure, which would spark sales competition between JCSAT and SCC. The management of JCSAT, along with the company's shareholders ITOCHU and Mitsui, began to feel that the business environment was difficult enough with just the two firms JCSAT and SCC, and that there was no room in the limited Japanese market for three satellite companies. The prospect of the emergence of a third company elicited a sense of crisis, and the JCSAT group felt that it had to find some way to collaborate with SAJAC. There was a further fear that SAJAC might partner with SCC. Against this backdrop, SAJAC moved forward with its satellite procurement. Meanwhile, JCSAT was considering on its own the launch of a third satellite with new, high-output transponders for the purpose of starting digital CS broadcasting service. But with a concern of an oversupply of transponders if both companies were to launch their own new satellites, JCSAT in the end changed its plans to launch its own satellite and, as part of a strategic partnership with this company, decided to secure the transponders on the spacecraft that SAJAC was preparing to launch.

JCSAT and SAJAC then began negotiations for an alliance in which the shareholders in both firms became involved. There were many obstacles to the plan, however, and negotiations became bogged down. The biggest hurdles to an alliance were a big difference in the perspective the two companies had about the current situation of the satellite communications market in Japan and the degree to which each felt a sense of crisis. However, JCSAT received the support of the Ministry of Posts and Telecommunications, which was also worried about oversupply, and as a result of persistent negotiations conducted between shareholding companies, by the end of 1992 the shareholders of SAJAC had begun to share JCSAT's view of the market and sense of crisis about oversupply. A breakthrough in the negotiations finally came when the two companies decided that rather than form an alliance they would merge together. The two companies had begun conducting due diligence by early the following year, and a team of members from both companies was formed to study the merger. Another team was established to handle procurement of the new satellite, as merger and satellite procurement process were conducted in parallel.

After nearly a year of negotiations, the two companies formerly merged on 17 August 1993 and formed a new company, Japan Satellite Systems, Inc. (JSAT). With this merger, Hughes, which had been a partner in JCSAT since its founding, sold its stake in the company, and Sumitomo Corporation and Nissho Iwai Corporation newly joined as shareholders. The result was that JSAT was a consortium of the four major trading companies in Japan, with the exception of Mitsubishi Corporation. Hughes, which had played such a pivotal role in the founding of JCSAT, became estranged from its partners ITOCHU and Mitsui and later approached SCC as part of efforts to launch DirecTV Japan.

JCSAT-3

PREPARATIONS

JCSAT had been making preparations for the procurement of JCSAT-3 since before the merger with SAJAC. There were three goals. The first was to secure high-output Ku-band transponders for CS digital broadcasting. The second was to offer Ku-band transponders with 36-MHz bandwidth that would cover all of Japan, including outlying islands, thereby resolving the transponder bandwidth and coverage issues that had kept JCSAT second to SCC in user acquisition. Finally, the third goal was to provide both Ku-band and C-band transponders that could cover the entire Asia-Pacific region and allow the company to begin offering an international telecommunications service for Asia and the Pacific. The third spacecraft was designed in anticipation of offering a full-scale international service, with the presumption that the company would receive approval for an international business license. At the time the company felt that eight Ku-band transponders would be sufficient for the CS digital broadcasting service it was planning in Japan; therefore, other Ku-band transponders remaining were intended to be sold outside Japan as an international service that could eliminate sales competition with SCC within Japan.

Around the spring of 1993, the JCSAT-3 procurement task force, which had been operating in parallel to the preparations for the merger of JCSAT and SAJAC, was making its final refinements to the satellite specifications, and on 30 August 1993, after completion of the merger, issued an ATP (authorization to proceed) to Hughes to begin manufacturing of the satellite, signing a formal contract in October that year. The inclusion in JCSAT-3 of C-band transponders intended for Asia raised many doubts among shareholders, but they were ultimately persuaded by the explanation that there would be little difference in the total cost of manufacturing and launching a satellite with an additional 12 C-band transponders along with the Ku-band units. This was the first stepping stone for JSAT's later entry into providing international services. In February 1995, JSAT became the first satellite operator in Japan to obtain an international telecommunications business license, and that year JSAT began transmitting an international television signal from Hawaii using the Hawaii beam, one of the international beams it had secretly had on JCSAT-1 from the beginning.

LAUNCH

JCSAT-3, like JCSAT-1 and JCSAT-2, was procured from Hughes (now Boeing). However, the satellite bus adopted for JCSAT-3, unlike the spin-stabilized Model HS-393 of the previous two satellites, was a three-axis-stabilized Model HS-601(now BS-601) Hughes had newly developed. Hughes held a patent on spin-stabilized satellites and had a preference for them but, unable to resist the trend toward larger and more powerful satellites, began developing three-axis-stabilized models, which offered numerous benefits in terms of power generation and flexibility in onboard antennas. At the time JSAT signed the contract, Hughes had yet to have such a spacecraft operating in orbit, but it had received several orders, including from the Australian company OPTUS, the Hughes-affiliated satellite companies HCI and DirecTV, and the U.S. government. The contract called for transfer of the satellite once it had been placed in orbit, and so Hughes selected the rocket, an ATLAS-IIAS from General Dynamics (later acquired by Lockheed Martin). The ATLAS rocket was originally developed for intercontinental ballistic missiles and had been widely used by NASA for spaceflights and other operations. Although the formal contract for JCSAT-3 was signed in October 1993, the actual program began with the submission of the ATP order at the end of August with the 24 months on-orbit delivery contract. So the manufacturing proceeded with a launch schedule for August 1995.

One of the main reasons behind procurement of JCSAT-3 was to end the struggling CS analog broadcast service (CS BAAN) as quickly as possible and establish a new multichannel broadcasting service on a digital format. One issue that arose after the placement of the order was procurement of a backup satellite. Should the launch fail it would be two years at the earliest before a replacement could be launched, which would cause a serious delay to the CS digital broadcasting service planned for commencement in late 1995 or early 1996 and would inflict considerable damage in terms of establishing the business.

JSAT approached its shareholders about procurement of a backup satellite for JCSAT-3, but with the considerable accumulated losses the company held, the shareholders responded that it was just barely possible financially to procure the third satellite and that a backup satellite was out of the question. However, the JCSAT-3 procurement team felt uneasy with the complete lack of a strategy to deal with the risk of a launch failure, and so consulted with the satellite manufacturer Hughes concerning measures to launch a replacement spacecraft as quickly as possible in the event of a launch failure. Their proposal was to preorder the long lead components. Should the satellite have to be remanufactured because of a launch failure, though work at the design phase could be abbreviated, procurement of some key components would take time, which would become the critical path that ultimately would prohibit any significant shortening of the manufacturing schedule. If JSAT were to commit to purchasing the long lead components (in the case of communications satellites these include the TWTAs, or traveling wave tube amplifiers, the branching and multiplexer filters), the manufacturing time could be shortened by around eight months.

JSAT appreciated Hughes' proposal and in December 1993 decided to procure the long lead components. Then a revised proposal was made to the shareholding companies regarding JCSAT-3's backup scheme, but they were not quick to give consent, and four months passed with no action taken. To make matters worse, since four months had passed, the cost to shorten the lead time by just the remaining four months had become too high. Things became even more confused after it was learned that the plan announced by SCC for Superbird-C included even higher-powered 100-W TWTA units, and the company became caught up in a debate about whether it could compete with 60-W TWTA units. JSAT went back to Hughes for further negotiations to obtain higher-powered TWTA units and revise other terms of the order. Having thus been able to get agreement from all shareholders, JSAT managed to place an order for the long lead components just under the wire on May 1. In fact, this order for long lead components made after long and furious debate later rescued JSAT from a crisis.

The manufacturing of JCSAT-3 proceeded smoothly, without any significant problems, and on 29 August 1995, exactly 24 months from the start of the program, the satellite was successfully launched from the U.S. Cape Canaveral Air Force Station in Florida onboard an ATLAS-IIAS rocket.

PROBLEMS

JCSAT-3 was operated smoothly after the launch. Sometime around September 14, about two weeks after the launch, the satellite was placed in geostationary orbit close to 128° E longitude as planned. However, during final confirmation tests and adjustments of the satellite bus prior to in-orbit testing (IOT) of communications systems scheduled to commence the following week, one of the solar wings was damaged by accident. A detailed review of the seriousness of this damage and its impact on future satellite operations was necessary at this point, causing substantial delays in the subsequent testing schedule and launch of services.

At the time, there were some in JSAT who viewed this problem very seriously. They took the view that the satellite should be entirely written off and replaced with the launch of a new one. However, these views eventually faded away in light of subsequent developments. An in-depth study by satellite manufacturer Hughes and JSAT's operations team concluded that the solar panels could still generate sufficient power and that there is a way to conduct maneuvers even with the damaged solar wing without any impairment in services. JSAT in the end was also able to take out in-orbit insurance.

To launch as quickly as possible multichannel digital broadcasting services (PerfecTV!), the primary goal of the JCSAT-3 launch, JSAT decided to begin services with JCSAT-3 on November 1. However, as part of risk management assuming the worst-case scenario, JSAT also decided to procure and launch JCSAT-4 as a backup satellite as quickly as possible.

In the end, JSAT's decisions to rapidly commence services using JCSAT-3 and launch JCSAT-4 as a backup would prove instrumental in establishing a competitive position for PerfecTV! and later SKY PerfecTV! service, which will be shown in the following pages. These decisions truly marked a turning point for JSAT's development to this day.

BACKUP SPACECRAFT JCSAT-4 (R)

Service on JCSAT-3 began in November 1995, and immediately thereafter work began on procurement of JCSAT-4 as a backup spacecraft. Because it would be a backup for JCSAT-3, the design would be basically the same, but because it had been decided to purchase 100-W TWTA units when ordering long lead components, the output power would be even greater. The procurement order for JCSAT-4(R) was made at the end of November 1995, with a very short delivery time of just 14 months, which was achievable because the long lead components had already been ordered.

While this satellite was being manufactured, a fuel leak was discovered on JCSAT-1 (discussed in the next section). To rescue the satellite, it was decided to temporarily operate JCSAT-4 at 150° E longitude after its launch, and therefore minor adjustments were made to the antenna pattern and Earth sensors so that it would have no difficulty operating at that orbital slot.

Although not part of the original scenario, the backup satellite JCSAT-4 (R) was manufactured on schedule in the exceedingly short period of 14 months and successfully launched on 17 February 1997 from Cape Canaveral in Florida in the United States onboard an ATLAS-IIAS rocket. It was immediately positioned at 150° E longitude to relieve JCSAT-1. Following the launch of JCSAT-5(1B) in December 1997, the original successor satellite to JCSAT-1, JCSAT-4(R) was next transferred to 124° E longitude at the outset of 1998, having already played an important role of backing up JCSAT-1. JCSAT-4(R) was temporally operated at 124° E slot for the JSkyB (which in fact merged to become SKY PerfecTV!) digital broadcasting service that began in spring that year. Thereafter, JCSAT-4 (R) was utilized in many other ways as JCSAT-R, filling the role of a dependable backup satellite that truly saved JSAT from a crisis.

DEORBIT

After the commencement of the service, JCSAT-3 continued to operate smoothly even with the damaged solar wing and, being the main spacecraft providing the SKY PerfecTV! service, has made the biggest contribution to JSAT's revenue. After JCSAT-3's successor satellite JCSAT-10 (3A), launched in August 2006, formerly took over its mission, JCSAT-3 was deorbited in the spring of 2007 at the end of its life. JCSAT-3 played an invaluable part in JSAT's business for more than a decade, and therefore I would like to say a farewell word, "Thank you, and goodbye JCSAT-3!".

FUEL LEAK ON JCSAT-1

From sometime around September 1995, amid the chaos caused by the problems following the insertion of JCSAT-3 into geostationary orbit,

JCSAT-1 began to show signs of anomalous behavior. Mr. Narita, a YSCC engineer, began reporting several anomalies revealed through data analysis. These included a "gradual slowdown in spin rate," "discrepancies in changes in satellite attitude," and "a constant deviation of satellite orbital position in the west direction compared with the calculated projections."

These signs suggested that a tiny force was acting on the satellite, possibly caused by a fuel leak. If so, this could be confirmed immediately from telemetry data showing a drop in fuel tank pressure. However, any drop in pressure caused by a fuel leak could not be confirmed directly. The pressure sensor had poor measuring precision, and the fuel tank pressure was highly dependent on the temperature, meaning that changes in pressure as a result of seasonal temperature changes would eclipse any other change in fuel tank pressure. JSAT immediately reported this issue to the satellite's manufacturer, Hughes, and requested a careful review. However, no progress was forthcoming for several months because everyone was busy dealing with the problems with JCSAT-3 at the time.

Mr. Narita, who grew impatient with this lack of progress at Hughes, continued to analyze the data using various methods. In January 1996, he finally obtained proof of a fuel leak. He carefully analyzed nutation data generated during maneuvers over the years. He found clear anomalous changes in the nutation period of the satellite beginning about September 1995, when a fuel leak was first suspected. Specifically, the data showed variations in the rate of change of the satellite's moment-of-inertia ratio (MOIR). This meant that the satellite's weight distribution was changing drastically caused not only by changes arising from ordinary fuel consumption, but also by a fuel leak.

Confronted with these data, Hughes finally recognized that there was a fuel leak and began specific analysis to locate the leak. Various tests were performed to this end. Based on these tests, it was believed that the leak came from the fuel-side thruster valve of the apogee motor that was no longer in use. In theory, a fuel leak from the thruster valve of the apogee motor was inconceivable because the thruster valve is downstream of the latch valve that serves as the main fuel supply valve. The latter was closed after the apogee motor firing operation completed for the insertion of the satellite into geostationary orbit. However, although this valve should have closed, it failed to do so because of a short circuit in the coil that actuates it. Tests showed that the valve had remained open all of the time. For more than seven years, fuel was suspected to have remained in contact with the thruster valve, causing the valve seal to fail and fuel to leak from the satellite.

ORDER AND LAUNCH OF JCSAT-5 (1B)

Subsequently, estimates of the rate of fuel leakage from JCSAT-1 forced JSAT to recognize the shorter life of the satellite to around August 1997. In light of this, in the spring of 1996 JSAT decided to procure JCSAT-5 (1B) as a successor to JCSAT-1 and in May that year rushed to placed an order with Hughes. This followed the quick decision to purchase JCSAT-4 in the previous year. To shorten the delivery time, JSAT used the same option from the JCSAT-3 contract it had for JCSAT-R and adopted a Model HS-601(Now BS-601) satellite bus. Because the earliest time this satellite could be manufactured and launched would be the end of 1997, it was decided to place JCSAT-4, which had already been ordered as a backup to JCSAT-3 and was scheduled for launch at the end of January 1997, temporarily at 150° E longitude and to migrate customers to it in preparation for the end of the useful operating life of JCSAT-1. It was a true tightrope scenario for being able to continue service at 150° E. This too was made possible because the long lead components had been ordered for a backup spacecraft, to prepare against a worst-case scenario of a failed launch of JCSAT-3. If these components had not been ordered, the launch of JCSAT-R would have been delayed to the end of 1997, and it would not have been ready in time to rescue JCSAT-1 customer from its unexpected shorter life caused by the fuel leak.

JCSAT-5 (1B), the satellite that had been ordered as a successor to JCSAT-1, was manufactured in a short 20 months and successfully launched on 3 December 1997 onboard an Ariane-IV rocket.

TRENDS IN DIGITAL MULTICHANNEL BROADCASTING SERVICES

Around 1992 and 1993, when the extraordinary demand JCSAT had enjoyed while aiding SCC customers after the in-orbit incident with Superbird-A was ending and the analog CS broadcasting business that had just begun was not going well, JSAT needed new strategies for growth. The company's attention turned to multichannel digital satellite broadcasting, which was just starting in the United States. DirecTV commenced services in 1993 in the United States. JSAT planned to be ahead of other companies to be the first to introduce multichannel digital broadcasting service in Japan. Analog satellite broadcasting already existed from Skyport and CS BAAN, and the consignment broadcasting system was in place to allow satellite broadcasting, but JSAT embarked on a bold challenge to introduce in Japan a new business model for digital multichannel broadcasting at a time when even the United States did not yet have a viable business model.

The consignment broadcasting system was an arrangement unique to Japan that provided for the separation of the hardware (broadcast equipment) from the software (programs) and allowed anyone, even those without a transmission facility, to easily enter the broadcasting business. The separation of hardware and software made it possible for businesses without the financial resources to build television towers or make other capital expenditures to become broadcasters. The system allowed anyone who created their own program to use the equipment of a telecommunications operator to easily become a broadcaster.

START OF PERFECTV! SERVICE

With the introduction of satellite digital broadcasting, in 1994 JSAT teamed up with its shareholder trading companies to establish the planning company DMC Planning, Inc., which in 1995 was incorporated as DMC Inc. (currently SKY Perfect Communications, Inc.). In October 1996, the first multichannel digital broadcasting in Japan, with about 50 channels, was offered under the PerfecTV! brand using the JCSAT-3 satellite launched in the previous year. PerfecTV! was a so-called "platform company" specializing in customer management and marketing. In effect it created a structure that relied on the programming companies to supply the content and the satellite operator to provide a highly reliable satellite link for broadcasting service.

The business model for PerfecTV! (now SKY PerfecTV!) has to be reviewed here for better understanding. The service was built on a three-company structure of the content provider that created the programs, the platform company, and the satellite operator that managed the satellite link. Each of them had their own role and shared the risk. In a sense they were partners in the same boat, making up the digital satellite multichannel broadcasting service PerfecTV! Subscribers paid viewing fees to the platform company PerfecTV!, which was funneled directly to the content providers. The funds remaining after paying fees to PerfecTV! and satellite link charges to the satellite operator were the programming company's earnings.

The multichannel satellite broadcasting market, which was as yet untapped in Japan, held an appeal for many companies. In September 1995, JCSAT's one-time partner Hughes established DirecTV Japan with SCC, forming a second platform company to PerfecTV! that began offering service in December 1997. This set the stage for the ever-widening competition between the two platform companies operating alongside each other.

RISE OF JSKYB, ITS OFFER OF COOPERATION WITH JSAT

In early July 1996, Masayoshi Son, the president of Softbank Corporation, approached JSAT to discuss the cooperation between the companies in developing another multichannel satellite broadcasting service platform in Japan, including the launch and operation of the new satellite for the purpose. At that time the JSAT group had just started the PerfecTV! venture, and SCC, in partnership with the U.S. company Hughes, was in the midst of procuring a satellite and making other preparations to set up DirecTV. Backing the Softbank initiative was News Corporation led by Rupert Murdoch, which had been operating the BSkyB network in the United Kingdom and had experience with satellite broadcasting. Both companies had acquired shares in TV Asahi Corporation and steadily made other preparations aimed at entering the broadcasting market in Japan. They were also planning to establish a satellite broadcasting business in Japan called JSkyB (Fig. 9.10).

Fig. 9.10   JSkyB is established in December 1996 (copyright held by Sky Perfect Communications Inc.).

The request from Softbank forced JSAT to make an extremely difficult decision. If the startup could not gain the cooperation of JSAT, it was possible that they would turn to SCC for help, and so the merits and demerits of refusing the offer required a lot of careful consideration. If JSAT turned away Softbank, on the one hand they might join with DirecTV Japan, while on the other hand JSkyB might also develop its business by procuring a satellite on its own, instigating a three-way struggle. PerfecTV!, despite its first-comer advantage, had yet to gain a solid foothold. Therefore JSAT wanted to avoid either case that the DirecTV Japan group would only get stronger, or PerfecTV! would face three-way competition. Then-JSAT President Mr. Yoshida felt that it would be best not to make Mr. Murdoch a competitor, and determined that if JSAT could build the kind of far-reaching cooperative structure Softbank President Son envisioned and was able to procure a satellite with little or no risk, then it would be in the best interest of the company to accept the offer.

Top-secret discussions were held immediately following the initial offer of the cooperation, and just one week later on 15 July 1996, News Corporation, Softbank, PerfecTV!, and JSAT reached a basic agreement. With the conclusion of the basic agreement, JSAT set to work putting together the form and content of the contract with JSkyB. The form of the contract was a reservation agreement for the use of transponders. Although Softbank's original plan was to use the satellite that News Corporation had already ordered Hughes in anticipation of implementing its ASkyB service in the United States, JSAT proposed to launch a new satellite, JCSAT-6 (4A), with 16 transponders out of 32 on this satellite provided for JSkyB. The framework also included the request that a deposit for the reservation be paid at the time the contract was signed. JSAT submitted a draft of this contract at the end of September, and negotiations were held three times, in Tokyo, New York, and Los Angeles, before a final agreement was reached in December. JSAT's counterpart in the negotiations shifted from Softbank to News Corporation, whose lineup included Mr. Clark and Mr. Pontual, who had been given full authority to negotiate by Mr. Murdoch, along with attorney Mr. Specter, while the JSAT team included Messrs. Akiyama, Suzuki, Kato, and Matsui, along with their legal council Mr. Kurosu from the law office of Graham and James.

Initially the proposal was to launch JCSAT-6 (4A) to provide service as already stated, but launching a new satellite would take more than two years. JSkyB, in order to compete with PerfecTV!, the first company to market, and with DirecTV Japan, hoped to begin service as quickly as possible. JSAT then made a proposal to temporarily use JCSAT-4, which was already being procured as a backup spacecraft for JCSAT-3. Instead of waiting for the launch of JCSAT-6 (4A) in early 1999, the idea would allow JSkyB to begin offering services in just one year, by early 1998, when the ground broadcast equipment would be ready. JSkyB appreciated and agreed with the plan; thus, JSAT was able to gain an advantage in the negotiations.

JSKYB AND PERFECTV! MERGE TO FORM SKY PERFECTV!

JSkyB's later increase in capital gave Sony and Fuji Television a stake in the project, and JSkyB became a company composed of four equal partners. By 1997 JSkyB had begun considering its broadcast system architecture. The JSkyB side proposed to JSAT and PerfecTV! the use of Simulcrypt, which would allow programming transmitted in either of the two platforms to be received with the same set-top box. But PerfecTV! was already broadcasting using Sony's CAS (conditional access system). And JSkyB planned to adopt CAS from News Data Systems, an affiliate of News Corporation. The idea was that the signal for both platforms' program would be transmitted with encrypted data compatible with these two types of CAS and could be viewed no matter which receiver was used or the program platform. In parallel with these discussions, the two sides held negotiations to try to work out some sort of alliance. The idea for a merger between PerfecTV! and JSkyB ultimately surfaced, and the two companies began moving in that direction. Three separate platforms were thought to be excessive for such a small market as Japan. As a result, the talk of Simulcrypt faded away, and it was decided that the JSkyB system would be designed for the Sony CAS system used for PerfecTV! In April 1998, JSkyB began using its transmission equipment to make digital broadcasting via JCSAT-4, but JSkyB never made a single broadcasting under its own name, as in May 1998 Japan Sky Broadcasting merged with Japan Digital Broadcasting, Inc., the company that ran PerfecTV!, giving birth to SKY Perfect Communications Inc. (SKY Perfect).

JCSAT-6 (4A), which had been planned for the JSkyB service, was successfully launched on 16 February 1999, using an ATLAS-IIAS rocket, and placed at 124° E longitude. With the new satellite in place, the digital broadcasting service using JCSAT-4 at 124° E (the former JSkyB service) was from June 1999 being provided by JCSAT-6 (4A). Freed from its relief role, JCSAT-4 was placed on stand-by at close to 128° E as JCSAT-R, an in-orbit backup spacecraft as originally planned.

DirecTV Japan Exits the Market Leaving SKY PerfecTV! as the Sole Platform

DirecTV Japan began offering a digital broadcasting service in December 1997, using the Superbird-C spacecraft launched by SCC in July that year. However, handicapped by having started its service nearly a year after PerfecTV!, and undeniably inferior in terms of number of channels and marketing capabilities compared to SKY PerfecTV!, formed by the merger between PerfecTV! and JSkyB, DirecTV Japan was forced to withdraw in September 2000. DirecTV Japan undertook measures to transfer its customers to SKY PerfecTV! and got some shares in SKY Perfect Communications. As a result, SKY PerfecTV! became the sole platform for digital multichannel broadcasting in Japan. High-profile content offerings such as all of the games of the 2002 FIFA World Cup and U.S. major league baseball games helped raise its appeal over the more limited terrestrial broadcasts and responded to the expectations of viewers. By 2006 SKY PerfecTV! had grown to be the world's leading digital multichannel broadcasting corporation, offering 290 channels, which originally was only 50 provided by eight transponders on JCSAT-3. The number of subscribers has also grown steadily to reach 4.18 million. SKY Perfect Communications Inc., listed on the Tokyo Stock Exchange's Mothers market in October 2000, moving to the TSE First Section in April 2004, and since September 2005 has been one of the Nikkei Average of 225 selected issues. As of 2006, 10 years after PerfecTV! began broadcasting in 1996, the broadcasting segment accounted for more than 60% of JSAT's revenue, making a significant contribution to the company's growth.

Looking back, the driver for JSAT's growth has been SKY PerfecTV!, which survived to be the sole platform for digital multichannel broadcasting in Japan. Such growth was possible because of the advantage of starting service a year earlier than DirecTV and the merger with JSkyB. This scenario became a reality because of JSAT's ability to provide service without delay even with the problem on JCSAT-3 after launch, the advanced purchase of long lead components as a backup prior to the launch of JCSAT-3, the immediate order for JCSAT-4 as a backup made at the same time as JCSAT-3 encountered the problem, and its procurement and launch in just 14 months, along with the success in proposing use of this backup satellite for JSkyB and integration of its service. This scenario was by no means something designed in advance, but it was the founding spirit of JCSAT to make preparations against risk, and the determination to make forward-looking resolutions when problems arise that led to its success.

LATER DEVELOPMENTS

The history of JSAT up through the birth of the SKY PerfecTV! service that underpinned its growth has been described. This is in a sense the climax of the JSAT story, but there have been an assortment of later developments. The following four topics are some of the more noteworthy, and a brief summary of each will be presented: 1) transfer of N-STARa and N-STARb to JSAT, 2) change of the corporate name to JSAT Corporation and listing on the First Section of the Tokyo Stock Exchange (TSE), 3) the Horizons Project, and 4) announcement of the business integration of JSAT and SKY Perfect.

NTT had been JSAT's biggest customer, using many Ku-band transponders on JCSAT satellites since the company's start in 1985, but at the same time it also had its own communications satellites, namely, CS-2 and CS-3, both of which were equipped with Ka-band and C-band transponders. It had also launched and operated successor spacecraft, N-STARa and N-STARb, in 1995 and 1996, respectively, with a mission that included mobile communications using the S-band by NTT DoCoMo. Some time around 1997, then-JSAT President Mr. Yoshida began advancing a strategy for transitioning NTT from a foremost customer to a foremost partner and began exploring ways to cooperate with the NTT Group. As a first step, in 1998 NTT Satellite Communications, Inc., providing an Internet service via satellite, was established jointly by JSAT and NTT Communications. Building on this trend, JSAT continued to push forward with its alliance strategy, and in March 2000 accepted the transfer of NTT Communications' ownership interest in N-STARa and N-STARb (Fig. 9.11). Under this deal JSAT would operate these satellites and provide the transponders to NTT Communications. Later, in 2002 and 2003, JSAT also acquired the interests held by NTT East, NTT West, and NTT DoCoMo, Inc. JSAT paid for the satellite transfer with stock through a capital increase. NTT Communications' interest in JSAT surpassed that of the four trading companies in a single bound, making it JSAT's biggest shareholder.

With the transfer of N-STARa and N-STARb, JSAT began receiving usage fee revenue from the NTT Group for the transponders on these satellites, which combined with an increase in income related to the SKY PerfecTV! service operated from JCSAT-3 and JCSAT-4 to provide JSAT with a significant year-on-year rise in revenue in fiscal 2000. For the first time since its founding,

Fig. 9.11   The N-STAR spacecraft. (See also color figure section at the back of the book.)

JSAT's revenue topped ¥30B, and within the year it had managed to eliminate its accumulated losses. Bolstered by such circumstances, in April 2000 the company changed its corporate name from Japan Satellite Systems (JSAT) to JSAT Corporation and in August 2000 achieved its longstanding goal of public listing (TSE First Section) for which it had been preparing. The offering price for shares was ¥700,000 but proved popular immediately after listing, at one point reaching a high of ¥1.5M.

Another major development was the joint venture with PanAmSat Corporation (now merged with Intelsat). Foreign-backed telecommunications services were permitted as part of the liberalization of Japan's telecommunications market, and PanAmSat had established an office in Tokyo and was conducting sales activities for a satellite telecommunications service in Japan and the Asia/Pacific region. In that sense, PanAmSat was one of JSAT's main competitors, particularly in the field of international communications. These competitors decided to join hands in a joint venture.

The business had its origin in the tightening of supply for Ku-band transponders in the United States. From 2000 through early 2001, PanAmSat had approached JSAT about jointly launching a Ku-band satellite to be placed in orbit at 127° W longitude ever since Japan (JSAT) had made a filing with the International Telecommunications Union (ITU) for a Ku-band satellite at that position. PanAmSat at the time owned and operated a C-band satellite at 127° W, and when it planned to launch the successor satellite wanted a hybrid satellite that also carried the Ku-band transponders that were increasingly in demand in the U.S. market. However, the right to use Ku-band at that orbital position had already been obtained by JSAT via the Japanese government through a filing with the ITU, so that JSAT's cooperation was essential. A proposal for a satellite jointly owned by operators in different countries was unprecedented, but JSAT, wanting to establish a strategic alliance with an overseas operator, as well as a beachhead in the United States for an international communications service, dedicated itself to the project with a team built around Messrs. Akiyama and Mizoguchi. Negotiations with the PanAmSat team led by the firm's Vice President Mr. Cuminale on overcoming legal obstacles in Japan and a joint ownership business scheme stretched on, with an agreement for a joint venture with PanAmSat signed in August 2001 (Fig. 9.12).

The framework for the agreement included a completely equal, 50-50 sharing of the costs for and revenue from the Ku-band transponders in the new satellite, while each company would compensate the other for satellite operating costs and fees for orbital rights and marketing incentives. Following this agreement, JSAT established JSAT International, Inc. (JII), a wholly owned subsidiary in the U.S. Horizons Satellite LLC, a procurement and holding company for the new satellite, which was also established through joint investment by JII and PanAmSat. The jointly owned satellite Horizons-1

Fig. 9.12   The signing ceremony with PanAmSat.

was successfully launched in October 2003 by Sea Launch with service commencing that same year. As of the end of 2006, Horizons-1's Ku-band transponders were operating at nearly full loading, and JSAT is pursuing a plan with Intelsat, which acquired PanAmSat, to launch a new satellite, Horizons-2, in 2007 above North America.

The final theme for the JSAT story is the issue of the business integration with SKY Perfect. As already mentioned, PerfecTV!, the predecessor of SKY PerfecTV!, began in 1994 as a planning company set up within JSAT along with its shareholding companies. JSAT remained a shareholder in PerfecTV! even after it became a business company, but since the merger with JSkyB that created SKY PerfecTV!, JSAT's share in the company has become minor, and the capital ties have become looser. During this time, the reigns of power over the SKY PerfecTV! business have been held by ITOCHU and Sumitomo—the original shareholders in JSAT—along with Sony and the terrestrial broadcaster Fuji Television, which were included at the time of the merger with JSkyB. In a certain sense, therefore, this merger is a return to the original owners.

There are two major factors behind the merger: a change in the business environment for broadcasting and telecommunications (more intense competition between media companies) and room for growth in the fee-based multichannel broadcasting market. JSAT's satellite communications service has revolved around the two main axles of telecommunications network services and broadcasting services. However, with the spread of optical fiber and growth of internet services that has accompanied it, JSAT is rapidly losing its cost competitiveness for telecommunications network services. Even bidirectional and multicasting network services linking multiple locations spread over a wide area, for which satellites used to have a very strong position, are being overwhelmed in terms of cost by IP-VPN services based on terrestrial fiber-optic lines and similar services. JSAT is gradually shifting the central axis of its network services from fixed services to mobile, and from domestic to international, but its dependence on revenue from the broadcasting business is high, and it has to be admitted that new subscriber acquisition for SKY PerfecTV! service is not proceeding well recently.

Currently, increasing the number of SKY PerfecTV! subscribers is a common and vital issue for both JSAT and SKY Perfect Communications. The rate of household penetration for CS digital multichannel broadcasting in Japan is under 10%, less than half the 20–25% rate in the United States. In March 2007, the two companies announced the medium-term management plan of the new holding company SKY Perfect JSAT Corporation, which is to roughly double the cumulative number of registered subscribers from the current approximate 4.2 million to somewhere around 8 million, with the goal of overcoming the competition for video transmission from CATV and Internet services. With analog terrestrial broadcasting scheduled to end in

Japan in 2011, high-definition broadcasting using terrestrial digital services is spreading rapidly. In line with this trend, CS digital broadcasting must also provide programming in HDTV format, utilize the set-top boxes for HDTV that are introduced to offer a variety of new enhanced features that will allow new types of services to satisfy customers, and adopt multimedia strategies that are not dependent on satellites alone. The decision for this business integration was made so that the capital, personnel, transponders, and other resources necessary to develop and implement these strategies could be invested in a timely manner and in sufficient quantities.

After the integration in April 2007, the benefits of this decision will become clear, and the further developments of both JSAT and SKY Perfect are highly expected.

CONCLUSION

This chapter has looked back at JSAT's more than 20-year journey from its founding to the present day. As shown through the story, the road leading to the present has not always been smooth. The efforts of those who preceded us to overcome a range of difficulties have brought JSAT to where it is today. It is often said that JSAT was "lucky," or that it was in some way fortunate. But reading this story, it might be understood that luck or good fortune had nothing to do with making JSAT what it is.

A spirit of challenge igniting the dreams and ideas that have been handed down unchanged since the company's founding, underpinned by carefully considered plans and strategies, company mergers made in a spirit of harmonious integration unconstrained by one's own company of origin, and perseverance and teamwork to overcome difficulties. These are the things that have quietly and progressively cultivated the JSAT spirit, culture, and traditions to shape JSAT into the company it is today.

There is no doubt that JSAT and SKY Perfect, as an integrated company, will have many challenging days ahead of it over the next 20 years. But so long as the positive traditions of JSAT remain, I am confident that the new company will be able to forge a new tomorrow.

Chapter 10

INSAT INITIATES COMMUNICATION REVOLUTION IN INDIA

U. R. RAO*

AND

A. BHASKARANARAYANA†

INTRODUCTION

The communication revolution ushered by the INSAT satellite network in the early 1980s has totally transformed India, by enabling people even in the remotest corners of the country to have access to modern communication facilities. In just under 25 years, INSAT system has grown into one of the largest communication networks in the world, providing access to over 90% of the country's population to nationwide TV broadcasting, a satellite-based radio networking system, a variety of fixed satellite services, quality educational broadcasts, regular meteorological forecasts, and locale-specific disaster warning facilities in the coastal districts of the country.

Indian Space Program made a modest beginning with the establishment of an Equatorial Sounding Rocket Launching Station in Thumba, near Tiruvananthapuram in 1963 for carrying out scientific experiments in aeronomy, upper atmosphere, equatorial ionosphere, and astronomy under the leadership of the Indian Space Research Organization (ISRO). However, realizing the vast potential of space technology for addressing a variety of socio-economic problems of the nation, particularly in the areas of communication, education, disaster management, and weather forecasting, ISRO soon focused its attention on developing a vibrant application-oriented space program on a totally self-reliant basis. The guiding vision for India's Space

Copyright © 2008 by the authors. Published by the American Institute of Aeronautics and Astronautics, Inc., with permission.

*Chairman; Physical Research Laboratory Council, Department of Space, Ahmedabad, India. Former Chairman, Indian Space Research Organization (ISRO), and Secretary, Department of Space.

†Scientific Secretary and Director, Satellite Communication Programmes and Frequency Management, Indian Space Research Organization, Bangalore, India.

Program was clearly enunciated by Vikram Sarabhai, the father of the Indian Space Program, as follows:

> There are some who question the relevance of space activities in a developing nation. To us, there is no ambiguity of purpose. We do not have the fantasy of competing with the economically advanced nations in the exploration of the Moon or the planet or manned spaceflight. But we are convinced that if we are to play a meaningful role nationally and in the Comity of Nations, we must be second to none in the application of advanced technologies to the real problems of the man and society.

It is this vision, which is relevant even now, that has guided ISRO in the development and application of space technology, sharply focused towards achieving rapid socioeconomic development of the country.

During the 1970s, Indian space efforts were primarily geared towards carrying out research and development in a variety of engineering disciplines of relevance to satellite and launch-vehicle technologies as well as carefully selected large-scale experiments in communication, education, and remote sensing of natural resources. During the 1980s and 1990s, the experimental initiatives undertaken earlier were progressively transformed into operational satellite systems for providing nationwide services in communication, navigation, broadcasting, education, weather forecasting, and management of natural resources. It was during this decade ISRO also succeeded in building indigenous capability in operational launch vehicles for launching both low Earth orbiting and geostationary satellites, enabling ISRO to become totally self-reliant. With its primary emphasis on large-scale application of space technology, on an end-to-end basis, towards national development, the Indian Space Program has distinguished itself as one of the most cost-effective and development-oriented space programs in the world.

Early Experiments

Based on the experience gained from the conduct of Krishi Darshan Experiment, which beamed educational programs to 80 villages around Delhi in 1967, ISRO was fully convinced that the use of nationwide TV broadcast, the most powerful medium of mass communication, was essential for tackling the massive problem of illiteracy in India [1]. To establish the immense potential of satellite technology for providing education, ISRO carried out the reputed world's largest sociological experiment called the Satellite Instructional Television Experiment (SITE) in 1975–1976, with the help NASA's ATS-6 satellite (Fig. 10.1). In the SITE experiment, specially tailored developmental video programs were broadcast for six hours on each day to community TV sets located in 2400 specially selected remote villages in six states for a period of one year, for imparting education in health, hygiene,

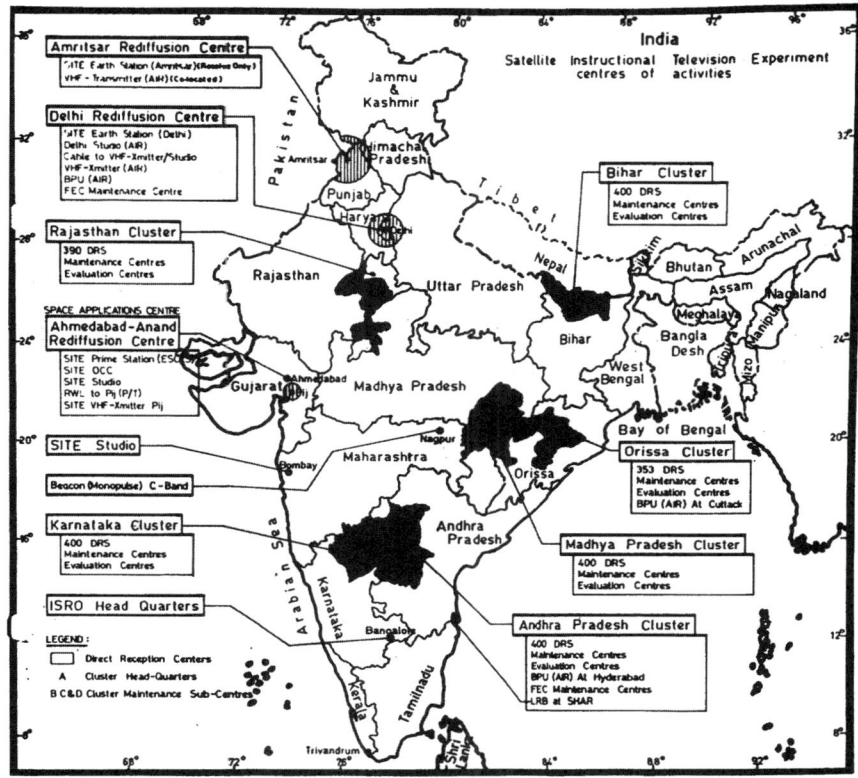

Fig. 10.1   A map of SITE.

environment, better agricultural practices, and family planning. The unique-ness of SITE [2] was that it became the first large-scale experiment to directly broadcast video programs to village community reception TV centers equipped with a carefully designed 10-ft chicken mesh antenna. The educational programs were specially prepared using the rich stock of mythological stories and cultural heritage of India as backdrops to make them entertaining and instructive.

Extensive evaluation of both hardware and software components of the year-long SITE experiment, conducted by a number of independent teams including those from outside the country, clearly demonstrated that SITE experiment had a very significant impact on rural population, thus firmly establishing the capability of satellite TV medium for rapidly transforming our rural society. The SITE experiment was followed by a hardware-oriented experiment called STEP conducted with the Franco-German satellite Symphony during 1978–1979, which enabled ISRO to gain experience in conducting communication experiments with indigenously designed and

TABLE 10.1   MAJOR EVENTS LEADING TO CONCEPTUALIZATION OF INSAT

No.	Experiment	Year(s)	Event
1.	Krishi Darshan	1967	Educational broadcast to villages around Delhi started
2.	ESCES Earth station	1967	Establishment of Earth station at Ahmedabad
3.	SITE	1975–1976	Educational broadcast to 2400 villages in India using ATS-6
4.	STEP	1976–1977	Communication experiments using symphony satellites
5.	APPLE	1981–1983	Experimental communication satellite launched by India

fabricated ground hardware, thus paving the way for ISRO to embark on its own communication satellite program. Table 10.1 gives the major events leading to conceptualization of INSAT.

CONCEPTUALIZATION OF INSAT

During the decade of 1960s, ISRO also initiated a number of studies, some of which were jointly conducted with external organizations such as Philco-Ford, Hughes, GE, and MIT-Lincoln Laboratory to conceptualize and arrive at an overall concept and design of a cost-effective satellite system, specially tailored to meet India's requirements. These studies clearly brought out the need for establishing a broadcasting system, consisting of direct reception of TV in remote areas to provide access even to the most remote areas of the country and through rebroadcast mode to all urban centers. The studies, in addition to providing broadcasting and long-distance telecommunication services, also stressed the importance of establishing a geostationary meteorological platform for regular weather forecasting and disaster monitoring, which are of paramount importance for a primarily agricultural country like India, totally dependent on the vagaries of the monsoon. To economize on cost, ISRO decided to employ a unique multipurpose satellite system combining communication and broadcasting with meteorological imaging on the same platform, unlike the conventional satellite system, which consisted of separate communication and meteorological satellites. With the growing pressure for establishing an operational communication and broadcasting satellite system, ISRO decided to procure the first-generation INSAT-1 series of multipurpose satellites from Ford Aerospace Communication Cooperation (FACC) because ISRO had just launched only its first experimental low-Earth-orbiting satellite Aryabhata in 1975 and had yet to develop the capability to build complex operational satellites.

INDIGENOUS TECHNOLOGY BUILDUP

In parallel ISRO decided to design, build, and operate an experimental geostationary satellite to gain experience required to undertake the design and fabrication of more complex, second-generation, and subsequent operational satellites, to follow the first-generation INSATs, during the decade of the 1990s. Towards achieving this objective, ISRO got a unique opportunity when its proposal for launching an experimental communication satellite, made in response to the offer of the European Space Agency (ESA) for a free launch in one of its early Ariane missions, was selected by ESA in 1977 among over 70 competing international proposals. A major milestone was thus achieved with the successful launch in June 1981 of the indigenously designed and fabricated three-axis-stabilized, geostationary communication satellite APPLE (Ariane Passenger Payload Experiment) weighing 670 kg and carrying two C-Band transponders (Fig. 10.2), using the third Ariane launch mission. APPLE, in addition to enabling ISRO to gain experience in building and launching an on-orbit operation of a communication satellite, also provided a continuous platform for over 27 months for broadcasting major national events and for

Fig. 10.2   The three-axis-stabilized, geostationary communication satellite APPLE. (See also color figure section at the back of the book.)

carrying out meaningful communication experiments. Large-scale experiments on time, frequency, and code division multiplex access system (TDMA, FDMA and CDMA), radio networking, computer interconnect, random access and packet switching with a variety of multiple access protocols such as ALOHA, slotted Aloha, and other protocols based on fixed and demand assignment were carried out using APPLE [3]. Spread spectrum multiple access system, data-compression techniques, and digital speech interpolation techniques were also experimented upon to gain experience in more efficient use of communication channels. APPLE was also used for conducting valuable experiments in education and training, telemedicine, and point-to-point and point-to-multipoint communication. Most importantly, APPLE experience enabled ISRO to indigenously undertake the design and fabrication of the second-generation and subsequent INSATs, which followed the first-generation INSAT's procured from FACC.

FIRST-GENERATION INSAT (INSAT-1) SPACE SEGMENT

The system configuration concept of the INSAT-1 satellites (Fig. 10.3), built by FACC to the Indian specifications, envisaged a space segment consisting of two multipurpose satellites—one as the primary satellite providing all services and the other as a major backup satellite providing additional

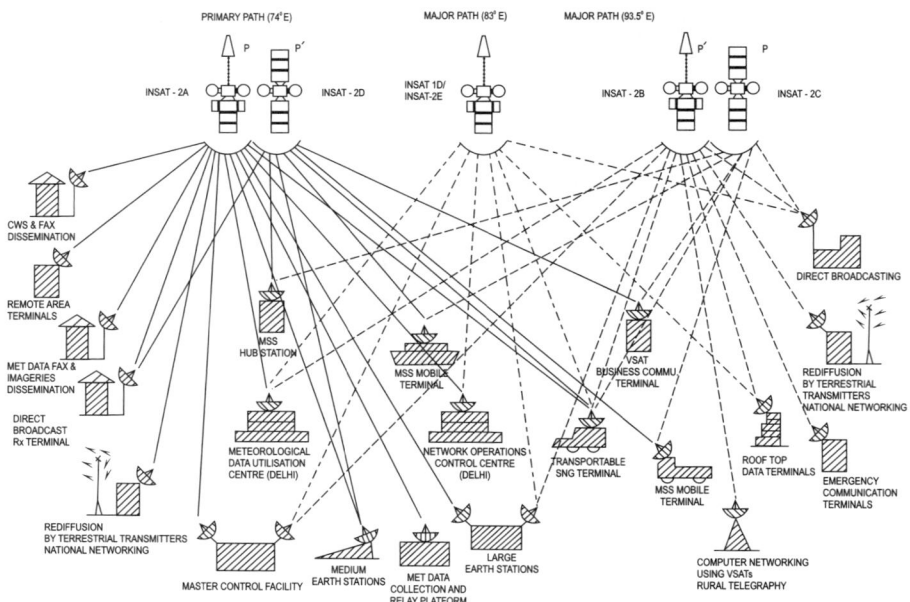

Fig. 10.3   The INSAT-1 system concept.

fixed satellite service utilization and also certain on-orbit backup capability. Each of the INSAT-1 satellites weighed about 1200 kg in the geostationary transfer orbit and about 650 kg in the geostationary orbit at the time of initial station acquisition [3,4]. A roughly 11.5-m^2 one-sided asymmetrical solar array of five panels, involving multiaxial deployment in orbit, provided 1185 W of electrical power at the beginning of life and about 900 W at the end of life, to meet the design requirement of seven years life.

The use of an asymmetrical solar array on INSAT satellites was to ensure an unobstructed clear field of view (FOV) into the cold space for the radiation cooler of the very high resolution radiometer (VHRR) Earth-imaging instrument. A deployed solar sail was used to offset solar pressure on account of the asymmetrical solar array. In its fully deployed configuration in the geostationary orbit, the length of INSAT-1 satellite from the tip of the solar sail to the extreme end of the solar array (Fig. 10.4) was 19.4 m. The three-axis-stabilized INSAT-1 satellites were equipped with a precision attitude control system for providing the high-accuracy stability required for the meteorological imaging mission. A magnetic torquer with a current coil placed around the periphery of the satellite body provided fine control. The satellites had a unified bipropellant propulsion system for both orbit raising from the transfer orbit to geostationary orbit and for station keeping and attitude maintenance during the entire seven-year life of the spacecraft mission.

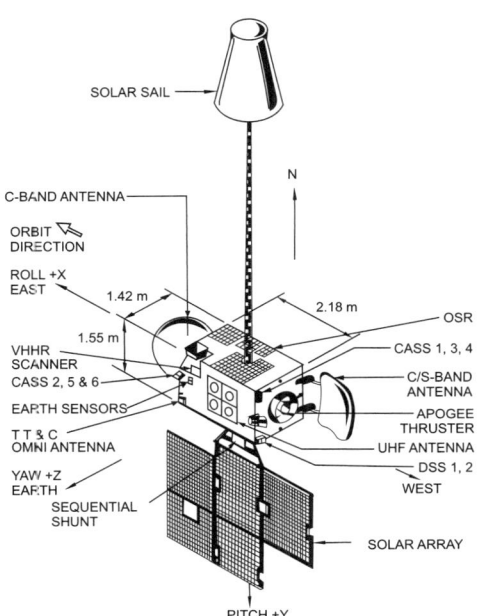

Fig. 10.4   The INSAT-1 on-orbit satellite configuration.

Each INSAT-1 satellite was designed to provide the following capabilities:

1) First is twelve national coverage C-band transponders of 36-MHz bandwidth each, operating in 5935–6425 MHz (Earth-to-satellite) and 3710–4200 MHz (satellite-to-Earth) frequency bands with 32 dBW (min) end-of-life EIRP over the primary coverage area.

2) The next capability is two high-power national coverage S-band TV broadcast transponders operating in 5855–5935 MHz (Earth-to-space) and 2555–2635 MHz (space-to-Earth) frequency bands, each capable of handling one direct broadcast (community reception) TV channel and five low-level carriers for services like radio program distribution, disaster warning, and dissemination of standard time and frequency signals with a 42 dBW (min) EIRP over the primary coverage area.

3) Third is a VHRR instrument for meteorological Earth imaging in the visible (0.55–75 μm) and infrared (10.5–12.5 μm) channels with resolutions of 2.75 and 11 km, respectively, with half-hourly full Earth coverage and 5-minute sector scan capability.

4) Last is a data relay transponder with global receive coverage with a 402.75-MHz Earth-to-satellite link for relaying meteorological, hydrological, and oceanographic data from unattended land- and ocean-based automatic data-collection platforms.

INSAT-1A launched in April 1982 on a Delta 3910 launch vehicle and failed after four months of functioning because of the initial nondeployment of solar sail, which culminated in a series of events resulting in the depletion of onboard fuel. Consequently, regular satellite communication service started only after the successful launch of INSAT-1B in August 1983, which functioned satisfactorily during the entire seven years of its design life to initiate the communication revolution in India. This was followed by the launch of INSAT-1C in July 1988 and INSAT-1D in June 1990 to provide nationwide services in communication, broadcasting, and meteorological services until the mid-1990s.

First-generation INSAT satellites were used extensively to provide over 7000 two-way voice or equivalent long-distance telecommunications and establishment of a number of captive business communication networks for meeting the requirements of major public sector organizations such as Oil and Natural Gas Commission, Indian Telephone Industries, National Thermal Power Corporation, Gas, Steel and Coal Corporations, and National Informatics Center, which connected all district centers with the National Centre at Delhi. A 50-station satellite-based TDM/TDMA rural telegraphy network was implemented in the Northeast Region.

Extensive use of meteorological data taken every half-hour and at a five-minute intervals during cyclone, hurricane, and flooded conditions was provided to the nation, which became the primary backbone of round-the-clock meteorological news dissemination (Fig. 10.5). The primary data from

Fig. 10.5   An example of meteorological data.

VHRR received by the Meteorological Department at Delhi were processed along with the data from the unattended data collection platforms located in inaccessible areas to derive products like cloud motion vectors, sea surface temperature, precipitation estimation, etc. The processed data along with cloud pictures were transmitted to over 40 secondary data utilization centers (SDUC) located in various parts of the country for further dissemination. A few hundred automatic meteorological data-collection platforms were installed in inaccessible regions including in Antarctica to provide meteorological information over such areas. A locale-specific disaster warning system (DWS) (Fig. 10.6), first of its kind in the world and which saved thousands of lives as well as livestock, with over 100 receivers located in selected cyclone-prone east coastal areas of the country was installed.

The two high-power S-band transponder channels onboard INSAT-1B were utilized by Doordarshan for nationally networked TV programs feed for a large number of low-power TV transmitters operating in the national TV system, or transmitting UGC-sponsored higher-education enrichment TV programs for school ETV program feed in selected states and for area-specific direct TV broadcast to augmented community TV receivers in selected rural

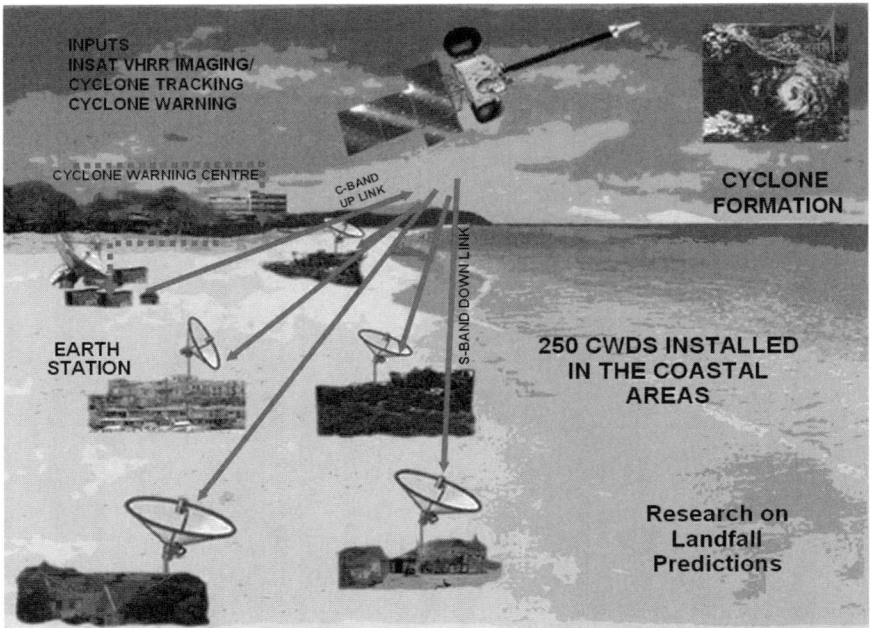

Fig. 10.6   A locale-specific DWS.

areas. These S-band transponders also supported a five-channel radio network-ing service, the cyclone disaster warning service, and a direct satellite retrans-mission facility for the processed INSAT meteorological (VHRR) images.

The most dramatic impact of INSAT was in the remarkable expansion of TV coverage in the country, which would not have been possible without the availability of satellite (INSAT) TV feed capability. In a period of just two years, over 200 TV stations were set up in addition to about 5000 direct reception sets (DRSs) in various parts of the country. The impetus provided by the first-generation INSATs has, over the last two decades, grown from just a handful of TV transmission serving a few million people to over 1200 transmitters covering over 90% of the country's population.

SECOND-GENERATION INSAT SATELLITES

With the experience gained from the successful design, fabrication, launch-ing, and operation of APPLE, ISRO took the bold decision to indigenously design and fabricate the second-generation three-axis-stabilized INSAT-2 series of satellites, weighing over 1900 kg and having much higher capacity compared to the first-generation INSATs (Fig. 10.7). These satellites carried 12 C-band and six extended C-band transponders, two high-power S-band transponders, a data relay transponder, and an improved VHRR with a

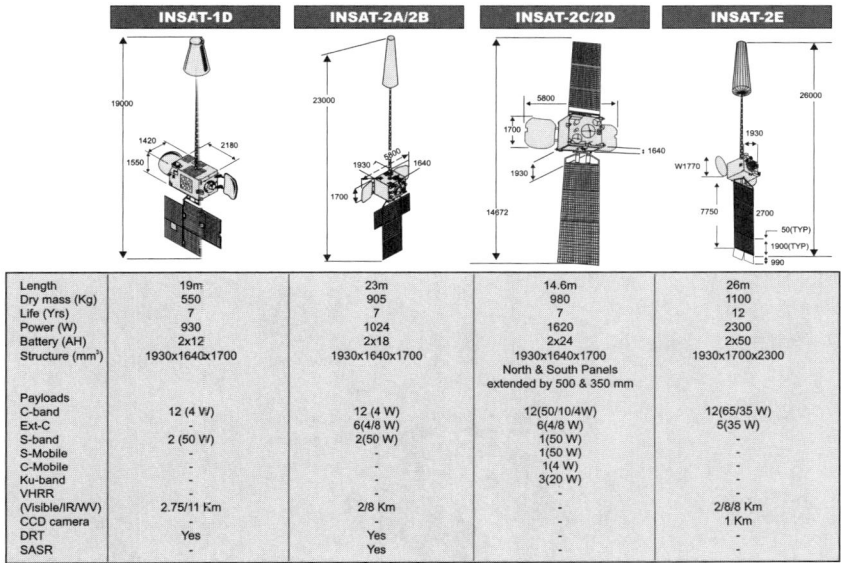

	INSAT-1D	INSAT-2A/2B	INSAT-2C/2D	INSAT-2E
Length	19m	23m	14.6m	26m
Dry mass (Kg)	550	905	980	1100
Life (Yrs)	7	7	7	12
Power (W)	930	1024	1620	2300
Battery (AH)	2x12	2x18	2x24	2x50
Structure (mm³)	1930x1640x1700	1930x1640x1700	1930x1640x1700 North & South Panels extended by 500 & 350 mm	1930x1700x2300
Payloads				
C-band	12 (4 W)	12 (4 W)	12(50/10/4W)	12(65/35 W)
Ext-C	-	6(4/8 W)	6(4/8 W)	5(35 W)
S-band	2 (50 W)	2(50 W)	1(50 W)	-
S-Mobile	-	-	1(50 W)	-
C-Mobile	-	-	1(4 W)	-
Ku-band	-	-	3(20 W)	-
VHRR				
(Visible/IR/WV)	2.75/11 Km	2/8 Km	-	2/8/8 Km
CCD camera	-	-	-	1 Km
DRT	Yes	Yes	-	-
SASR	-	Yes	-	-

Fig. 10.7   INSAT-1–INSAT-2 comparison is given. (See also color figure section at the back of the book.)

resolution of 2 km in the visible and 8 km in the infrared. To compensate for the higher mass of the asymmetrical satellite, the length of the solar boom and sail had to be suitably increased, which resulted in increasing the length of INSAT-2A and 2B satellites from the tip of the solar sail to the extreme end of the solar array to about 23 m (Fig. 10.8).

With the successful launch of INSAT-2A in July 1992 and INSAT-2B in July 1993 on Ariane-4 and concurrent availability of meteorological imaging from INSAT-1D and two INSAT-2 satellites, ISRO decided to delete VHRR and go for conventional configuration for INSAT-2C and 2D satellites with two-sided solar arrays. INSAT-2C and 2D weighing over 2000 kg launched in December 1995 and June 1997, respectively, also carried 12 C-band and six extended C-band transponders, more than half of which operated at a much higher power level, with two of them covering a large geographic region from West to Central to South East Asia. Deletion of VHRR enabled ISRO to introduce two new services, a mobile communication service by reconfiguring one of the high-power S-band transponder, and fixed satellite service in Ku-band using three Ku-band transponders.

INSAT-2D and INSAT-2E, launched in 1997 and 1999 respectively, incorporated many new technologies in spacecraft design including corrugated cylinder to improve the efficiency of the structure, nickel hydrogen batteries, high-efficiency GaAs solar cells, use of flaps with shaped memory alloys for solar pressure compensation, ASIC-based TTC and AOCS systems and

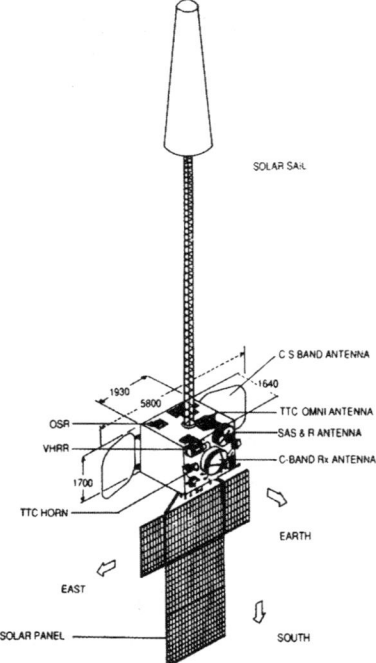

Fig. 10.8   The INSAT-2 satellite.

shaped beam and dual-gridded antennas to increase payload capacity. INSAT-2E also included a water vapor channel having 8-km resolution and a charge-coupled-device (CCD) camera operating in the visible, near infrared, and short-wave infrared at 1-km resolution for augmenting the meteorological capability of the satellite. INSAT-2E has been operating since its launch with eleven 36-MHz equivalent C-band transponder capacity, which has been leased to Intelsat, an international customer.

THIRD- AND FOURTH-GENERATION INSAT SATELLITES

Rapid expansion of VSAT services and growing demand from communication and broadcasting services necessitated the development of third-generation INSATs, which began with the launching of INSAT-3B, which carried 12 extended C-band and three Ku-band transponders in March 2000. Since then, INSAT-3A, 3C, and 3E, each weighing in excess of 2700 kg, have been launched to provide extensive communication capability in C-, extended C-, and Ku-bands and also to enhance the sensitivity of VHRR. As the communication payloads grew in capacity and capabilities, it was decided to separate communication and meteorological spacecrafts to avoid mutual electromagnetic interference. Accordingly, a dedicated meteorological satellite weighing 1 ton and carrying VHRR payload with water vapor channels and a data-collection transponder was launched from Satish Dhawan Space Centre, Sriharikota (SDSC) on 12 September 2002 by PSLV, which was very cost effective. Subsequent to the launch, the satellite was named Kalpana, in honor of the Indian-born American astronaut Kalpana Chawla.

The Indian Geostationary Launch Vehicle (GSLV) had its first experimental flight in 2001, and this was followed by the second and third launches to launch GSAT-2 and GSAT-3 (Edusat) satellites carrying C- and Ku-band transponders. The liftoff mass capability of GSLV was gradually increased from 1550 to 1950 kg during this period.

INSAT-4A, the first satellite of INSAT-4 series, is the heaviest and most powerful communication satellite launched so far by ISRO. The satellite weighing 3100 kg at liftoff was launched on 22 December 2005. It is designed to meet the direct-to-home (DTH) broadcast requirements of India. The satellite carries 12 Ku-band transponders, each with an EIRP of 52 dBW, and 12 C-band transponders, each with an EIRP of 39 dBW.

IMPACT OF INSAT

India, as of now, has one of the largest domestic communication satellite systems with 175 communication transponders on nine satellites—INSAT-2E, INSAT-3A, 3B, 3C, 3E, GSAT-2, Edusat, (GSAT-3), INSAT-4A, and Kalpana—providing a variety of communication and meteorological services to the country. Table 10.2 provides the details of all of the communication satellites launched by India, beginning with APPLE. With over 10,070 two-way speech circuits covering about 492 routes and linking 704 Earth stations of various sizes, the vast reach of the unique multipurpose INSAT satellites has been advantageously used for providing nationwide radio networking; administrative, business, and computer communication; VSAT networking; and emergency communication services. More than 54,000 VSAT terminals, including those installed by the National Informatics Centre (NICNET), are operating today to cater to the fast-growing requirements of a large number of both public and private closed user groups. Many important users like NSC, BSC, and bank ATMs use VSAT links as their communication backbone.

The INSAT system has become the primary backbone of round-the-clock, regular, half-hourly meteorological imaging and continuous weather forecasting. The primary data from VHRR received by the Meteorological Department at Delhi are processed along with the data from automated unattended data-collection platforms located in inaccessible areas to derive products like cloud motion vectors, sea surface temperature, precipitation estimation, etc. The processed data along with cloud pictures are transmitted

TABLE 10.2   COMMUNICATION SATELLITES OF INDIA

No.	Spacecraft	Launch date	Launched	Mass, kg	Position	Major payloads	Status
1.	APPLE	19 June 1981	Ariane-1	670	GSO 102°E	2 C-Band	Mission completed
2.	INSAT-1A[a]	10 April 1982	Delta	1150	GSO 74°E	12-C, 1-S, VHRR, IR 8 km, 2.5 km	Failed after 5 months
3.	INSAT-1B[a]	30 Aug 1983	Space shuttle	1190	GSO 74°E	12-C, 1-S, VHRR, IR 8 km, 2.5 km	Mission completed
4.	INSAT-1C[a]	21 July 1988	Ariane-4	1190	GSO 93.5°E	12-C, 1-S, VHRR, IR 8 km, 2.5 km	Failed after 2.5 years
5.	INSAT-1D[a]	12 June 1990	Delta	1292	GSO 83° E	12-C, 1-S, VHRR, IR 8 km, 2.5 km	Mission completed
6.	INSAT-2A	10 July 1992	Ariane-4	1905	GSO 74°E	12-C, 6-Ext.C, 2-S, VHRR, DRT, SASR	Mission completed
7.	INSAT-2B	23 July 1993	Ariane-4	1932	GSO 93.5°E	12-C, 6-Ext.C, 2-S, VHRR, DRT, SASR	Mission completed
8.	INSAT-2C	7 Dec 1995	Ariane-4	2020	GSO 93.5°E	12-C, 6-Ext.C, 1-S, MSS, 3-Ku	Mission completed
9.	INSAT-2D	4 June 1997	Ariane-4	2070	GSO 74°E	12-C, 6-Ext.C, 1-S, MSS, 3-Ku	Failed after 4 months
10.	INSAT-2E	3 April 1999	Ariane-4	2550	GSO 83° E	12-C, 5-Ext.C, VHRR, CCD	In operation
11.	INSAT-3B	22 Mar 2000	Ariane-4	2070	GSO 83° E	12-Ext.C, 3-Ku, MSS	In operation
12.	GSAT-1	18 April 2001	GSLV-D1	1540	Inclined orbit	2-C, 2-S, 1-C steerable	Mission completed
13.	INSAT-3C	24 Jan 2002	Ariane-4	2650	GSO 74°E	24 -C, 6-Ext.C, 2-BSS, MSS	In operation
14.	KALPANA-1	12 Sept 2002	PSLV-C4	1060	GSO	VHRR, DRT	In operation
15.	INSAT-3A	4 April 2003	Ariane-5	2950	GSO	12-C, 6-Ku, VHRR, CCD, DRT, SASR	In operation
16.	GSAT-2	28 Sept 2003	GSLV-D2	1540	GSO	4-C, 2-Ku, MSS	In operation
17.	INSAT-3E	28 Sept 2003	Ariane-5	2775	GSO	24 -C, 12-Ext.C	In operation
18.	EDUSAT	20 Sept 2004	GSLV-F1	1950	GSO	6-Ku, 6-Ext.C	In operation
19.	INSAT-4A	22 Dec 2005	Ariane-5	3086	GSO	12-Ku, 12-C	In operation

[a] Satellites procured from outside India, in this case from Ford Aerospace Communication Corporation (FACC), U.S.

for further dissemination to about 40 secondary data utilization centers (SDUC) located in various parts of the country. During cyclones and other emergency situations, INSAT VHRR is capable of imaging every five minutes in the sector scan mode. Beginning with the INSAT-2 series, a search and rescue payload was also added, which complements the efforts of COSPAS-SARSAT system and provides real-time detection of 406-MHz distress alerts within the Indian Ocean Region. At present, INSAT-3A located at 93.5° E is providing service with search and rescue payload onboard. The payload picks up and relays alert signals originating from distress beacons of maritime, aviation, and land users. Since its establishment in 1991, the satellite's Aided Search and Rescue Program has helped in saving more than 1500 lives.

The most dramatic impact of INSAT has been the rapid expansion of TV dissemination in India through the installation of about 1400 transmitters, providing access to 90% of India's population to national as well as regional services through 60 channels of telecasting. Use of transportable Earth stations and satellite news gathering vehicles now allows extensive real-time coverage of important events anywhere in the country. Six developmental communication channels are being operated to feed over 2000 distant education/training classrooms spread across India. Recognizing the importance of the interactive communication system, a number of experiments were conducted for imparting developmental education to target audiences of different types both in the rural and urban areas. Encouraged by the success of these, a few large-scale experiments have been undertaken, a typical example of which is the Jhabua Developmental Communication Programme in Madhya Pradesh that was implemented with more than 1500 receive terminals for imparting developmental education in the predominantly tribal areas.

TELE-EDUCATION

Space communication is now being very effectively used for distance-education applications in the country. Transmission of educational programs has been one of the high-priority areas for Doordarshan. In this direction an exclusive 24-hour Educational TV, Gyandarshan was implemented by Doordarshan in January 2000. Gyandarshan-III (Ekalavya) channel dedicated to technical education was started on 26 January 2003. Six digital channels are earmarked for specific aspects of educational/developmental programs in technical education, agriculture, vocational, training, secondary education, and distance education. Curriculum-based programs are produced with the active involvement of state educational administrators and teachers. Satellite-based enrichment programs for school children are produced by several State Institutes of Educational Technology (SIET). Education media research centers and audio-visual research centers at different places produce programs

for university students. These enrichment programs provide quality education within the reach of students in small towns and villages.

With the mushrooming growth of educational institutions in the country
and acute shortage of qualified teachers, the demand for the use of latest
technology to support the process of education has been steadily growing. To
support this challenge, ISRO launched "Edusat" (Fig. 10.9), a state-of-the-art
geostationary communication satellite having a coverage spread over the
entire country and specially designed for imparting wide-ranging educational
programs including higher education, vocational training, primary school
education, and teachers' training. Edusat has been specially configured with
multiple beams covering different regions of India, operating in Ku-band,
with an EIRP of about 55 dbW.

The expansion of the Edusat program to cover the entire country is planned
in a phased manner. The Edusat pilot project, conducted before the launch of
Edusat using INSAT-3B, covered university education in three states, namely,
Karnataka, Maharastra, and Madhya Pradesh. In the current semi-operational
phase of the Edusat program, Karnataka Primary Education Project under
Sarva Shiksha Abhiyan covering 885 primary schools was made operational
using the southern regional beam of Edusat. Similar networks have been set
up in 850 schools in Hindi-speaking states of Madhya Pradesh, Bihar, Uttar
Pradesh, and Chattisgarh; in 885 schools of Gulbarga in Karnataka; and in

Payload	EIRP	G/T
5 Ku -Regional beam	53 dBW	7 dB/K
1 Ku - National beam	50 dBW	3 dB/K
6 Ext-C-National beam	37 dBW	-1 dB/K

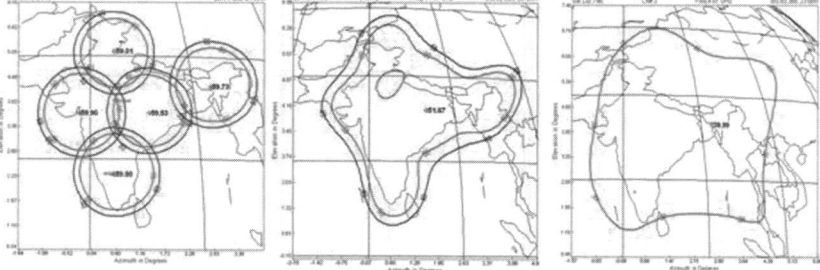

Fig. 10.9   The Edusat satellite. (See also color figure section at the back of the book.)

400 schools in Mallapuram, Kerala. At present, under the Edusat program, a total of 38 networks connecting 7600 classrooms, some with interactive terminals and others with receive-only terminals, are under implementation, 60% of which are already operational. The Edusat program expected to become fully operational by 2007–2008.

TELEMEDICINE NETWORKING

India has just about one qualified medical doctor for every 2500 people, compared to one for every 250 people in the United States, over 90% of whom practice in urban and semi-urban centers, leaving less than 10% of qualified doctors to take care of about 65% of country's population. In spite of the efforts of the government to establish public health centers in rural areas through about 25,000 primary health centers and about 150,000 subcenters, the medical assistance available in rural areas, particularly for treating serious medical complications, is sadly very meager. With the emergence of satellite communication technology as a very powerful tool, it is now possible to provide a very wide spectrum of medical and healthcare services, ranging from dissemination of basic healthcare information to assisting rural doctors in medical dispensation, through interaction with identified specialists in urban areas. Under disaster situations, telemedicine can become a lifesaving program for mitigating the sufferings of affected people and dealing with postdisaster rehabilitation effort.

The use of INSATs for providing specialized medical facilities to even the remote rural areas of the country, by connecting them with medical experts located at specialty hospitals in major cities, is fast emerging as an important application of the satellite communication system. Telemedicine consists of customized medical software integrated with computer hardware, along with medical diagnostic instruments connected to a commercial VSAT at each rural location, which, in turn, is linked to a superspecialty hospital through INSAT. The medical history of the patient including x-ray, ECG, and other diagnostic records are transmitted to the specialist doctors, who can diagnose and advise on the course of treatment through video conference with the paramedic at the patient's end. The specialist doctor can even guide the doctor at the patient's end during a surgery, thus enabling patients in distant and rural areas to have access to the best medical treatment.

ISRO's telemedicine network today covers 173 hospitals consisting of 139 remote-area hospitals in several states such as Karnataka, Kerala, Jammu and Kashmir, Lakshadweep, and North Eastern States networked with 34 superspeciality hospitals. More than 100,000 patients have already taken advantage of the teleconsultation and treatment since the establishment of the telemedicine network 18 months ago. With encouraging results from the experience of the present network, the remaining states in the country are now preparing to establish similar networks in their states. Karnataka State has already initiated the establishment of SatCom-based telemedicine facility in all of their district hospitals and a few trust hospitals by connecting them to speciality hospitals in Bangalore and Mysore.

VILLAGE RESOURCE CENTER

The INSAT system has demonstrated its capabilities over the years to provide services related not only to communication but also healthcare, education, weather information, disaster management, etc. To take the benefits of space technology to rural areas in a big way, ISRO has initiated a program for setting up INSAT-based village resource centers throughout the country, in partnership with the concerned state, central agencies, institutions, universities, research organizations, self-help groups, and nongovernmental organizations. The VRC program is meant for reaching the benefits of space technology directly to the communities at the grass roots level through a single window delivery of need-based services in the areas of education, health, nutrition, weather, environment, agriculture, market information, and other alternate livelihood opportunities to improve the overall quality of rural life. The VRCs are designed to handle both dynamic and generic information to empower rural communities through spatial information infrastructure, which will help in enhancing ecological and livelihood security of rural people and enable them to realize substantial value addition to their farm outputs. Space-enabled applications such as telemedicine, tele-education, and a natural resource database with resource management advisories will be made available in the VRCs.

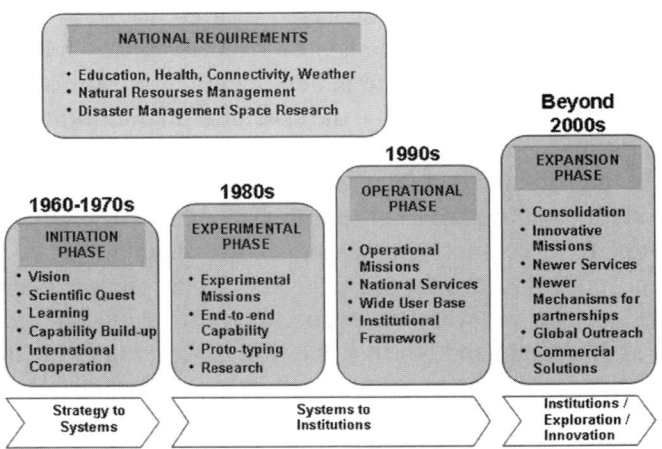

Fig. 10.10   The evolution of INSAT.

CONCLUSION

The INSAT system, which was dedicated to the nation by Indira Gandhi on 11 February 1984 has grown beyond all expectations, evolving through four generations of steady development and initiating a total communication revolution in this process (see Fig 10.10). With its unique, innovative, and very cost-effective design, a modular structure with a built-in capability to be able to provide a variety of communication, broadcasting, and meteorological services, and its emphasis on providing value-based services particularly to the remote rural-area population, INSAT has been able to make a very significant impact on the lives of Indian population. The socioeconomic impact of the INSAT system today scans across a wide variety of services related to communication, broadcasting, meteorological services, disaster warning, search and rescue operation, telemedicine, and developmental training for the benefit of the entire population of India, both urban and rural. Being the only geostationary meteorological satellite over the Indian ocean, INSAT has become an indispensable component of world weather watch. With over 175 transponders from nine geostationary spacecraft, INSAT is no doubt one of the largest domestic communication satellite systems in operation.

REFERENCES

[1] Sarabhai, V., et al., "A National Satellite for Television and Telecommunications," Proceedings of the National Conference on Electronics, 1979.

[2] Rao, U. R., Kasturirangan, K., Sridhara Murthy, K. R., and Pal, S., *Perspectives in Communication*, World Scientific, Singapore, 1987.

[3] Rao, U. R., Pant. N., Narayanan, K., Ramachandran, P., and Singh, J. P., *The Indian National Satellite System – INSAT: Space Communication and Broadcasting*, North-Holland, Amsterdam, 1987, p. 5.

[4] Rao, U. R., *Space Technology for Sustainable Development*, Tata McGraw-Hill, Delhi, India, 1996.

Chapter 11

THE DBS DIMENSION: HOW U.S. DBS SUCCESS WAS ACHIEVED

JIMMY SCHAEFFLER*

AND

LLOYD COVENS†

The story of the successful global evolution of the direct broadcast satellite (DBS) industry is in fact made up of many stories of daring and risk-taking pioneers and companies. In earlier chapters, we have looked at the evolution of direct satellite broadcasting in many parts of the world. Here, we will focus largely on the United States. We start by highlighting the fundamental tenets that made it all work.

TENETS OF DBS' SUCCESS

Nine specific tenants of success stand out in today's DBS world. These include 1) the ability, before launch and regularly thereafter, to raise substantial hundreds of millions, if not billions, of dollars to support the DBS venture; 2) a solid technological infrastructure, including first-rate conditional access and digital video compression systems; 3) a strong collection of sophisticated and savvy executive managers; 4) strong marketing and industry communications; 5) good partners on the hardware, software, retail, installation, and operational sides; 6) a population base whose demographics amply support the notion of a growing and thriving telecom venture; 7) attractive content, hardware, and pricing that resonates with those potential subscribers and viewers; 8) the support of the U.S. government, especially as it relates to early and fair access to content by a nascent video business; and 9) a surprising ability to prove naysayer after naysayer wrong as it relates to

*Chairman, The Carmel Group, Carmel-at-the-Sea, California.
†President, LTen, Com, Denver, Colorado.

the performance of the U.S. DBS industry. Yet, it remains one of the great ironies of the DBS industry in North America that when Stanley S. Hubbard used to travel the country trying to sell investors on his nascent DBS dream, called U.S. Satellite Broadcasting (USSB), some in the audience would deride his vision, saying that DBS really stood for "Don't Be Stupid." How little did they know.

EARLY HISTORY

Although the DBS industry is a relatively new industry, its true roots go back decades, to the time of World War II and German rockets that bombarded England. The mathematician, government auditor, and science fiction author, Arthur C. Clarke, first wrote about three fixed orbital satellites, equal distances apart above the Earth, more than 60 years ago, in his October 1945 article entitled "Extra-Terrestrial Relays." Clarke, who was then a Royal Air Force radar officer, U.S. Air Force captain, and prominent member of the British Interplanetary Society, summarized his theory in his now-famous "Letter to the Editor" in the British publication, *Wireless World*:

> An 'artificial satellite' at the correct distance from the earth would make one revolution every twenty four hours; i.e., it would remain stationary above the same spot and would be within optical range of nearly half the world's surface. Three repeater stations, 120 degrees apart in the correct orbit, could give television and microwave coverage to the entire planet.

Clarke's "geostationary theory" was an amazing conceptual development for its time because it set the foundation for something that mankind knew very little about, but which would revolutionize the concepts of telecommunications within Sir Arthur's lifetime. (As a tribute to his accomplishments, Clarke was knighted by Queen Elizabeth of England in 1998. Unmarried and childless, he lived and worked on a hill above Colombo, in the Indian Ocean island nation of Sri Lanka, until his death.)

Clarke later built his greatest fame as the author of the now-famous novel *2001: A Space Odyssey.* The novel's literary predecessor, a Clarke work entitled *The Sentinel*, was the original basis for collaboration on the famous movie, *2001: A Space Odyssey*, involving Clarke and the renowned movie director and producer, Stanley Kubrick, during the late 1960s. Sir Arthur's truest brilliance lies in the fact that he was, like many of the great inventors and conceptualists who preceded him, able to derive solutions to challenges involving satellites, with few modern-day technological devises or theorems to aid him.

Indeed, it was Arthur C. Clarke who famously observed that "... any sufficiently advanced technology is indistinguishable from magic." This often-quoted phrase, mixing rare common sense with even rarer brilliance and technological know-how, is the perfect description of many elements from today's DBS infrastructure.

EARLY U.S. DBS HISTORY

Following Sir Arthur's midcentury geostationary satellite proposal, the first great step toward the implementation of audiovisual satellites was the founding of the United States' National Aeronautics and Space Administration (NASA), in 1958. In 1962, two events of significance occurred. First, the U.S. Congress approved the U.S. Communications Satellite Act, which was the precursor to creation of the U.S. version of the international space consortium, called Communications Satellite Corporation (COMSAT). In the same year, the U.S. satellite, Telstar I, carried the first TV transmission by satellite, an eight-minute experimental transmission from France to the United States.

The most formidable early achievement in the field of communications satellites was the launch of the first Early Bird orbiter in 1965. Soon after, in 1972, the "Open Skies" policy was enacted, allowing much in the way of private company and private individual uses of direct-to-home (DTH) satellites. The resultant tinkering aided the satellite industry by providing much of the trial and error that was necessary to help get commercial audio-visual satellites launched and operational.

In the late 1970s, the Federal Communications Commission (FCC) began licensing private dishes, and in 1982 it authorized the first DBS licensees. In 1983, the first satellite dedicated solely to the distribution of cable TV was launched. This year was also an important because of the four additional DBS-only orbital slots that were authorized by the FCC, following allocation during international telecom meetings conducted and sponsored by the Geneva, Switzerland-based International Telecommunications Union (ITU). By doing this, the FCC was, in essence, conveying to the world that the first U.S. DBS satellite launch was expected to be a success, and it was further encouraging the new DBS industry to move forward. In effect, the DBS industry was finally allowed to flourish.

Since those early days of the U.S. DBS industry, DBS satellites have become more and more an important part of our society and everyday life. Also, in the early years, for many savvy investors the DBS industry became something of an investor's dream, which resulted from DBS's almost exponential growth rates during the late 1990s. More and more subscribers were clearly measured, in part, by way of increased shareholder values (and vice versa), which was made most clear in the rise of EchoStar's stock during the 1997–2000 time frame.

DBS DOZEN IN THE DBS DECADE

In mid-2004, at the end of 10 years of U.S. availability, the 20 millionth U.S. home installed its high-power 18-inch diameter dish antenna and began downloading hundreds of channels of digital sound and pictures. This chapter in many ways is a salute to so-called "DBS Dozen" executives and entrepreneurs who have helped make DBS in the U.S. today what it most clearly is, i.e., a strong and dynamic snapshot of a successful American business. Indeed, a look at these executives' successes (and failures) is also a good way to measure how U.S. DBS, as an industry, has itself succeeded. We now summarize the achievements of these pioneers in alphabetical order.

JOE CLAYTON

Joe Clayton was until recently chairman of the board for Sirius Satellite Radio. Before that, he was Sirius' chief executive officer, and before that he was the key executive at RCA, when it successfully bid for and became the dominant step-top box vendor for DirecTV (as that satellite operator first launched in the 1993–1994 time frame). Clayton's vision was in recognizing and implementing the importance of DBS as a communications product and service and in helping his parent company at the time, France-based Thomson, to make an extremely successful business venture from supplying critical hardware to the burgeoning DBS business in the United States.

CHARLES DOLAN

"Chuck" Dolan is the founder, patriarch, and chairman of the board of a Long Island, NY-based telecommunications company, Cablevision Systems. Dolan, along with key players from News Corporation, General Motors Hughes, and NBC, was one of the initial investors in and developers of a precompression era DBS venture called Sky Cable. In 1991, Sky Cable failed in its effort to deploy DBS to Americans, but it set the stage for what would come later because of the implementation of quality digital video compression, which allowed more content to be pushed to consumers using the same basic hardware. Sky Cable also whetted Dolan's appetite for another DBS venture (focused on the delivery of HDTV), which also faltered in 2005, called Voom. Nonetheless, Dolan's courageous visions of what could be, as well as his efforts to actually succeed in DBS, clearly qualify him among the original "DBS Dozen."

CHARLIE ERGEN

As noted in the section specifically discussing his company, EchoStar, Charlie Ergen is remarkable because he is the only one of the early and key U.S. DBS players that is still working the business and has consistently done so since the true implementation of the industry in the early 1990s. Ergen is presently the president, chairman. and CEO of EchoStar, which is the parent company of the DISH Network, as well as other units that implement technology and other DBS-related businesses. Ergen, together with early partners, Jim DeFranco and Ergen's wife, Cantey, founded EchoStar as a C-Band (big dish) business in the 1980s and has since grown EchoStar to the point where DISH Network is the third largest multichannel pay TV operator in the U.S. (after cable operator Comcast and DirecTV). EchoStar's current subscriber count exceeds 13 million, it offers hundreds of video and dozens of audio channels, it also owns Sling Media, and it has interests in many other technology- and telecommunications-based operations.

AL GORE, JR.

In early 1981, fourth-term Tennessee Congressman, Albert Gore, Jr., told a gathering of big-dish (i.e., C-Band) dealers in Nashville, Tennessee, "You are the only protection the American consumer can expect from the giant cable industry. We must prevail upon Congress to make sure your viability as an option is assured." Ten years later, overriding a veto by the first President George Bush in August 1991, Gore once again was at the right place and the right time, passing federal legislation that literally blew life into the DBS industry. That law he championed required all programmers who were owned by cable TV operators (i.e., notably Time Warner, TCI, Comcast, and Cablevision) to provide those same signals on the basis of ". . . equal and fair access" terms and conditions to other multichannel pay TV providers, such as the rival DBS operators. With that, a key DBS component was in place, that is, that of supportive legislation.

EDDY HARTENSTEIN

It is far from hyperbole to label Eddy Hartenstein "The Father of U.S. DBS," if for no other reason than that he was the first top-level executive to actually take the concept of DBS satellites relaying audio and video signals directly back to consumer dishes on the ground and make it happen in America. Together with his key technologist at the time, Bill Butterworth, Hartenstein took the nine key elements noted at the beginning of this chapter and actually formed them into the United States' first successful high-power DBS operator, named DirecTV. As chairman and CEO of DirecTV from the late 1980s to 2003, Hartenstein traveled the globe tirelessly on its behalf and ultimately became the chief voice for the nascent industry, which is a role he served until his retirement from DirecTV in 2005. Today, DirecTV serves approximately 16 million U.S. subscribers, and its core ownership is transitioning from that of Rupert Murdoch and his News Corporation to an original multichannel operator from a previous cable era, Liberty Media's John Malone.

BOB PHILLIPS

Bob Phillips is the current CEO for the National Rural Telecommunications Cooperative (NRTC), located in Herndon, Virginia, just outside of Washington, D.C. Phillips is instrumental in marshalling the substantial forces of the NRTC toward the investment of $125 million in the early development and later growth of DirecTV. The NRTC has, within the 2004–2006 time frame, divested itself of the core of its DirecTV-related assets. Although like just about all of the others on this list, his DBS involvement has waned considerably since its peak; Phillips and his NRTC remain deeply connected to satellite offerings in the form today of a major investment in the satellite broadband provider, WildBlue. WildBlue competes in the satellite delivery of broadband with DirecTV spin-off, Germantown, VA-headquartered HughesNet.

STANLEY S. HUBBARD

The son of a pioneer radio and over-the-air broadcaster, Stanley S. Hubbard was one of the original entrepreneurs to bid on a license for a Ku-Band small dish satellite system in the early 1980s. With license in hand, the senior Hubbard traveled the nation in search of money, ultimately raising approximately $100 million, much of which went to help DirecTV launch its first satellite on 17 December 1993. Hubbard was able to corral the nation's core premium channel assets, that is, HBO and Showtime, to exclusive DirecTV DBS agreements under the umbrella of his St. Paul, Minnesota-based company, USSB (see Fig. 11.1). USSB then became a key partner with DirecTV, both financially and in the operation of the DirecTV platform, which USSB later sold to DirecTV for over one billion dollars. Once operational, the senior Hubbard turned the day-to-day operational aspects over to his son, Stanley E. Hubbard, in the early 1990s.

STANLEY E. HUBBARD

Stanley E. Hubbard acted as operational head of USSB from the time of its launch as a DBS system, in mid-1994, and remained in that role until the sale of USSB to DirecTV in 1999. As such, he was, by many accounts, one of the most respected executives in the DBS business, in large measure because of his follow-through and respect for others. Moreover, together with his father and siblings, especially brother Robb, Stanley E. was able to step into his father's rather large shoes at a relatively young age and hold his own with some of the most sophisticated and shrewd managers in the DBS and related businesses. Stanley E. has since gone on to head another Hubbard Broadcasting startup, the movie-based informational and entertainment Reelz Channel.

Fig. 11.1   Stanley E. Hubbard, Robert W. Hubbard, and Stanley S. Hubbard with a USSB display.

JOHN MALONE

Working with Bob Magness, the CEO of cable giant TCI, its president John Malone cut his teeth early in the U.S. telecom business. TCI, at the time of its sale to AT&T in 2001, was the largest cable entity in the United States. From then until late 2006, Malone kept his eye on the multichannel business while amassing various lucrative content assets through his Liberty Media organization. In late 2006, Malone was able to work a tax-free strategy that involved a key trade of stock he held in Rupert Murdoch's News Corporation in exchange for a controlling stake in what once was Murdoch's "Crown Jewel," that is, DirecTV. What Malone does with the purchase of DirecTV is a question of great excitement for many U.S. DBS and broader telecom industry observers.

RUPERT MURDOCH

Australian-born Rupert Murdoch, like the Hubbards, also comes from a long line of media executives. His earliest ventures in the DBS field include a controlling stake in the U.K.-based BSkyB DBS service and the aforementioned SkyCable effort. Indeed, during the dozen or so years between SkyCable and the purchase of DirecTV from GM Hughes, in 2004 Murdoch tried unsuccessfully to acquire a large DBS U.S. presence no less than six separate times. His success culminated in the acquisition of the controlling ownership in DirecTV, which lasted approximately three years, at which point Murdoch felt compelled to improve his holdings in his parent company, News Corporation, by trading News Corporation shares owned by Liberty Media for Murdoch's shares of DirecTV.

TED TURNER

R.E. "Ted" Turner did much to develop the home-delivery side of satellite TV. He believed that home satellite could be a functioning complement to cable TV. Putting his money where his mouth was, Turner developed his own Cable News Network (CNN), delivering signals directly from satellites into his own and millions of other C-band satellite dishes, beginning in 1981. Many in the advertising business (who supported these fledgling networks) looked at the then-booming big dish business as a major source of new viewers, who were often well-heeled viewers. They installed dishes in the hinterlands, a place where an often expensive cable company was never likely to string a coax wire. In this manner, the "Man from Georgia" helped drive the ultimate success of the concept known as direct-to-home (DTH) satellite service.

CARL VOGEL

Carl Vogel is important to the history and development of DBS in the United States for three key reasons. Early on, as he moved from Jones Intercable to the role of Charlie Ergen's first president at EchoStar, Vogel's financial acumen was instrumental in helping the fledgling telecommunications company establish itself as a future major player. Vogel next became important to the history of U.S. DBS because of his successful executive leadership at several other cable- and DBS-related organizations in Canada and the United States. Until early 2008, Vogel set that track yet again as the president of EchoStar, in a career move that began for him in 2005. He retired from the presidency of EchoStar in February 2008.

FIVE KEY DBS COMPANIES

The following five companies ultimately launched their own DBS systems in the skies above the United States. Today only two of these, DirecTV and EchoStar, are still in operation.

DIRECTV

DirecTV is the perfect product of the nine tenets introduced in the opening paragraphs of this chapter. Although not all things were done perfectly from start to finish, enough were done well to sustain a business that will likely operate many scores of years, if not decades, from today.

At its outset during the late 1980s and early 1990s, DirecTV's former CEO and chairman, Eddy Hartenstein, along with Hughes' technical guru, Bill Butterworth, pushed the idea of a DirecTV system through the General Motors (GM) and GM Hughes (GMH) boards of directors. Hartenstein's and Butterworth's combined business and technical prowess, together with over one billion dollars in backing from the combination of GM, Hubbard Broadcasting's USSB, and the NRTC, were the glue that held the parts together for this new competitor to U.S. cable.

Also critical to the early growth of DirecTV was the deal it entered into during the 1992–1993 time frame to have the first million DirecTV set-top boxes (and satellite dishes and remote control units) supplied by France's Thomson Consumer Electronics. While Thomson was then doing business in the U.S. under the brand names General Electric (GE), Proscan and RCA, within today's U.S. DBS space, it is focused solely on the trade name RCA. The combination of solid content, operational, and hardware partnerships were instrumental in the longer-term success of the premier U.S. DBS operator.

Sony Electronics of America was the second major manufacturer to join DirecTV in the distribution of STBs. Sony became the second set-top box and related DBS hardware supplier in 1995, after Thomson had ended its exclusive sales term, having sold one million set-top boxes by mid-1995. Once the DirecTV/USSB/NRTC/Thomson combination launched its first satellite in December 1993, that event paved the way for subsequent launches of new DBS services by rivals PrimeStar (in March 1994), EchoStar (in March 1996), and AlphaStar (in mid-1996). PrimeStar and USSB were later purchased by DirecTV in 1999, and AlphaStar closed its doors in 1997.

A specific factor worth noting that was behind the success of DirecTV, as already alluded to, was the power of digital video compression to not only deliver huge volume of bits for services like HDTV, but also the ability of digital video compression to take the same transponder that would deliver one channel before and turn that into a transponder that would instead deliver six, eight, ten, or more channels in the same bandwidth. This was the scientific magic that made the DirecTV DBS business model—and later the followers—actually work.

ECHOSTAR

Turning to the other major remaining player in today's U.S. DBS market place, one cannot help but admire the success of another great risk taker, EchoStar's Charlie Ergen. Never one to shy away from doing battle with major corporations, the poker-loving native of Oak Ridge, Tennessee—known around his Denver, Colorado-based EchoStar Corporation as "Elvis"—was in the thick of the home satellite business in the late 1980s and early 1990s, as the money, technology, and a Congressional lift were all moving to buttress the promise of long-term DBS for many millions of viewers (which Ergen's earlier large dish, C-band business had tried, but failed, to deliver).

As chronicled by *Denver Post* editor, Stephen Keating, in his insider's cable/ satellite business publication, *Cutthroat: High Stakes and Killers Moves on the Electronic Frontier* (Johnson Publishing, Boulder, Colorado, 1999) during the years since EchoStar had opened its doors as a C-band satellite retailer, it had also become a major equipment distributor and designer. As such, it would match wits with a slew of larger entities. Among them were TCI and its chief John Malone; General Motors' subsidiary, Hughes Communications, Inc.; telephone giant MCI; and Rupert Murdoch's News Corporation. Also battling EchoStar would later be PrimeStar Partners (run by cable's biggest guns, i.e., Comcast, Time Warner, TCI-TSAT, Cox Communications, Cablevision, Newhouse, Continental/MediaOnce, and GE Americom); General Instrument Corporation; Hughes Network Systems; and RCA.

As an example of Ergen's business and gambling prowess, sometimes he would force the opponent to spend millions of dollars as a cost of entry into the satellite world, typically by bidding higher and higher against them and then dropping out of the auction at the last minute. At other times, he would outgun bigger companies with simple, but elegant, answers to what consumers wanted in their television options. These included options such as an early and continuous support for simple and reliable digital video recorders (DVRs). And in this latter example, the fact that he would later be challenged in courts by companies like TiVo, which had what appeared to be earlier DVR patent claims, was yet another business risk he appeared quite willing to take.

Interestingly enough, the first satellite in the EchoStar fleet was launched by a Chinese Long March rocket in December 1995. Yet the true risk taker in Ergen is perhaps best exemplified by what happened before and after that launch. The Long March satellites sent up right before and just after that EchoStar-1 satellite launch failed; indeed, one launch met a fiery demise, failing to reach orbit, and instead crashed back onto the ground, killing several Chinese peasants.

In 1995, Carl Vogel a time-tested, up-and-coming cable executive with the cable company, Jones Intercable, went across Interstate 25 in Denver to give Ergen a big assist as president and the one generally credited with doing the early major fundraising. "Without me, they (Wall Street) would have never touched him," Vogel once noted. Like he had with his lead technologists, Mike Kelly, Mike Dugan, and Mark Jackson, Ergen was able to draw strong efforts from his people.

For years Ergen lead the Washington, D.C. push to have legislative approval for "local-into-local" DBS carriage of broadcaster signals. In December 2000, President Clinton signed into law the enacting legislation that allowed DISH Network and DirecTV to begin selling the local channels of network and independent affiliates across the United States. At the end of 2006, hundreds of local channels were being refed from local markets and rebeamed into DBS homes at an average $5.99 per month/subscriber. Elsewhere, DISH Network has been particularly active in building interactive services into its consumer offerings. DISH Network also has instituted a big commitment toward ethnic/cultural and other specialized feeds to very small niche viewing groups.

Returning to the consolidation theme, at an early point, Ergen was, ironically, the lead distributor for the roll-out of DirecTV hardware. And twice he has dabbled with a deal to combine forces with Rupert Murdoch. Moreover, in the 2001–2002 time frame Ergen looked again at the books of rival DirecTV for an acquisition, while the government decided that a single, monster DBS company was not yet in the public interest.

PRIMESTAR

PrimeStar, somewhat ironically, was the first U.S. DBS system to actually launch, going operational in March 1994, three months before DirecTV. The irony lay in the fact that PrimeStar, from the beginning, was owned and operated by several major cable operators, whose core business, that is, cable TV, was the only real competitor to DBS when the latter service launched in mid-year 1994. TCI, Time Warner, Cox, and Comcast were all early owners of the PrimeStar medium-powered, medium-sized dish DBS service located in Bala Cynwyd, Pennsylvania. When PrimeStar's medium-power Ku-band owners tried to purchase assets that would have allowed the PrimeStar service to truly compete with the technologically superior high-power DBS providers, the U.S. government, in the form of the Department of Justice's Anti-trust Task Force, blocked the sale. As a result, the owners of PrimeStar were forced to sell its 2.2 million subscribers and other assets to rival DirecTV in 1999. This resulted in the termination of PrimeStar as a going U.S. DBS concern.

USSB

USSB was the brainchild of the broadcasting family headquartered in St. Paul, Minnesota—the Hubbards. Stanley S. Hubbard first applied for a high-powered Ku-band satellite broadcast license in the early 1980s, and then parlayed that into a partnership with Thomson Consumer Electronics, DirecTV, and the NRTC in the early 1990s. He also was able to obtain exclusive DirecTV platform agreements with premium channel service providers HBO and Showtime, which turned out to be a part of the DirecTV service that DirecTV could not live without. Indeed, in 1999, DirecTV purchased the premium services and hardware assets of USSB for well more than a billion dollars. During its years of operation, from 1994–1999, USSB was ably run by Stanely S.'s son, Stanley E. In the sale to DirecTV, the Hubbards retained full ownership of one satellite transponder, which later aided the launch of Hubbard Broadcasting's new movie entertainment and informational venture, the Reelz Channel.

ALPHASTAR

AlphaStar was a U.S. DBS service for the United States market developed by Canadian firm Tee-Comm Electronics. It launched in mid-1996, under CEO Murray Klippenstein and discontinued operations with 40,000 subscribers in September 1997, when the company entered bankruptcy proceedings. It was the first U.S. DBS service to use the internationally accepted DVB-S broadcasting standard. Because it was a medium-power instead of a high-power service, the weaker satellite signal meant larger dishes were required; AlphaStar employed a 39-inch satellite dish receiver. Tee-Comm, the parent company of AlphaStar, had originally cofounded the partnership that created Canada's Bell ExpressVu DBS service, where Tee-Comm served as a technology supplier, but later divested all interest in that venture to ExpressVu. Looking back, most would say the combination of insufficient investment, strong competition, and superior competitive technology were three of the key ingredients that spelled the demise of AlpaStar.

KEY U.S. DBS CHALLENGES

Nevertheless, despite the many successes, the U.S. DBS industry has had its challenges. In the early days, these primarily included struggles with technology, regulations, and various naysayers, including those from competitive industries, especially cable. As already noted, during the early 1990s, when DBS pioneers like Hubbard Broadcasting's Stanley S. Hubbard were trying to raise funding for their DBS businesses, prominent cable executives poked fun at the DBS industry, often joking that the acronym "DBS" actually stood for "Don't Be Stupid."

All along, technology challenges have been substantial. Particularly troublesome has been the ability of thieves, also known as pirates or hackers, to manipulate the technology inside the set-tops boxes in a way that allows them to obtain DBS signals either without paying or without paying their fair share. This promises to continue as a true dilemma to the future success of U.S. DBS. It has already cost the key operators billions in stolen revenues. Yet another challenge is being able to carry sufficient bits, or data points, through the channels that deliver signals to DBS consumers. Digital video compression has shown great promise to help alleviate this concern (for all multichannel operators, including DBS, cable, and telco). One more key technology dilemma involves the delay of a signal going up from the ground to a satellite and coming back down to the ground. This "lag time" of about a second reeks havoc with live two-way transmissions, such as those required of telephone calls and certain forms of two-way interactive TV.

Other more recent challenges have involved a vivid response to DBS successes during the past 10 years. In the United States and Canada, for example, many billions of dollars have been spent by cable and telco competitors to grow and enter the multichannel video marketplace. Indeed, the new wires into the homes of almost 120 million North American TV households (TVHH) is big and robust and capable of adequately delivering signals of video, broadband, and voice services, amounting to a so-called "triple-play" service (which the surviving North American DBS players DirecTV and EchoStar and Canada's Star Choice and ExpressVu cannot, by themselves, offer as competitive packages). Indeed, the telcos go one step further by offering the so-called "quadruple play," which adds cell phone service to the three services just mentioned.

Today, as the DBS industry in the United States approaches 30 million subscribers in the United States alone (and approximately 100 million globally), the industry and its patrons can say that these critics were sorely mistaken. In fact, despite some recent set-backs in the United States, DBS continues on a record pace to successfully gobble up wired subscribers and compete with cable like no competitor has ever done before. Indeed, today, even for the naysayers, DBS is clearly an investment and entertainment choice of the future.

GLOBAL DBS TRENDS

One of the clear trends worldwide is consolidation. For the DBS industry in Canada, for example, the narrowing has gone from five in the early to mid-1990s to just two main service operators today. Similarly, in the United States, the original licensees numbered 31, yet today there are no more than a handful of total licensees remaining, and only two main ones are operational. A third one, Dominion (operator of the "Sky Angel" Christian network), has determined it will switch from the delivery of its signals via satellite to an Internet-delivered service sometime in 2008.

To carry the consolidation theme one step further, the two true remaining U.S. DBS players, EchoStar and DirecTV, have already, in 2001–2002, pushed the consolidation envelope, this time in the form of a $26 billion failed attempt to merge their two companies. Looking forward, both DirecTV and EchoStar are expected by many, in the years ahead, to attempt to merge again, once the telephone company (telco) video competitors are more established, and resultant antitrust and anticompetitive concerns are alleviated.

Moreover, when one looks across the global transom for additional examples in Japan, the United Kingdom, and South America, one constant theme stands out: in general, it is very difficult for any large political or geographical region to simultaneously support more than one remarkably cost-intensive DBS venture. In short, in country after country and region after region, consolidation becomes an almost inevitable road toward long-term success. In the end, there almost always are strong arguments backing the efficiencies and economies of scale that support a successful conglomeration.

The first instance of major two-or-three-players-into-one consolidation is seen in the United Kingdom, where BSkyB emerged the victor in a two-operator market notable for its substantial lack of a competitive cable or telco video services. BSkyB, as of late 2007, services an estimated 8.5 million subscribers.

In Japan, dating back to the late 1980s and lasting right up to the turn of the millennium, several DBS players vied for supremacy in the Land of the Rising Sun. Like the United Kingdom before it, however, one final system operator, SkyPerfecTV, emerged victorious. SkyPerfecTV today services roughly 4.2 million subscribers.

In greater South America, for more than a decade, one system owned by the global telecom operator, Rupert Murdoch, vied against a second competitor in the Latin American DBS realm, which was controlled by the U.S.-based DirecTV. Murdoch's system, Sky Latin America, gobbled up DirecTV Latin America in 2004, when Murdoch acquired a 34% controlling interest in the DirecTV parent company for $6.6 billion. The new Latin American satellite company that is a combination of Sky Latin America and DirecTV Latin America was transferred in mid-2008, along with the DirecTV parent company, to John Malone's Liberty organization. This follows Murdoch's 2006 sale of DirecTV in order to solidify his ownership the parent company, News Corporation.

CONCLUSIONS

Tracking the movement of these companies and momentum toward consolidation in each of these preceding examples are various combinations of the "success" dynamics noted in the first section of this chapter. In short, whether elsewhere in the world, or somewhere in North America, the answer to the question of "How DBS Success Was Achieved" can be reduced to the proper mixture of the following ingredients: finance, technology, population dynamics, executive talent, regulation, partnerships, marketing and communications, as well as consolidation.

Chapter 12

GLOBAL POSITIONING SYSTEM: ORIGINS, EARLY CONCEPTS, DEVELOPMENT, AND DESIGN SUCCESS

KEITH D. MCDONALD*

INTRODUCTION

This is an account of the origins and development of the U.S. Navstar Global Positioning System (GPS), an enterprise in which I was privileged to play a significant engineering and management role. It traces GPS development from the pioneer work and precursor system concepts of the 1960s through the years from 1968 to 1974 when the director of Defense Research and Engineering chartered the Navigation Satellite Executive Steering Group (NAVSEG) to develop GPS into a Department of Defense (DoD) Four Service Program. It describes how NAVSEG managed the complex GPS investigations, analyses, coordination, and system designs, which in early 1973 culminated in the successful completion of NAVSEG's Development Concept Paper (DCP) for GPS. The DCP described the system concept, design, benefits, and cost and was prepared as the necessary DoD program approval document. It included all of the essential characteristics for the GPS as well as other information that all four military services had agreed to.

The initial Air Force plan to obtain program approval for a navigation satellite system was not successful. The U.S. Air Force later used NAVSEG's DCP design characteristics in their submission to successfully gain approval for the GPS program from the Defense System Acquisition Review Council (DSARC) in December 1973. The Air Force subsequently was designated by the DSARC as executive agent for GPS and carried out the system's

Copyright © 2008 by Keith D. McDonald. Published by the American Institute of Aeronautics and Astronautics, Inc., with permission.
*Chairman and Technical Director, NavtechGPS, Springfield, Virginia; Former Scientific Director, U.S. Department of Defense Navigation Satellite Program; Executive Secretary, Four Service Navigation Satellite Executive Steering Group; Chairman, Navigation Satellite Management Office.

implementation and deployment stages. GPS achieved operational status in 1993.

Navstar was originally proposed as the name for the U.S. GPS by John Walsh, deputy assistant secretary of the Air Force. The TRW Corporation, which had used and possibly trademarked the term, did not object to DoD's use of it. Apparently an acronym for "navigation system (or satellites) using timing and ranging," Navstar now primarily identifies spacecraft in the GPS constellation. As for the name, the global positioning system, Thomas F. Rogers, a vice president of the Mitre Corporation and a previous deputy director of defense research and engineering, promoted this title. A leading advocate for navigation satellite technology, Rogers worried that an unsuitable name might damage a large development program. In his view, the system's name should not refer directly to the system's satellite character because at the time satellite systems were costly and largely untried. After considerable discussion, DoD and others accepted the global positioning system name. This name, by focusing on the system's function, proved an excellent choice that has held up well.

From its initiation in 1968, the development of the GPS Defense Department Program was a daunting challenge as a potentially large and costly program that the military departments' operational leadership either seriously questioned or opposed. Most defense planners at the time considered a space-based navigation and positioning system as technologically premature, economically unsound, and fulfilling no clear military need.

In the decade following the mid-1960s, many new civil and military technology investigations and concept studies, such as those leading to GPS, had difficulty in obtaining funding and receiving acceptance. At the time, some considered navigation satellite systems to be "solutions looking for a problem."

The technology base of the 1960s set the stage for the rapid growth and evolution of satellite system developments and applications, especially in the communication, surveillance, and navigation areas. I describe the salient events of the period as well as details of the principal satellite-based navigation system concepts that government, industry, and university researchers had under development at the time. This material provides an important perspective on many of the precursor investigations. These studies and experiments established the sound scientific and engineering foundation for GPS operation. Moreover, it provided both the analytical and practical design bases for the system elements required to deploy a successful satellite-based navigation system. The significance of these ground-breaking efforts has been generally much underestimated.

I present a description of the DoD actions taken to form a four-service organization, NAVSEG, which was charged with determining the practicality, viability, usefulness, initial design, schedule, and cost of a navigation satellite system for defense applications as well as addressing other related issues. It worked energetically for about four years to develop and coordinate its findings with the military departments. It was assisted by the Navigation Satellite Management Office, a group established by the NAVSEG Four Service Charter to provide technical expertise, carry out engineering investigations, prepare appropriate technical papers, assist in the coordination process, and administratively support the NAVSEG.

The system development activities of the NavSat Management Office were substantial, and after technology briefings, meetings, discussions, and a period of interservice coordination, they succeeded in fulfilling NAVSEG's mission by producing an initial design for the GPS in 1972. This design provided essentially all of the principal characteristics and capabilities of the later developmental and operational GPS ground- and space-based elements. With this system design, other data, and the support of the Office of the Director, Defense Research and Engineering, NAVSEG prepared a DCP in 1973 for the Defense System Acquisition Review Council (DSARC) to approve GPS as a major development acquisition.

Finally, I shall explain how, after the joint services NAVSEG completed its GPS DCP in late 1972, the Defense Systems Acquisition Review Council (DSARC) indefinitely delayed meeting to approve the GPS program. In this hiatus, an unusual chain of events allowed the U.S. Air Force Space and Missile Systems Organization (SAMSO) to bypass NAVSEG and appropriate the product of NAVSEG's concept development effort. SAMSO then used the NAVSEG DCP system architecture to obtain Defense Systems Acquisition Review Council approval for the Air Force's GPS Joint Program Office to contract and implement GPS.

This chapter offers an accurate and documented description of NAVSEG's four-service coordinated activities from 1969 to 1973. It was these activities that originated and successfully configured the GPS architecture and ultimately led to the program's acceptance. In the 1970–1973 period, NAVSEG created the basic GPS system configuration and prepared a detailed DCP on the system for DSARC approval. From the outset, operational GPS has in fact employed the same architectural elements and configuration that NAVSEG established in 1972 and early 1973. Only after more than 20 years, in the mid-1990s, were these NAVSEG design features reviewed for significant revision or modernization [1–7].

In the conclusion, I discuss the USAF Joint Program Office (JPO), now the USAF GPS Wing, and the final decision process that emerged for contracting and implementing the GPS Engineering Development Program. As someone who was closely involved with nearly all aspects of the early program, I try to provide a balanced and fair exposition of the entire development of GPS. I've based this account on a number of sources: my recollection and notes, the comments and records of others, and interviews with many of those involved in the GPS development program in those years.

This account will demonstrate the outstanding scientific and engineering capabilities that NAVSEG concentrated on the navigation satellite system development. It will also display the admirable cooperation among nearly all of the technical and management experts who supported the military departments' joint activities in the early years of GPS development. This sizable cooperative group brought to fruition a useful, needed, and state-of-the-art technology in the navigation and timing fields. The development of GPS has dramatically changed not only DoD operations but, in many ways, the world as well.

Perhaps the most gratifying result for those of us who worked in the early formulation and development of GPS is the pervasive acceptance and widespread applications of the system by an enormous number of users in an incredible variety of uses. Although position, velocity, and time have always been important in navigation, they have now become extraordinarily accurate, available, inexpensive, and essential to hundreds of millions of users. GPS is arguably one of the most capable as well as technically and economically useful developments in the history of navigation and technology.

FORMATION OF THE DOD JOINT NAVIGATION SATELLITE PROGRAM

We start with the formation of the DoD Joint NAVSEG. Following this, we present the key early technology efforts that laid the foundation for the final system. We then present the development and evolution of the system itself with some comments on its management structure.

In late 1968, largely as a result of the diverse but promising government, industry, and university studies, the military departments and the DoD's Director of Defense Research and Engineering (DDR&E) decided "in the interest of forming a reasonable basis for a decision on a Four-Service Navigation Satellite System" to establish a charter for NAVSEG with an associated Navigation Satellite Management Office. The charter for the NAVSEG and the NavSat Management Office was prepared at the direction of the DDR&E by the offices of the assistant secretaries for research and development of the military departments. It was dated 19 November 1968 and was signed by all members of the NAVSEG and coordinated with their military departments and their research and development organizations. The charter clearly assigns the responsibility to the NAVSEG in Article 1.1 "to provide the overall management of the NAVSAT Program" and in Article 1.2 "to establish program objectives and provide guidance for and coordination of all tasks related to NAVSAT research and development." A scientific director for the program was also selected to serve as the principal scientific and technical advisor and as executive secretary of the NAVSEG. The scientific director was also to serve as chairman of the management office. The main responsibility of the NAVSEG was to assist the military departments, the DDR&E, and other DoD planning activities in coordinating, selecting, and recommending an agreed-upon course of action.

Its charter required NAVSEG to address the following issues:

1) Is a space-based navigation system technically feasible, and if so, what are its characteristics and capabilities?

2) Can the system, compared to existing systems, provide substantial navigation performance enhancements to fulfill military missions and support other applications?

3) Can the system be implemented and phased into use in a timely manner?

4) Is the system economically viable, considering development, implementation, and operational costs?

5) Does a navigation satellite system provide such significant benefits to all military departments that the system can be developed and implemented as a joint service capability?

The NAVSEG was formed as a four-service activity whose members included Army, Navy, and Air Force civilian officials at the secretariat level; military officer representatives from the Army, Navy, Air Force, and Marine Corps; and a civilian scientific director. The NAVSEG Chairmanship rotated annually among the civilian officials representing the military departments. The NAVSEG was patterned after a similar group established in the mid-1960s, the joint Communications Satellite Executive Steering Group. This group's task was to evaluate the feasibility and viability of a system of military satellites for communication. This model appeared appropriate for the navigation satellite area. The make-up of the NAVSEG and many of its accomplishments are shown in Fig. 12.1.

The program also incorporated some inherent and perhaps inevitable interservice rivalries, primarily for funding and program leadership, which were largely unproductive and occasionally stymied coordination efforts. The establishment of a joint program was difficult, but over time a cooperative atmosphere emerged. The group's members accepted the need to work together to configure a system and program that would take advantage of the benefits of new technology.

The joint service NAVSEG substantially determined the success of the DoD Navigation Satellite System program. Many in military leadership roles, however, were unaware of or complacent about the potential for navigation satellite systems. In some cases, military planners proved to be wary, threatened, and distrustful of the activity. Space technology activities were recognized as costly and of uncertain success, with a well-earned reputation for large cost overruns. There were nevertheless a number of factors that

• Active program now for over 35 years

• **GPS Program Created by DoD NavSat Executive Steering Group (NAVSEG[1])**
Members (8): 3 Mil. Dep't Secretariat[2] level civilians, 4 Military officers, 1 Civilian Scientific Dir./Exec. Sec'y (started 1970) Also, an ODDR&E[3] observer.

• **NAVSEG activities and accomplishments during 1970–1972:**
- identified and analyzed GPS technical options and operational needs - coordinated activities with: Military Dept's, Joint Chiefs of Staff, industry… - developed viable GPS navigation satellite system concepts - accomplished initial GPS system design - established GPS perf. benefits, cost estimates (with Institute for Def.Analysis…)

• **NAVSEG accomplishments by 1973:**

- prepared a GPS DCP[4] for consideration by DSARC[5]. This provided a basis for the approval by DoD/DDR&E of the GPS and in funding, contract awards for:
    1. spacecraft, boosters, control system and user equipment fabrication,
    2. developmental system final design and implementation
    3. engineering tests of GPS feasibility, performance and user equipment

1. NAVSEG initiated by DoD/DDR&E Charter in November of 1968. Most active from 1970 through1973.
2. These were Deputy Assistant Secretary or Special Assistant levelDoD executives reporting to their
*   respective Assistant Secretaries for Research and Development.*
3. ODDR&E - Office of the Director, Defense Research and Engineering - who reported to SecDef.
4. Development Concept Paper   5. Defense System Acquisition Review Committee

Fig. 12.1   GPS program and development background.

encouraged the decision to move forward to investigate the feasibility of a Defense Navigation Satellite System.

In late 1969, I became the scientific director of the DoD Navigation Satellite Program. By NAVSEG Charter I also served as the executive secretary of the NAVSEG and chairman of the newly formed Navigation Satellite (NavSat) Management Office. The management office's staff consisted of a civilian chairman, seven military officers, and two civilian administrative personnel.

As chairman of the NavSat Management Office, I was responsible for a wide range of activities that included performing technical analyses, organizing meetings, providing presentations, recording minutes for NAVSEG meetings and related activities, and generally keeping the community—including the military departments, research and design laboratories and supporting contractors—informed of the group's status and progress.

I was also responsible for coordinating navigation satellite program activities with the Office of the Director of Defense Research and Engineering (ODDR&E), industry, the Federal Contract Research Centers concerned, the civil community (notably, NASA, and the U.S. Department of Transportation), and some parts of the academic community.

The NAVSEG held regular meetings for its members about every two to three weeks from late 1969 through mid-1973, when meetings became somewhat less frequent. I arranged for the group to receive frequent briefings of navigation satellite developments by investigators from government agencies, laboratories, academic institutions, and industry.

The NavSat Management Office in coordination with NAVSEG concurrently sponsored a series of meetings on the development and configuration of a Defense Navigation Satellite System that could satisfy the requirements placed on the system by the military departments and the Requirements Group of the Office of the Joint Chiefs of Staff. Meeting regularly for about three years, this body was known as the NavSat Management Office System Development Group and at times the System Configuration Group.

In 1970, as scientific director of the program and on behalf of the NAVSEG, I made an invited presentation to the President's Science Advisory Committee (PSAC), then chaired by Richard Garwin of IBM. This presentation described the technology and potential of the Defense Navigation Satellite System Program and the issues that were of concern at the time. The advisory committee received the briefing with great interest and favorably commented on the activities to date. The navigation satellite program received the PSAC's encouragement and approval.

EARLY TECHNOLOGY EFFORTS AND THE DEFENSE ENVIRONMENT

The U.S. commitment to the Vietnam War in the late 1960s presented special difficulties for the decision to create NAVSEG to develop GPS. The U.S. military departments first began to seriously consider the satellite-based navigation system concepts just as the United States was attempting to withdraw from the Vietnam war. After the North Vietnamese Tet offensive of January 1968, antiwar sentiment mounted in the United States and abroad. New efforts to draw down the American commitment, first by President Lyndon Johnson and then by his successor Richard Nixon, produced increasing pressure from the American public and Congress for substantial reductions in the defense budget. This brought cuts in both military personnel levels and defense technology program funding.

The late 1960s was thus an extraordinarily inauspicious time to initiate a costly and unproven high-technology program. Such military technology programs had poor track records: they frequently fell years behind schedule, incurred high cost overruns, and often resulted in failure. Most military leaders and planners found it questionable whether or not a projected untested navigation service based on a satellite system could improve the operational effectiveness of their forces, have any useful impact on an adversary, or provide a significant military advantage. In a time of declining resources, defense planners naturally tended to focus principally on weapons platforms that were well known, accepted, and proven such as conventional tanks, ships, and aircraft.

It had nonetheless been only a decade or so since the Soviet Union had won the "space race" in early October of 1957 [8] by placing the first space vehicle Sputnik I, shown in Fig. 12.2, into Earth orbit. Space launches by the

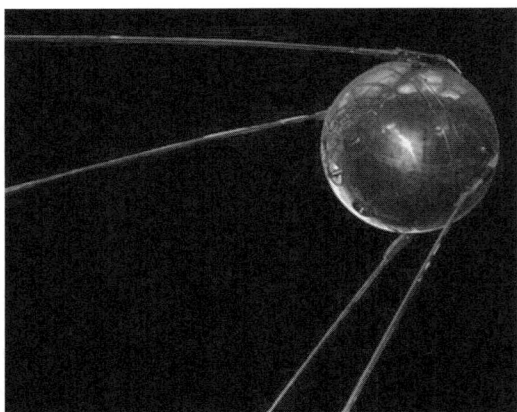

Fig. 12.2   The Russian satellite Sputnik I.

United States and other nations rapidly and aggressively extended activities in this area. The promise of satellite technology was widely recognized, and plans were in place for solidifying its future applications.

By 1967–1968, some members of the military departments and the Office of the Secretary of Defense (OSD) had become interested in the performance and potential of space-based navigation and positioning systems. This was the result of two factors. First, there was an increasing concern for the serious limitations and deficiencies of the U.S. military's navigation and position determination systems and the resulting high cost in men and material. Secondly, there was a growing awareness that recent developments in satellite technology offered the means that could provide the new and improved capabilities needed.

It was clear from the military experience in Southeast Asia that U.S. navigation and position determination capabilities were not adequate to perform many required missions successfully. Frequent large errors in aircraft position and velocity as well as errors in ground location affected troop operations and weapons delivery performance, especially in bad weather or in hostile conditions. Not only was more accurate targeting data needed, but the precise position, velocity, and acceleration components acting on the aircraft and the weapon being delivered were found critical to a successful mission.

When it was necessary to attack a "hard" target, such as a fortified steel and concrete bridge, aircraft navigation and target location errors frequently caused the ordnance to miss the more vulnerable parts of the objective. This required additional aircraft sorties and placed the aircraft and flight crew at much higher risk with attendant air crew casualties, loss of aircraft, and, at times, collateral damage. These serious concerns required attention and correction. Satellite technology appeared to offer solutions. The new capabilities and improved performance of navigation satellite systems appeared to provide the potential for substantially improving military operations and mitigating the serious and widespread deficiencies of the current systems.

Moreover, for many years serious military—mainly naval—communications problems had contributed to the loss or damage of U.S. naval vessels, imprisonment of military personnel, and related issues that affected the morale and defense standing of the United States. These incidents had resulted in serious concerns as well as diplomatic and military embarrassments for the United States.

As the promise of satellite technology was increasingly recognized, programs were in place or solidifying for its future applications. U.S. organizations that had begun to explore satellite applications included NASA, DoD, the intelligence community, many segments of industry, and a number of universities. Satellite technology applications that were considered included civil and military communications, Earth observation and weather prediction, geodesy and space sciences, air traffic management, and missile tracking and surveillance. A description of the principal initiatives in satellite navigation follows.

U.S. NAVY NAVIGATION SATELLITE SYSTEM: TRANSIT

The Navy Navigation Satellite System (NNSS) was a satellite-based navigation system employing user receiver measurements of the Doppler change or "shift" of the carrier signals transmitted from satellites at about 1100 km (600 n miles) above the Earth. The NNSS, or Transit system, provided passive users with data on the orbital elements (ephemeris parameters) of the spacecraft, which, with the Doppler measurements, made position determination feasible [9–11]. The original NNSS Transit Oscar spacecraft is shown in Fig. 12.3. The later Transit TRIAD spacecraft is shown in Fig. 12.4. Transit receivers were primarily used by submarine and other marine vessels for navigation or on stationary platforms for survey and geodesy. The system was developed under a Navy contract by the Applied Physics Laboratory (APL) of Johns Hopkins University. Transit became operational in the early 1960s and was phased out in favor of GPS at the end of 1996. The principal investigator for the program was Richard Kerschner of APL.

During 1960–1989, the developmental and empirical efforts in the NNSS program provided excellent information for subsequent satellite-based navigation investigations. This information included 1) data on the Earth's gravity field and the effects of high-altitude atmospheric drag; 2) empirical data on the space environment, especially the effects of the Van Allen radiation belt on solar panels; and 3) methods for the design and fabrication of reliable hardware for operation in space.

Fig. 12.3   The NNSS Transit satellite on the Oscar spacecraft. (See also color figure section at the back of the book.)

NASA STUDIES

From 1963 to 1969, NASA and its contractors carried out a number of studies at NASA Headquarters in Washington D.C. at the Goddard Space Flight Center and at contractor facilities. I managed and participated in several of these studies, which addressed the various methods by which space-based systems could be used to determine position, velocity, direction, trajectory, and other characteristics of a moving craft. Some of this work defined the basic measurement techniques that were feasible with satellite systems and estimated performance characteristics. This was useful and of some significance to the development of GPS.

Under a NASA contract, Communication and Systems, Inc., investigated ranging and range difference techniques, determined errors and performed error analyses, and evaluated system accuracy and related performance factors. As principal investigator for the program, I focused the effort on the fundamental measurement parameters and their performance in satellite-based navigation systems. NASA also sponsored a number of early satellite-based ranging experiments, propagation measurements, and related work

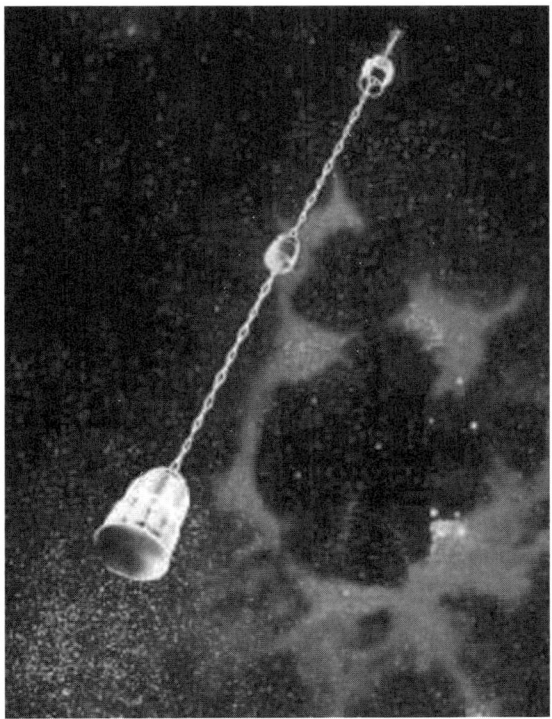

Fig. 12.4 The Transit TRIAD satellite configuration in orbit (courtesy of Johns Hopkins University Applied Physics Laboratory). (See also color figure section at the back of the book.)

carried out by Roy Anderson and investigators at the General Electric Company in Schenectady, New York (1958–1965).

DEPARTMENT OF THE ARMY ELECTRONICS COMMAND: TacNavSat SYSTEM

From 1966 to 1968, the Department of the Army's Electronics Command (ECOM) at Ft. Monmouth, New Jersey, sponsored a contract with Communication and Systems, Inc. (CSI) of Falls Church, Virginia, to investigate techniques, preferably using existing communication satellites and other means, to provide navigation information to military aircraft and other military uses. I was the program manager for this study, which developed and analyzed a number of techniques for providing space-based navigation capabilities to aircraft and other military uses. The final recommended system configuration [12] employed digital PRN-coded signals for ranging, atomic standard clocks for precise timing, and the transmission of a data message that included ephemeris, clock and region-based calibration, or differential, corrections. (PRN-coded is the abbreviation for pseudo random noise coded digital modulation.) As project manager, I found its characteristics to be ideal for use in a satellite ranging system because of its autocorrelation properties that provided excellent accuracy of range measurement, its low power requirements, and its fundamental excellent characteristics against most jamming signals. It is the basic signal modulation now used in GPS, the Russian GLONASS, and the planned European Galileo system.

This early work employed the principal system elements and technology upon which GPS was later configured; however, this study had some significant constraints relating to the available spacecraft. The study nevertheless comprehensively evaluated the most capable techniques for providing position, velocity, and time using satellite-based technology. This program's analysis and recommendation for the signal structure, power requirements, and the use of digital signal correlation receiver technology in 1967 and 1968 forecast very closely the types of digital signals and power levels recommended several years later in the Air Force 621B studies (1970–1972) and about six years later (1973–1975) in the final design for GPS.

DEPARTMENT OF THE NAVY'S NAVAL RESEARCH LABORATORY: TIMATION SYSTEM

From about 1965 to 1974, the Department of the Navy at the Naval Research Laboratory (NRL) investigated a navigation satellite system concept called "timation." This program proposed a medium-altitude constellation of satellites, an analog signaling technique known as side-tone ranging, precise quartz clocks, and later atomic standard clocks onboard the spacecraft. Although the program was primarily a study program, two spacecraft were built and launched in the program's latter stages. The first, Timation I, was launched with stable crystal clocks, and the second, Timation IIA (Navigation Technology Satellite-1), was launched in 1974 carrying two cesium atomic clocks. The Timation side-tone ranging is a modulation technique that uses a number of modulating frequencies on a selected carrier signal. The modulating "tones" are selected so as to provide accurate phase measurement at the highest frequency tone with the other tones providing for the resolution of the phase wavelength (cycle) ambiguity. This technique was used in the Goddard range and range rate (RARR) satellite tracking system with considerable success. This NRL investigation, which established system performance guidelines and highlighted the capabilities and applications of highly stable, space-borne clocks, played an important role in the development of GPS. Figure 12.5 shows an NRL Timation satellite with its gravity gradient stabilization "boom" pointing toward Earth.

Fig. 12.5   The NRL Timation satellite (courtesy of U.S. Navy Naval Research Laboratory). (See also color figure section at the back of the book.)

DEPARTMENT OF THE AIR FORCE'S SPACE AND MISSILE SYSTEMS ORGANIZATION AND THE AEROSPACE CORPORATION PROGRAM 621B

From 1967 to 1973, the Department of the Air Force at the Space and Missile Systems Organization (SAMSO) in El Segundo, California, supported by the Aerospace Corporation, investigated the use of satellites for navigation. This investigation focused on the use of Earth synchronous satellite constellations consisting of a combination of inclined, eccentric, and geostationary spacecraft using PRN-coded ranging signals. This work provided important contributions to the GPS program, especially in the signal structure area.

OTHER INITIATIVES.   There were also several other initiatives in the late 1960s that were of interest but ultimately proved to be of limited application to GPS development. These included the following:

1) From 1967 to 1969 the Department of Transportation Air Traffic Control Advisory Committee (ATCAC) met in Washington, D.C., to investigate ground- and space-based techniques for navigation and air traffic control. I participated in this work for about a year with the MITRE Corporation in

McLean, Virginia. The effort involved many investigators and consultants from government, industry, and universities.

2) From 1966 to 1968 the U.S. Air Force used Thompson Ramo Wooldridge Corporation (TRW) of Los Angeles to investigate the use of satellites for navigation.

3) During the mid- to late-1960s, university laboratories (including the University of Pennsylvania, the Massachusetts Institute of Technology, the University of California at Los Angeles, the Ohio State University) and others investigated the use of satellites for position and velocity determination.

4) From 1968 to 1970 the U.S. Air Force contracted with the Hughes Aircraft Company to investigate techniques for using satellites to obtain position information.

5) From 1966 to 1968 the U.S. Air Force had TRW of Los Angeles investigate the use of satellites for navigation.

PRECURSOR SYSTEM CHARACTERISTICS AND CONTRIBUTIONS TO GPS

There has been a good deal of confusion over the years about how the final concept and preliminary design for GPS were developed. The NAVSEG configured the GPS in essentially all of its important technical aspects by using the capabilities of the NavSat Management Office combined with the assistance of DoD laboratories. As chairman of the management office, as well as the GPS Program's scientific director, I directed the development group's activities in the synthesis of a satellite-based system that would fulfill the military departments' requirements and mission needs. This produced frequent and reasonably successful consensus of the NavSat Management Office sponsored development group participants who represented the military departments.

The navigation satellite system that was configured by the NAVSEG and its management office was not a modified version of either the Navy Timation system, the Air Force 621B, or the Army TacNavSat. The resulting NAVSEG initial GPS configuration was a new system that used the best available features, some from precursor systems and some originated by NAVSEG members or their organizations. The system configuration was determined, designed, and established by the NAVSEG, the organization specifically reponsible for and chartered by the 1968 four services charter to accomplish this work.

As chairman of the NAVSEG Management Office, I organized regular system development meetings that involved science and engineering experts from the military departments and their laboratories, industry contractors, university specialists, Federal Contract Research Center representatives, and consultants. From 1970 through early 1973, this work actively addressed and resolved most of the critical technical issues in the development of a navigation satellite system for defense applications. These critical issues included the frequency and signal selection; position, velocity, and time (pvt) measurement techniques; error correction methods; orbit selection; constellation arrangement; ground control system operations; and other factors. This continuing series of system development meetings, combined with prior investigations, formed the important technical basis for most aspects of the initial system design for GPS.

Of the many activities over the years that related to the development of space-based navigation systems, the five principal investigations briefly summarized in the previous section emerged as those most directly relevant to the development of a GPS design. The first order of business for the system development meetings was for the NAVSEG members and the system development team to become familiar with the technology and techniques used in those precursor program activities that could influence the design of GPS.

The NAVSEG provided a competent group and support staff for evaluating the technology and system concepts available and also worked to establish a system that could meet the specified defense requirements. By early 1973 the group had completed the GPS DCP containing a description of the recommended system configuration and the NAVSEG requested a meeting with the DSARC.

The preliminary system configuration for GPS was established and received general agreement by the NavSat System Development Team and the NAVSEG by mid-1972. It was carried forward in the DCP. I shall now describe in some detail how the NAVSEG evaluated and incorporated some elements of the five principal precursor programs that are shown in Fig. 12.6.

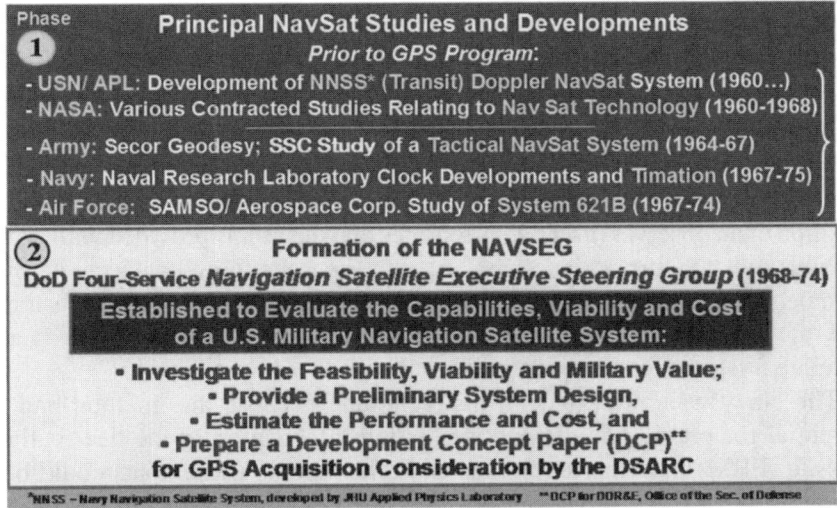

Fig. 12.6   The phases of the early navigation satellite developments leading to GPS.

NAVY NAVIGATION SATELLITE SYSTEM (NNSS OR TRANSIT)

DOPPLER MEASUREMENTS OF THE RUSSIAN SPUTNIK. In October 1957, two U.S. scientists, George Weiffenbach and William Guier, with the Johns Hopkins University Applied Physics Laboratory (APL), made observations and measurements of the Doppler shifts of the transmissions from Sputnik I. They also developed a technique for determining the satellite's orbital parameters from these measurements [11]. These investigators used the physical principle that changes in the relative motion between a spacecraft carrying a stable oscillator (or "clock") and the observer had the effect of shifting the signal frequency (or causing a frequency change) at the observer. This frequency shift, or Doppler shift, of the transmitted signal frequency from a spacecraft is especially useful if it can be taken over a substantial part of the transit of the spacecraft's observable path above the Earth.

Measurement of the Doppler frequency component provides an excellent record of the spacecraft-user relative motion caused by the motion of the spacecraft in its orbit. The relative motion in this case refers to the relative motion between the system user and the various spacecraft involved in the determination of position, velocity, and time. This could be as many as 12 or so spacecraft for a 24-spacecraft constellation. Each user-spacecraft combination will have its own relative motion with its own Doppler component. Although there is a large separation distance between the user and the spacecraft, most of the relative motion variations of concern are caused by the user. The spacecraft orbital motion is well known, has a predictable geometry, and can be readily accounted for. If the position (and motion) of the observing antenna involved in the measurements is known, it was shown that the orbital path and the orbital elements of the spacecraft could be analytically determined.

U.S. NAVY INTEREST AND APPLIED PHYSICS LABORATORY EFFORTS. The converse of the Sputnik orbit determination process was of interest to the U.S. Navy, especially for their submarine fleet. The Doppler principle can be configured with cooperative spacecraft moving in known orbits that could be broadcast to users. The Doppler measurements provide the information needed to determine the position of an observer (user). This requires that the user is provided with orbital path data, that is, the orbital elements of the spacecraft, and an accurate measurement record of the Doppler shifts of the signals during the satellite's transit over the observer. This was the basic premise upon which the Navy Navigation Satellite System (Transit) program was based.

The Navy was interested in this technique because this method had the potential for providing accurate navigation information to the fleet ballistic missile (FBM) nuclear submarines and other users. These boats could be at sea for several months without position updates, which could result in their navigation errors accumulating to unacceptable levels.

To investigate this technology, the Navy arranged for APL to analyze the technical approach and implement a developmental system. The program required the placement into orbit of a number of experimental NNSS (Transit) spacecraft and the transmission of their projected orbital elements frequently to the users. Two versions of the APL Transit spacecraft are shown; Fig. 12.3 shows the NNSS Transit Oscar Satellite and Fig. 12.4 the Transit TRIAD Satellite.

The orbital elements were determined by a ground-based tracking network and up-loaded periodically (every few hours) to the spacecraft. The location of a user craft, typically a ship, could be determined by the difference and the rates of change between the expected carrier signal frequency and the observed frequency that had been influenced by the satellite motion. This provided Doppler measurements for the craft that, with the Transit ephemeris data, could be integrated to obtain a craft position solution. The Doppler measurements as well as the received orbital ephemeris data for the spacecraft required excellent precision for accurate position determinations.

The system was configured, built, and demonstrated rapidly, in about two years from contract approval until the first spacecraft was in orbit [13]. The developmental Transit spacecraft successfully demonstrated the concept. The NNSS became operational in 1964 and provided excellent marine- and land-based services for over 32 years. The system was phased out in favor of GPS in 1996 [14].

Other significant contributions of the NNSS program to the GPS program are as follows: 1) accurate determination of the Earth's force field and the orbital parameters influencing satellite motion, especially for low-altitude satellites, including the effects of the Earth rotation resonance terms; 2) experimental determination of the Van Allen radiation belts' effects on solar panels and the design and fabrication of new solar-cell devices to mitigate this effect; 3) development of a data message containing the principal force field components and a simplified method for representing gravitational effects in a combined manner; and 4) development of spacecraft reliability and long life-time fabrication methods. These pioneering activities contributed to excellent reliability for the Transit (and subsequent) electronic and electromechanical subsystems and substantially improved spacecraft lifetimes in orbit.

NASA STUDIES; NAVIGATION SATELLITE SYSTEM ALTERNATIVES

System design considerations and capabilities for over 20 system concepts for satellite-based navigation systems were configured and analyzed. This provided a reasonable comparison of the performance capabilities of the principal system alternatives. A few of these had been investigated to some extent system by others. These concepts and the analysis of alternatives provided an excellent foundation of satellite-based navigation system con-figurations and performance information. A number of studies analyzed capabilities, limitations, and tradeoffs for these systems. I was the project manager and principal investigator (PI) for several of these investigations. The results of much of this work will now be summarized.

MEASUREMENT TECHNIQUES AND SYSTEM CONCEPTS INVESTIGATED. The techniques and concepts investigated are as follows: 1) single and multiple satellite Doppler systems, 2) combined multiple satellite Doppler and ranging systems, 3) multiple satellite range difference (hyperboloidal) systems, 4) multiple satellite ranging systems (multilateration), 5) multiple satellite range and range rate systems (multilateration with Doppler measurements), 6) two satellite range difference hyperboloidal systems with user baro altimetry, 7) single satellite crossed baseline interferometric (angle) measurement systems combined with user barometric altimetry or ranging, 8) combinations of the preceding, 9) combinations of some of the preceding with barometric or radar altimetry, and 10) other techniques including the use of space-borne directional (and monopulse) antenna systems combined with ranging capabilities.

Conceptual implementation of these techniques included both active and passive user systems. In the active case, the user was required to transmit a signal to a spacecraft to obtain a measurement. This was typically mechanized by measuring the response time of a transponded reply triggered by the spacecraft's receipt of the user transmission. The passive application was more desirable because the system capacity became unlimited, and further, the receivers would not be saturated by user transmissions. The cost and complexity of passive user equipment appeared more reasonable. Additionally, passive user operation was strongly preferred in military applications.

PRACTICAL CONSIDERATIONS RELATED TO THE IMPLEMENTATION OF NAVIGATION SATELLITE SYSTEM CONCEPTS. These considerations included the following: 1) optimizing the number of satellites in the satellite constellation consistent with cost and providing adequate signal capabilities in the desired area of coverage; 2) spacecraft constellation altitude selection for Earth visibility as well as spacecraft and launch-vehicle cost considerations and spacecraft vulnerability; 3) spacecraft orbital configuration, or constellation arrangement, for acceptable geometric performance with users; 4) optimization of the performance capabilities with other constraints, such as lifetime and replenishment; and 5) minimizing system cost consistent with performance, reliability, time to deploy, and other factors.

SATELLITE RANGING AND ACTIVE POSITIONING STUDIES AND EXPERIMENTS. This early experimental work on 1) satellite ranging techniques, errors, accuracy and propagation effects and 2) navigation satellite positioning techniques was performed by investigators at the General Electric Company (GE) in Schenectady, New York, from 1959 to 1973. This work was done under contract to the Air Force Cambridge Research Laboratory (AFCRL) and to NASA.

Measurements were made of the Doppler signals from Sputnik by a group headed by Roy Anderson of the GE General Engineering Laboratory in Schenectady, New York. These measurements, similar to those observed by the investigators at APL, were then used to establish the orbital parameters and other characteristics for the Sputnik spacecraft. The success of this work resulted in the group tracking each new satellite placed into orbit for 48 hours after launch, as requested by AFCRL.

A program initially funded by GE measured the phase and frequency instability of satellite transmissions through the ionosphere [15]. A knowledge of the ionosphere and its propagation group (time) delay characteristics was important to establish the accuracy of range and range difference measurements from spacecraft (S/C).

The group at the laboratory also investigated the capabilities and accuracy of range and range difference measurements from spacecraft to the Earth. To accomplish this, a tone-code ranging signal was configured that provided a range accuracy of about 0.1 microsecond (about 100 feet or 30 meters). The group developed a satellite system concept to determine position by the use of either active transmissions or passive signal receptions.

In this navigation satellite system concept, a ground station transmits to one satellite a short interrogation signal, modulated by an audio frequency tone and the digital address of an individual user craft, each of which has a unique address. (This description is based on *General Electric Schenectady Space Programs 1957–1983*, GE Co., Schenectady, NY, p. 20. This is an abridged version of a historical record prepared for the History Center of the Institute of Electrical and Electronics Engineers, Brunswick, New Jersey, and the Hall of Electrical History, Schenectady Museum, Schenectady, New York.) The satellite repeats the transmission, and it is received by all of the user craft and the ground station(s). The one craft that is uniquely addressed automatically retransmits the signal to two or more satellites, including the one that sent the interrogation signal.

Two or more satellites repeat the signal, and the ground station receives the response from each satellite. The ground station measures the time from its interrogation to the first return from the interrogating satellite and then the time of the craft's response that is sent back through all satellites. The time measurements are converted to range measurements, yielding the range from each satellite to the craft. Because the ground station knows the location of each satellite, it can calculate the location of the craft.

The concept was tested using NASA spacecraft transponders to obtain ranging accuracy data and, later, position determinations. This work was done using NASA advanced technology satellite ATS-1, ATS-3, and ATS-6 spacecraft and involved corrections for the instrumentation errors associated with the receiving equipment. The experiments demonstrated position accuracies of 0.25 n miles using VHF band frequencies with their large ionospheric uncertainties and about 0.1 n mile using the L-band transmissions of ATS-6 [16].

This work was important because it provided important data and empirical information on the ranging capabilities of narrowband signals operating in a realistic space-to-Earth environment. The research focus on the delay characteristics of the ionosphere as a significant error contributor was important. The early concept work on a navigation satellite system was also useful and instructive even though it was not a passive system and incorporated an analog ranging technique.

In a February 1964 report for NASA, the GE team proposed a navigation satellite system consisting of 24 satellites in four orbit planes inclined 51 degrees at an altitude of 5000 n miles. The number of satellites needed for four-S/C-in-view coverage of the Earth was known to be approximately this number, and the other parameters selected were also reasonable system design values.

The contributions that were of importance to the GPS program were 1) the experimental evaluation of the ranging accuracy from a narrowband signal propagating through a realistic satellite-to-Earth environment and 2) the early development of navigation satellite and surveillance satellite system concepts with examples of implementation techniques and performance capabilities.

U.S. ARMY ELECTRONICS COMMAND DEVELOPMENT AND DESIGN OF A TACTICAL NAVIGATION SATELLITE SYSTEM

This study was contracted to Communications and Systems, Inc. by the U.S. Army Electronics Command (ECOM) at Fort Monmouth, New Jersey in 1966. An investigation of alternatives as well as a specific navigation satellite system design was requested in late-1965 by ECOM. The design desired was for a system to be used primarily with military aircraft but also by others in a large tactical area. The system design had a number of constraints, including 1) the user navigation capabilities were to be passive, real time, and accurate; 2) the system was to use, to the extent possible, existing spacecraft; and 3) performance was to be of higher quality than available from alternative navigation capabilities.

The investigation was done during the period 1965 to 1968, the period of the U.S. Army's massive Vietnam war buildup, which peaked just before the North Vietnamese Tet offensive. The Army was well aware of the navigation shortcomings in their tactical (and other) operations. Some on the Army Research and Development engineering staff who had been following the navigation satellite system studies contracted by NASA and others were convinced that satellite technology offered a possible near-term solution. They hoped that existing civil and military spacecraft already in orbit (deployed primarily for communications) could provide adequate platforms for establishing a useful, albeit rudimentary, navigation satellite system for regional use.

At the time, ECOM staff recognized that to use existing satellites with their frequency and power limitations to achieve a substantially improved navigation capability was optimistic and probably not feasible. The staff nevertheless believed that it was an appropriate area for investigation to determine the characteristics for near-term feasible navigation satellite system techniques. In late 1965, the Army at Fort Monmouth awarded the contract for this investigation to Communication and Systems, Inc. (CSI) of Falls Church, Virginia, a subsidiary of the International Telephone and Telegraph Company (ITT). I was the designated program manager and the principal investigator for the effort.

ORBITAL EPHEMERIS AND LAUNCH-VEHICLE CONCERNS. In the design of the Army tactical navigation satellite system, it became clear that accurate orbital elements, or ephemeris parameters, for the position with time of the utilized spacecraft (S/C) in orbit would be required operationally for any system candidate selected. These data would normally be determined by the ground segment and transmitted by uplink sites to the spacecraft that would store the data in memory. The spacecraft would then either simply relay the data to the users or store the data onboard and periodically transmit, or broadcast, the data for the users.

The ephemeris data would provide the user receivers with the accurate position of each spacecraft as a function of time. The various spacecraft form the reference "locations" or positions (the S/C transmit antenna phase centers) from which ranges, range differences, or angles are measured. Therefore, it was essential to establish and communicate accurate data on the spacecraft positions to the users. As discussed later, if accurate data on the spacecraft positions are not known, then relative measurement, or differential, techniques can be used based upon measurements made in a local region of interest. This early application of differential techniques to mitigate the effects of spacecraft position and atmospheric errors was necessary and important. The spacecraft error was typically very large (>100 m) and the ionspheric errors were significant.

SPACECRAFT AND CONSTELLATION ALTITUDE SELECTION. In order to address future NacSat systems determination and prediction for extended periods (days, weeks, or longer), it appeared necessary to employ higher-altitude orbits, for example, near circular orbits at altitudes of about 10,000 to 30,000 km above the Earth's mean sea level. Spacecraft in these higher orbits had larger regions of coverage indicating the need for a smaller number of spacecraft to cover a given region. Higher orbits also have reduced levels of atmospheric drag, substantially lower gravitational effects and much smaller variations in these and other effects. The high-altitude orbits are therefore much more stable and predictable than low Earth orbits, for example, orbit altitudes of 200 to 5000 km. This indicated that medium- to high-altitude orbits (10,000–30,000 km altitudes) were of principal interest. At it turned out, the final orbit altitude selected for GPS was 20,186 km, well within the recommended range. This medium altitude orbit was a compromise, to provide excellent global coverage and to provide a desirable, repeating ground track arrangement for the initial GPS test program at the White Sands Missile Test Range. Later, some minor disadvantages of the half siderial day (11 hours 58 minutes) repeating ground trace orbits were found.

The geostationary orbit was not appropriate for a number of reasons. First, the spacecraft coverage, reaching to only 70–75 degrees of latitude, did not cover the polar regions and therefore did not meet the requirement for global coverage. Second, geostationary spacecraft would all be in the equatorial plane, an undesirable coplanar geometric arrangement that would degrade accuracy in the equatorial region. Some variations were attempted, including cases where same spacecraft were geosynchronous but with high inclinations and eccentricity, such as the Air Force Aerospace 621B's constellations, especially those configured later in the program. Other concerns, however, eliminated these satellite arrangements from consideration.

Establishing an "optimum" altitude and satellite constellation configuration for a navigation satellite system was difficult. In addition to the preceding considerations, the increased range to the Earth was initially of concern. However, the increased space loss is normally compensated for by higher-gain spacecraft antennas. Higher-altitude orbits, however, required larger, higher-energy, and more capable launch vehicles. The higher-altitude requirements resulted in larger vehicles and substantially greater launch costs. Figure 12.7 illustrates the medium-altitude baseline constellation with six orbit planes that was selected for the operational GPS configuration.

COVERAGE CONSIDERATIONS. The DoD frequently needed critical navigation system coverage over selected global geographical areas, but with little predictability or warning as to the location of many of these areas. Because a given region might be of interest for navigation system coverage for only a limited time, that is, a few months to a few years, the use of communications, weather, or surveillance spacecraft as platforms for navigation capabilities was not appropriate. Using the navigation function for only a fraction of a spacecraft's lifetime appeared difficult to justify. If a small set of spacecraft were dedicated to navigation and could be moved in longitude when and where they were needed, then the use of a combination of geostationary and other spacecraft might be appropriate for some applications. Readiness and time delay considerations, however, were a difficult concern to overcome except by the use of a dedicated constellation.

Other system compromises, especially in the cost arena, were required. In many areas, however, such as satellite fabrication and test as well as

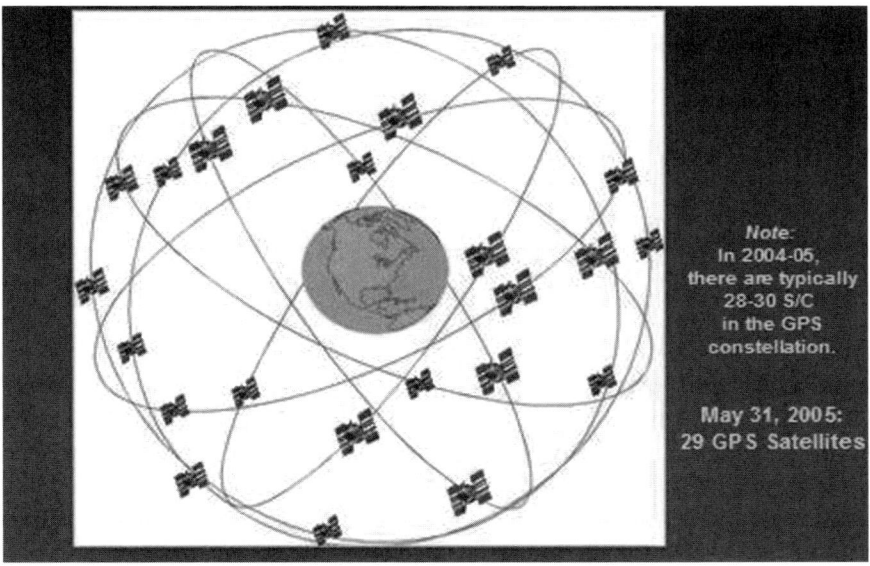

Fig. 12.7   The Navstar GPS operational constellation (24-satellite baseline configuration). (See also color figure section at the back of the book.)

launch-vehicle production and integration, costs were poorly known in the early 1970s. Only coarse estimates, which did not provide the basis for a meaningful optimization process, were typically available. The principal activity was therefore directed to investigate the alternatives available and to select the candidates that were found under analysis to provide the most acceptable system performance in light of the measurement parameters and navigation solution techniques employed.

The navigation system candidates developed and configured for the U.S. Army tactical navigation satellite system used the NASA investigations and other investigations that were available. The digital (PRN) signal structure that we configured and analyzed for the Army "TacNavSat" investigation is a precursor to the signal structure employed later in the Air Force 621B system. The Air Force-Aerospace Corporation 621B Request for Proposals listed the U.S. Army ECOM Tactical Navigation Satellite Study of late 1967 by K.D. McDonald et al. of CSI as a reference. It is quite possible that both investigations arrived at the same conclusions independently even though the work on the ECOM study predates the Aerospace Corporation results. The Aerospace Corporation had capable professionals familiar with digital spread spectrum techniques on staff, including Phil Diamond, James Woodford, Walt Melton, and on contract, James Spilker, Charles Cahn, and others.

TacNavSat Survey of Available Spacecraft in Orbit.   A review of the existing spacecraft in orbit found that they were not appropriate for the configuration of an accurate active or passive navigation satellite system. The spacecraft were in nearly all cases communication satellites operating in geostationary orbits with little available power for the transmission of navigation signals. As mentioned, analysis showed that the geometric performance of two or more geostationary ranging spacecraft degraded significantly near the equatorial region, a region of substantial interest.

The frequencies of operation for the available spacecraft were at C-band (4–6 GHz) for civil spacecraft, and, in the case of military spacecraft, at X-band (8–12 GHz). The power densities at or near the Earth's surface available from the available spacecraft were found to be inconsistent with a practical navigation satellite system for mobile users. This assumed that mobile users would have low-gain antennas.

We also found from NASA orbital determination and prediction experts that spacecraft location information, or orbital ephemeris data, was not typically known to better than a few hundred meters at that time (~1966). Considering these factors, we recommended and the Army ECOM management agreed to focus the investigation on the characteristics of the best-candidate navigation satellite system elements for possible future implementation.

Analyses were then accomplished for a considerable number of candidate NavSat systems, including many of the candidates mentioned earlier. We determined estimates for the accuracy of the candidate measurement parameters and selected placement positions for the candidate spacecraft in orbit. I completed, with my project group, covariance analyses of the system candidate errors including the spacecraft geometric aspects (GDOP, PDOP, HDOP, and VDOP) for a variety of systems. [The geometric quality of the position and velocity determinations available from a satellite constellation is related to their dilution-of-precision (DOP) characteristics. GDOP is geometric dilution of precision; PDOP is position DOP; HDOP is horizontal DOP, and VDOP is vertical DOP.] We then prepared plots of the capabilities for the principal candidates and evaluated these against the latitude and longitude regions of interest to the users.

SIGNAL STRUCTURE CANDIDATES EVALUATION.   During the ECOM study, we also analyzed the signal structure candidates, including side-tone ranging, swept FM modulation, pulse and pulse position modulation, and PRN spread spectrum techniques. We believed that the digital techniques had significant advantages over the analog techniques. Many of the new developments and advances in technology were directed to digital systems. We selected a particular set of direct sequence PRN spread spectrum signals for the Army TacNavSat system and configured a correlation receiver capable of operating with the signals to determine position, velocity, and time. (Many of the details of this correlation receiver and the PRN-coded spread spectrum signal structure used in the ECOM-CSI study are given in the third quarterly report for this effort [12].)

In summary, the following were the principal technical design results and anticipated operational capabilities, based upon the analytical studies:

1) **Range and range rate measurements** between the satellites and the users were found to provide the best accuracy in position and velocity. An extensive analysis of the capabilities of various measurement parameters was performed in the context of the satellite geometry and the basic measurement accuracies that could be expected from various measurement techniques [17].

2) **Atomic time standards were selected** to establish accurate time atomic scale clocks. They were specified for use in the ground control stations and recommended for use onboard the navigation spacecraft when feasible. Atomic time standard specialists at Ft. Monmouth indicated that these stable devices should be available for use in space in the near term.

3) **Spread spectrum signals using code division multiple access (CDMA) techniques** were established for the system. These PRN-coded signals were found to provide excellent performance with the modest power available onboard the spacecraft in this time period. The user could obtain an accurate ranging signal using receiver signal correlation technology.

4) **Use of L-band frequencies (1–2GHz)** to obtain adequate antenna effective area for low-gain hemispherical coverage mobile (aircraft) antennas was recommended. C-band with higher power levels was also considered but discarded in favor of L-band. If military X-band signals were to be used, a substantial increase in spacecraft power appeared necessary and was impractical at the time. L-band operation provided a reasonable compromise between the receiving antenna effective area and the effects of ionospheric group delay.

5) **Use of reference stations near the tactical area** compared the system position data measured by a receiver with known (or surveyed) positions to "calibrate" or provide relative difference (or "differential") corrections to the users. These corrections would eliminate nearly all of the bias errors in the user position caused by the large uncertainties in the spacecraft orbital positions as well as corrections for the troposphere and the ionosphere. This was necessary because the ground-based determinations of spacecraft positions were not sufficiently accurate at the time and inconsistent with the accuracy required by the user. Discussions were held with NASA experts in this area, including Joseph Siry of NASA Goddard Space Flight Center. Siry indicated that their best determinations at the time (1966) of spacecraft position error were about 200–250 meters. Further, no measurements were normally performed on the available communication satellites of interest. A technique was therefore needed in the TacNavSat program to correct for the spacecraft position errors. A local region "calibration," or differential correction, technique was configured for the TacNavSat system to provide this needed user corrections. Differential corrections came to be used extensively in navigation satellite technology about 10 years later. The ground station measurements had

significant ephemeris errors that, with other errors, required periodic "calibration" (or differential correction) by a deployed set of reference stations. A data link was configured to provide these corrections to users in near real time, preferably at a low data rate through the satellites. The reference station correction technique was a precursor (by about 10 years) to the use in GPS of similar differential corrections, now widespread, albeit for other purpose.

6) **Use of two frequencies for ionospheric correction** was based upon the well-known dispersive characteristics of the ionosphere. Although recognized as important for a global system, it was not employed in the Army tactical NavSat system design because the ionospheric error components were included in the reference station differential correction data for the relatively limited geographic region considered.

7) **The system used a digitally modulated spread spectrum signal** formed from a family of direct sequence codes using carrier modulation by BPSK (binary phase shift keying). These signals were demodulated in a receiver by correlation processing using internally generated replica codes (of the transmitted codes) and would be searched over time for code correlation with the demodulated signals received from the satellites. Searching over frequency for the received carrier signal was also necessary to compensate for the Doppler component related to the relative motion effects between spacecraft and user. This process provided information on the Doppler shifts of the received signals that the user receiver used to determine three-dimensional velocity.

8) **Data message information to the spacecraft** was provided from the ground control stations for satellite retransmission to the users. These messages included ephemeris data and clock correction data for the spacecraft; corrections based upon "actual minus observed" or similar measurements of position differential corrections; time corrections on a similar basis; and other data associated with the spacecraft and the ground control segment.

9) **The system would provide spacecraft and users with accurate time determinations (to 20–50 nanoseconds or better).** This would have desirable aspects in time division or code division techniques, such as communications, that might rely on the spacecraft transmissions.

10) **An estimated three-dimensional accuracy of 10–20 meters (or better) in the horizontal and a velocity accuracy of 0.3 feet per second (10 centimeters/second)** appeared feasible with a tactical navigation satellite system with the characteristics established. A principal concern was the availability of (or cost of launching) spacecraft that were inclined to the equatorial plane so as to provide users with good geometry for the range and range rate determinations of position and velocity.

PRESENTATIONS AND REVIEWS. I briefed the Deputy Assistant Secretary of the Army, Howard P. Gates and staff in 1968, on the TacNavSat Study results. Gates, a very capable and experienced engineer and technology manager, took great interest in the study and supported its further discussion and dissemination to others in the DoD. Gates later chaired (1968–1969) the NAVSEG.

In mid-1967, Phillip Diamond and Peter Soule, representing the Aerospace Corporation, the principal "in-house" supporting contractor to SAMSO, visited me and requested a briefing on the U.S. Army tactical navigation satellite program that I was directing. The Aerospace representatives discussed the engineering aspects of the effort in detail with me and took notes on the system design and performance. Diamond and Soule were apparently preparing data relating to a request for proposal (RFP) for a navigation satellite study program, designated Program 621B, that the Air Force was undertaking at SAMSO with Aerospace Corporation support.

When the RFP for the initial System 621B program study was published, it included the Final Report for the Army Navigation Satellite System investigation as a source document.

U.S. NAVY'S NAVAL RESEARCH LABORATORY INVESTIGATION OF TIMATION

The U.S. Navy had provided the national time standard for many years. This responsibility dated back many decades to the activities of the Naval Observatory (NObs) in Washington, D.C. Measurements of astronomical time (UT2) involving the rotational motion of the Earth had been accurately recorded against the celestial background by NObs for over 100 years. Although the technology for time standard measurements has changed dramatically over the years (e.g., transitioning to atomic time standards), NObs with the support of the Naval Research Laboratory (NRL) continues to provide the precise time standard for the United States and maintains and upgrades its expertise.

An NRL group, headed at the time by Roger Easton, had been investigating techniques for using satellites to transfer precise time over long distances. This involved developing accurate and stable clocks, or signal generators, that would perform well in a space environment and that were capable of operating reliably in space for long periods. When the Navy research and development community initiated their investigations of navigation satellite technology, the NRL precise timing group under Roger Easton became the focal point for the Navy investigation. The accurate measurement of time intervals (representing propagation distances) and the capability for maintaining stable, accurate time over long periods in orbiting spacecraft were key system requirements. The NRL navigation satellite system became known as the Timation system because of its origins in the precise and stable timing area.

The early prototype space-borne clocks were placed in thermally controlled multiple ovens, each inside another to provide excellent thermal stability for the innermost crystal oscillator, or "clock." This resulted in good quality frequency stability for the clocks. Figure 12.5 shows the Timation spacecraft, and Fig. 12.8 shows the Timation IIA/NDS spacecraft under test. Figure 12.9

Fig. 12.8   The NRL Timation IIA satellite under test.

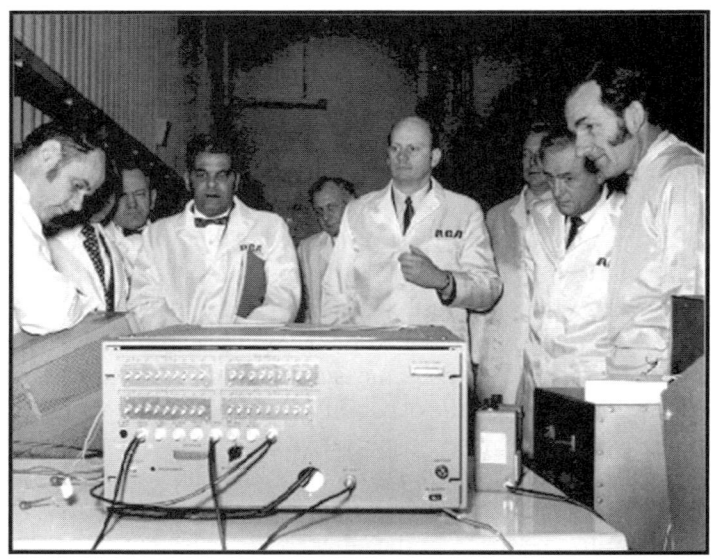

**Fig. 12.9   A Timation tour group, including the author (hand visible), at RCA Astro
Electronics (circa 1970).**

TABLE 12.1 CHARACTERISTICS OF THE NRL TIMATION NAVIGATION SATELLITE SYSTEM CONSTELLATION

Characteristic	Value
Coverage	Global
Number of spacecraft N	27
Orbit altitude a	~7500 n.mi., or 14,000 km (above the geoid)
Orbital eccentricity e	0 (circular)
Orbit planes	3
Spacing of right ascensions	120 deg (symmetrical)
Spacecraft per plane	9
S/C phasing between planes	13.33 and 26.67 deg
Inclination i of planes	~60 deg
Signal modulation at S/C	Multiple side tone
Modulation type	Analog
User measurement	Multiple side-tone phase measurement
Carrier frequencies	L1: 1575.42 MHz
	L2: 1227.6 MHz
Range measurement precision	~2–5 m (1σ)
Horizontal position accuracy (2 frequency)	~10 m (1σ)
Vertical position accuracy (2 frequency)	~15 m (1σ)
Principal error contributors	Ionospheric group delay, spacecraft ephemeredes, and troposphere delay and uncertainties

shows a Timation tour group at the RCA Astro-Electronics Spacecraft Production Facility that fabricated the spacecraft.

By about early 1971, the NRL Timation navigation satellite system was configured to have the principal characteristics shown in Table 12.1 [18].

The NRL investigations into the Timation navigation satellite system made the following salient contributions to the initial development of the global positioning system: 1) they demonstrated the viability and capabilities of space-borne stable clocks for use in navigation satellite systems, and 2) they configured medium-altitude satellite constellations for worldwide navigation coverage using spacecraft in circular, moderate inclination orbits. Some contributions were found advantageous and recommended for inclusion in the global positioning system. Others were found inappropriate and were not employed.

The Air Force Space and Missile Systems Organization (SAMSO, now the Air Force Space Command) in El Segundo, California, funded the Aerospace Corporation to study a navigation satellite system program designated 621B. Although the Air Force had also contracted with the Hughes Corporation and others for substantial efforts in this area, the principal continuing work in the development of the 621B concept was accomplished by the Aerospace Corporation engineering staff and their consultants.

By late 1969 and into early 1970, when the DoD Navigation Satellite Executive Steering Group received briefings on the U.S. Air Force System 621B system concept, the principal characteristics were established as shown in Table 12.2. Later characteristics are also given in the table.

The investigations into the Air Force 621B navigation satellite system made the following contributions to the development of the global positioning system: 1) they configured and recommended the use of digital PRN-coded signal modulation for the measurements determining user position and velocity; 2) they configured synchronous altitude satellite constellations for worldwide navigation coverage using inclined and circular orbits and eccentric and equatorial constellations (this was interesting but found to be deficient in some respects); and 3) they analyzed the viability and capabilities of space-borne atomic standard clocks for use in navigation satellite systems. Aerospace worked with NRL in this area.

As in the case of the Navy Timation investigations, these contributions and further work were later briefed, reviewed, and considered in the context of a viable defense navigation satellite system (DNSS). Some efforts were found advantageous and recommended for inclusion in the global positioning system; others were not.

INITIAL NAVIGATION SATELLITE SYSTEM EVALUATION, CHARACTERISTICS SELECTION, AND DEVELOPMENT

As detailed in the second section, the founding of the NAVSEG and NavSat Management Office were established in part to investigate the feasibility and military usefulness of a navigation satellite system. The principal secretariat members of the initial NAVSEG and their terms as NAVSEG chairman were as follows: Howard Gates, deputy assistant secretary of the Army for Development, who served during 1968–1969 as the first NAVSEG chairman; Harry Sonnemann, special assistant to the assistant secretary of the Navy for Research and Development (Robert Frosch), who served during 1969–1970 and 1972–1973 as chairman; Michael Yaromovych, deputy assistant secretary of the Air Force for Research and Development, who served during 1970–1971 as chairman; John Walsh, deputy assistant secretary of the Air Force, who

TABLE 12.2   PRINCIPAL AIR FORCE/AEROSPACE 621B SYSTEM CONCEPT
CHARACTERISTICS

Characteristic	Value
Coverage	Global to about 70° latitude (North and South)
No. of spacecraft N (2 cases)	15 (3 rotating X constellations) or 16 (4 rotating Y constellations
Orbit altitude	Spacecraft in geosynchronous elliptical orbits combined with geostationary orbits
Orbital eccentricity e	0.22 for elliptical orbits and 0 (equatorial) for geostationary orbits
No. orbit planes	X configuration 4 (3 elliptical and 1 equatorial) or Y configuration (4 elliptical and 1 equatorial)
Spacecraft per plane	X configuration 4 in elliptical planes and 3 in geo, or Y configuration 3 in elliptical planes and 4 in geo
Spacing of planes	X configuration 120 deg (symmetrical) or Y configuration 90 deg (symmetrical)
Inclination i of inclined orbits	30 deg for initial constellations
	At later time: 60 deg for "egg beater" constellation (to obtain global coverage)
Signal modulation	Pseudorandom noise coding (PRN)
Modulation type	Digital (BPSK)
Carrier frequencies	L1: 1575.42 MHz
	L2: 1227.6 MHz
Range measurement precision	~2.0–5 m (1σ)
Horizontal position accuracy (2 frequency)	~10 m (1σ)
Vertical position accuracy (2 frequency)	~15 m (1σ)
Principal error contributors	Spacecraft ephemerides, troposphere delay, and ionospheric group delay and uncertainties

served the latter part of the 1970–1971 term; and Frank Ross, of the Office of the Secretary of the Air Force, served as chairman during 1973–1974. Active deliberations extended just over four years, from late 1969 to early 1974. I served as the NAVSEG's civilian executive secretary as well as the chairman of the NavSat Management Office and Scientific Director of the DoD Navigation Satellite Program. [I was brought in from industry (The MITRE Corporation) and reported to the chairman of the NAVSEG. My civil service position (Public Law 313/GS-16) was supported under agreement with the Navy (OP-76).] A representative from the Office of the Director of Defense Research and Engineering (ODDR&E) was also directly involved in the

NAVSEG. George Salton, a civilian staff engineering and management representative, assisted substantially in the preparation of the DCP.

After the initial briefings to the NAVSEG, the management office organized a group of concerned and capable engineers and scientists to both structure and evaluate acceptable candidate navigation satellite system configurations and evaluate their capabilities. This evolved into a nearly three year series of DNSS development meetings. The Army, Navy, and Air Force NavSat system developments and preferred system configurations were briefed with issues addressed by the NAVSEG and system configuration group members.

The options and associated analyses were presented, discussed, and debated vigorously. The basic technical differences and originator biases relating to the system candidates became very apparent. The strong points and weak areas of the candidate systems resulted in substantial revisions and improvements of the proposed system configurations over an extended period.

In my role as scientific director of the DoD Navigation Satellite Program and as chairman of the NavSat Management Office, I chaired the extended series of system configuration meetings that developed evolving navigation satellite system candidates to satisfy the DoD's rigorous demands. The NAVSEG and others regularly considered the candidate configurations and status of the systems. This process was established to lead to an agreed-upon system configuration, which would be analyzed for technical feasibility, cost, performance, and defense suitability. The NAVSEG would then prepare a DCP to assist DoD management in the acquisition decision process.

Initially, the military departments established the required performance characteristics for the defense navigation satellite system. Later, at a higher level, a requirements group (J6) within the Office of the Joint Chiefs of Staff (JCS) consolidated and defined these performance characteristics for all normal military operations. We will not address these detailed requirements because to a large extent they evolved simultaneously with the early activities of the management office system development meetings. We shall cover these requirements as they relate to capabilities of the system candidates.

At the time, a DCP was the conventional way to structure a new defense system that was expected to exceed in cost about $100M. The DCP was to be reviewed by the secretariat-level officials who made or strongly influenced the decision process. The DCP was to be concise, comprehensive, and limited to a total of 20 pages describing the system characteristics, capabilities, performance, and cost. It was the principal source of program information placed before the Defense Systems Acquisition Review Council (DSARC), which would meet to consider the program as a major defense system acquisition. When it met, DSARC could decide to approve, reject, or defer judgment on

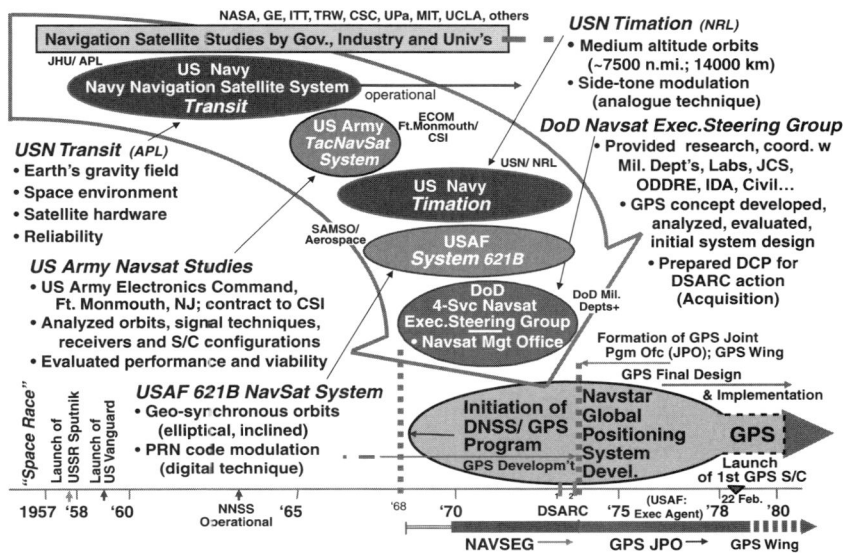

Fig. 12.10 The GPS program initiation and evolution.

the program to a later time. Figure 12.10 shows the principal steps in GPS development.

Because each military department had performed investigations of satellite-based navigation systems, each had its own preferred system approach. For this reason there was little agreement at the outset among the military department representatives, who revealed considerable competition and disagreement. Although these parochial views were never completely overcome, the group succeeded in reaching the accommodations needed.

As chairman, with the assistance of the DDR&E representative and the NAVSEG's civilian and military members, it was in fact possible to come to reasonable decisions by early 1972 on most of the technical characteristics for a DoD navigation satellite system. These characteristics were then incorporated into successive drafts of the development concept paper, which by early 1973 resulted in a final DCP for consideration by the NAVSEG and ultimately by the DSARC.

From the beginning of the management office's system development meetings, the candidate systems were separated into the systems that each military department's research and development activities proposed and supported. These included variations and generations of these systems that occurred as the meetings proceeded over about three-and-one-half years. The group provided comments to explain why they considered certain proposed system characteristics acceptable, desirable, or consistent, or not, with the capabilities required for the system.

NAVIGATION SATELLITE SYSTEM PRINCIPAL CANDIDATES

The candidates presented to the Management Office System Configuration Selection Group for consideration included the systems discussed in the following four subsections. (The NASA work is not presented here because many of its concepts and developments were incorporated in the U.S. Army ECOM TacNavSat and other investigations.)

ALTERNATIVE CONFIGURATIONS OF THE NAVY NAVIGATION SATELLITE SYSTEM

Alternative configurations of the NNSS, commonly known as the Transit system, were developed under the sponsorship of the Navy Special Projects Office by APL of Johns Hopkins University. The U.S. Navy wanted to use this system with the nuclear fleet ballistic missile (FBM) submarines and other Navy platforms. Richard Kerschner and his staff at APL promoted a number of clever higher-altitude extensions of the Transit system, most of which also included ranging capabilities, for consideration by the selection group. One of these involved a constellation of 154 spacecraft to provide worldwide coverage and reliable four-satellite visibility.

ARMY TACTICAL NAVIGATION SATELLITE SYSTEM

The Army Tactical Navigation Satellite System was developed under the sponsorship of ECOM by Communication and Systems Incorporated for use with tactical aircraft, with land and marine vehicles and possibly manpacks in large tactical areas. As program manager and principal designer for this system, I discussed its characteristics, capabilities, and features in detail with ECOM officials and with Howard Gates, deputy assistant secretary of the Army at the time. The system's final results were presented later to members of NAVSEG after it was formed with Gates as the first chairman.

NAVY TIMATION NAVIGATION SATELLITE SYSTEM

This program was developed at NRL under U.S. Navy sponsorship. The program addressed the timing and clock technology capabilities required for operation of spacecraft providing signals for position determination and time transfer. The Navy Timation system was configured as a medium-altitude orbit candidate for a satellite-based worldwide, accurate, continuously available navigation satellite system. Roger Easton of NRL, program manager for Timation, was articulate in his support of the system.

AIR FORCE 621B PROGRAM OF SAMSO AND THE AEROSPACE CORPORATION

SAMSO funded the study and development of a U.S. Air Force navigation satellite system, and the Aerospace Corporation in El Segundo, California, primarily accomplished the work. The Aerospace Corporation began its work on the 621B navigation satellite development program in about 1967. The

621B activity became a significant Aerospace program involving a sizable staff under the leadership of Walter Melton and Phillip Diamond of Aerospace. The program was apparently a favorite of the Aerospace Corporation's President, Ivan Getting, who obtained regular, substantial Air Force funding and support for the program effort. (Ivan Getting was a corecipient of the Draper Memorial Prize by the AIAA in 2002 for his "management skills in obtaining and maintaining the funding and other resources for the USAF/ Aerospace 621B Program which contributed to the development of the Global Positioning System.")

DEVELOPMENT OF THE GLOBAL POSITIONING SYSTEM

The four main activities, or phases, that were planned and executed to develop GPS included the following: 1) configure a representative system and evaluate it as a military capability, 2) present system for DoD program approval and obtain funding, 3) arrange for the military department to serve as executive agent for the program's development, and 4) arrange for the executive agent to complete development, test, and system implementation. Based on the relative merits of the candidate systems, the NAVSEG and the NavSat Management Office developed the system configuration shown in Table 12.3.

Spacecraft constellations that were most desirable from a coverage and stability perspective were with moderate to highly inclined orbits (55–65 deg). The constellation was to provide global coverage and global visibility of 4–7 spacecraft continuously to an arbitrary user with position dilution of precision (PDOP) values of six or better (less) at a minimum elevation angle to the horizon of approximately 5–7.5 deg.

The preliminary engineering development activity in the second phase was necessary to provide a representative navigation satellite system configuration in enough detail so that its characteristics, capabilities, and costs could be evaluated. This was critical to sufficiently determine its viability as a defense system and to demonstrate the capabilities and potential of GPS to all military departments.

A substantial amount of effort was devoted first to developing the basic GPS configuration, which the NavSat Management Office and the military department technical experts accomplished. (These primarily include the Army ECOM TacNavSat-sponsored efforts at Fort Monmouth, the Naval Research Laboratory Timation efforts, and the SAMSO Aerospace Corporation Program 621B efforts.) Second, effort went into assessing the value of the military capabilities of GPS, an effort in which the Institute of Defense Analysis (IDA) substantially assisted. Much of this evaluation was accomplished by the IDA in Arlington, Virginia, at the request of the NAVSEG and DDR&E. Dr. Flax, the president of IDA at the time, strongly supported this

TABLE 12.3   GPS INITIAL CONFIGURATION

Salient Characteristic	Description
Basic navigation measurement technique	Passive, 4 or more satellite pseudoranging, combined with Carrier Doppler measurements
Carrier frequencies	2 separated L-band frequencies for iono correction
	L1: 1575.42 MHz
	L2: 1227.6 MHz
Modulation method	Digital binary phase shift keying (BPSK)
Codes employed	1) Short sequence PRN for acquisition ~0.5–2 Kbits (TBD)
	Nominal ~1 Mbps rate; Length ~1 kbit
	2) Long sequence PRN; for precision; encrypted (NSA)
	Estimate ~10 Mbps rate; Length ~long; days to weeks (TBD)
Receiver operating technique	Digital PRN correlation range and range rate tracking receiver using internal PRN replica codes; 4+ S/C observables for autonomous position, velocity, time
	Receiver clock bias and clock drift solutions
	Short PRN sequence for rapid acquisition
	Long PRN sequence for precision
Accuracy, rms	1) Short sequence: 10–20 m (horiz.); 15–30 m (vert.)
	2) Long sequence: ~5–10 m or better
Constellation	1) 3 orbit planes; 21–27 + S/C; high-altitude circular orbits; nominal: 24 S/C in medium-altitude circular orbits; 8 S/C per plane
	2) Inclination: ~63 deg (not critical; 55–65 deg acceptable)
	3) Separation of ascending nodes: 120 deg for 3 orbit planes
	4) S/C spacing in orbit plane: 45 deg
	5) Relative phasing for S/C in orbit between planes: 15 deg
Spacecraft timing	1) Cesium beam and rubidium atomic standards, multiple for reliability
	2) Stability of clocks: 1 pp 10 exp 12 to 1 pp 10 exp 13; mod. term

(Continued)

TABLE 12.3   GPS INITIAL CONFIGURATION (CONT)

Salient Characteristic	Description
Control segment	1) 3–5 + Colocated uplink and monitor stations, distributed
	2) Master station at central location: Colorado Springs, other
Received power into a 0-dB upper-hemispheric coverage user antenna	1) At L1, L2: (low rate code, ~1 Kbps) ~160 dBW
	2) At L1, L2: (high rate code, ~10 Mbps) −163 to −166 dBW
	3) Launch vehicle: expendable—Atlas, Delta, other
	4) S/C antenna: phased array—Earth coverage
	5) TT&C antenna: conventional horn or helix
Launch sites	1) Vandenberg AFB
	2) Kennedy Space Center

work, which was done primarily under the guidance of Harold Chielek, an IDA senior member of the technical staff. Harry Sonnemann, NAVSEG chairman, other members of the NAVSEG, and I met with Flax and with Chielek frequently during this work. An IDA final report was published.

The NAVSEG characteristics for the system provided the data needed for further evaluation of the GPS initial configuration by the NavSat Management Office, the NAVSEG, and the IDA. These analyses proved to be useful and very positive.

The NAVSEG and the NavSat Management Office in the second phase prepared a DCP based upon 1) results of the various analyses and supporting engineering work; 2) NAVSEG deliberations and the development group recommendations; 3) military departments' interest in the capabilities and potential of GPS; and 4) cost, military benefits, and value of GPS as assessed by IDA.

The DCP was produced with the support of the Office of the DDR&E and the combined efforts of George Salton of ODDR&E, the NavSat Management Office, and the NAVSEG. A number of DCP drafts were circulated and commented upon as supporting data were developed and as work progressed.

The program's third phase consisted of the preparations for the DSARC and the DSARC activity. The DCP was completed at the end of 1972 and ready for presentation to the DSARC in early 1973. Unfortunately, a DSARC review of the program could not be scheduled at this time. After the November

1972 national election that returned Richard Nixon to a second term, the president made sweeping changes in a host of presidential appointments. Many positions in the Executive Branch, especially in the DoD, became vacant or had lame duck officials until well into 1973. Although a number of officials who had served in the first Nixon administration were still in office during this period, many could make only very limited decisions because they might soon be replaced. For this reason and possibly other considerations, NAVSEG found that its presentation to the DSARC had been delayed indefinitely.

After the second Nixon administration had been in place for several months, John Foster, the DDR&E for the previous eight years, resigned and left government service. [Foster became a vice president of Thomson Ramo Wooldridge (TRW) Corporation.] Malcolm Currie was appointed as the new DDR&E.

As NAVSEG was completing the DCP, which described the developmental configuration for the GPS program, it was necessary to coordinate with the military departments to determine which service might take on the executive agent responsibility. This was a significant commitment because the executive agent department would need to agree to significant funding of the program each year over the program period as well as a substantial personnel commitment. The NAVSEG estimated that an initial commitment of $100M appeared appropriate to start the final engineering development and contract efforts that needed to be accomplished in a timely manner.

In late 1972, the chairman of the NAVSEG, Harry Sonnemann, and I arranged to discuss the acceptance of the primary program responsibility with the secretariat-level decision makers in the Air Force, the Navy, and the Army. We believed it important to have an indication of which of the military departments would be willing to accept the executive agent role in the development and implementation of the system. The DDR&E would have a strong influence on the decision process, but we believed it useful to determine the interest and concerns of the departments at this time.

The Department of the Army was under intense financial pressure with their continuing major involvement in the Vietnam War. The Army indicated that the program was not a feasible option for them and declined consideration as lead agency for the activity.

When Sonnemann and I met with the Air Force secretariat officials, we proposed that the Air Force consider sponsoring the GPS program and act as the executive agent. In our judgment the program was clearly within Air Force assigned mission areas. We recognized, however, that recent cuts in the acquisition and development of Air Force aircraft and other programs were a very significant concern to the Air Force at this time. Air Force officials, including the secretary of the Air Force, John McLucas, agreed to consider our proposal, but they were not optimistic because of the department's budgetary stresses. A few days later, the Air Force informed us that the department could not find any way in their planned budget to fund and support the GPS program. The Air Force at this time declined to become the executive agent for GPS.

Sonnemann and I then met with the Assistant Secretary of the Navy for Research and Development Robert Frosch, his staff, and with others to see if the Department of the Navy would sponsor the GPS Program. The Navy indicated that it was not a possibility at the desired funding level of $100M. However, we were informed that some alternative arrangement might be feasible. Some time later, we were contacted, and the Navy made a proposal to fund the GPS program at an initial level of approximately $55–$60M. The Navy had its own budgetary constraints and was being forced to reduce the size of its fleet substantially. This offer by the Navy, although financially modest for the program requirements, was at least a path for initiating the GPS program.

We advised the NAVSEG members of the results of these meetings, contacts, and responses. There was some surprise that the Air Force had declined and some minor disappointment with the funding level offered by the Navy. There was nevertheless considerable satisfaction in knowing that the program was now in a good position to proceed to the engineering development stage. Although this commitment was not essential, it placed the program in a good position for later consideration by the DSARC.

Not long after this, I found that the two Air Force officers on my staff at the NavSat Management Office were busily putting data together for meetings with the Air Staff and the Office of the Secretary of the Air Force. When I inquired as to the reason for this, they informed me (with some reserve) that there was Air Force interest and activity relating to a development program and that they would inform me fully about this later. I did not press with further questions or delay their efforts. I surmised that it was NAVSEG's development program, the GPS program, that was now under further Air Force review.

AIR FORCE RECONSIDERATION OF THE GPS PROGRAM

A short time later, I was informed by one of the Air Force officers of my staff that the Air Staff and Air Force Secretariat were seriously concerned that the Navy was to become the executive agent for the GPS program. At the time the Navy was managing a large communications satellite program, known as Fleetsat, with the Air Force at SAMSO in the Los Angeles area. Because the Air Force had the space technology mission area, the Navy used the SAMSO facilities and stationed naval officers there to manage the program. Air Force officers could see that for the Navy to manage the GPS program, it could considerably expand the number of naval officers at SAMSO and increase Navy–Air Force tensions. In their view, this could create a serious institutional, personnel, and morale problem. The result was a high-level Air Force serious reconsideration of their earlier refusal to undertake the GPS program. The Air Force recognized that to secure the program they would now have to compete with the Navy.

A short time later, Air Force officials informed the DDR&E of their interest in the GPS program and proposed more substantial funding than the Navy's earlier offer. The Air Force's stated plan was to staff and locate the program at SAMSO. Without officially informing NAVSEG, the Air Force had now agreed to sponsor and fund the program. The Air Force now offered what the NAVSEG was striving for, a strong and committed GPS executive agent candidate.

Following their decision to commit resources to the GPS program, the Air Force apparently wanted to guarantee that they would receive the program sponsorship. The SAMSO management might have been unwilling to risk an open competition with the Navy for DSARC selection of the GPS program. In mid-1973, the NAVSEG found that in late 1972 the Air Force had quietly established an office for navigation satellite systems at SAMSO and had designated a director for the office. The new director, Col. Bradford Parkinson, referring to the navigation satellite system concepts espoused by the Army, the Navy, and the Air Force told the Aerospace Corporation *Crosslink* magazine [19], "I entered the picture when those three concepts were in a death struggle—none of them was going anywhere." In fact, in late 1972 when Parkinson entered the picture, the NAVSEG had configured and essentially completed a Four Services navigation satellite system (GPS). NAVSEG was then incorporating this system into a final DCP. Indeed, by late 1972, the NAVSEG's three military departments' representatives had generally agreed to the joint (Army–Navy–Air Force) GPS configuration that NAVSEG had coordinated and designed. This system is essentially the GPS configuration that we have now and have had for the past 25 years.

DSARC CONSIDERATION OF THE GPS PROGRAM

DSARC-1 ON AUGUST 7, 1973; AIR FORCE/DDR&E MEETING

Shortly before August of 1973, when the DSARC on the defense navigation satellite program was scheduled, I asked Harry Sonnemann, chairman of the NAVSEG, if he had received notice of the meeting. Sonnemann was a special assistant to Robert Frosch, the assistant secretary of the Navy for Research and Development [OASN(R&D)]. Sonnermann indicated that he had not. Other NAVSEG members, for example, Victor Friedrich, OASN (R&D) were not invited, nor was I as the scientific director of the DNSS (GPS) program. (Victor Friedrich was the immediate past chairman of the NAVSEG and the special assistant to the assistant secretary of the Army for Research and Development.) Later, we found that the DSARC had been scheduled as an event limited to Air Force and DDR&E personnel.

At this unusual DSARC meeting, the Air Force Program 621B was apparently presented as the sole navigation satellite system for consideration. NAVSEG had earlier evaluated the 621B program and found it could not meet the established JCS and other military requirements. Records indicate that the DSARC declined to approve the 621B program at the meeting. Those in attendance at the DSARC reported that the new director of Defence Research and Engineering (DDR&E), Malcolm Currie, requested the Air Force to change the GPS program to provide a system that would meet the needs of all of the military departments and that all departments could support.

The Air Force at SAMSO appears to have subsequently changed their navigation satellite system to match in all essential respects the configuration that the NAVSEG had established about a year earlier. There would have been little difficulty in changing the Air Force 621B program to the NAVSEG GPS configuration, a system that representatives of the research and development organizations of the military departments already were familiar with, had coordinated with the NAVSEG, and in principle agreed to. The NAVSEG configuration had employed the spread spectrum PRN-coded signal initially developed for 621B by the Aerospace Corporation. The main changes required were to employ the NAVSEG circular-orbit satellite constellation arrangement with medium-altitude satellites and a few other details. It is unfortunate that there was a four-month delay (from August until December) in the presentation of this configuration to DSARC-2. When the Air Force presented this changed program to DSARC at a second meeting on 17 December 1973, the program was readily approved.

In late 1973, the Air Force thus unilaterally took over the joint services GPS system architecture that the NAVSEG had developed. In the process, the Air Force bypassed the commitment by all military departments to the NAVSEG 1968 coordination agreement. These activities cut short in its final stage an extraordinarily successful and timely coordinated effort by the Four Services NAVSEG to implement the GPS program.

A few comments on the capabilities of the Air Force 621B program appear in order. In the previous year, the NavSat Management Office with its development group had evaluated the performance characteristics of System 621B and had found a number of serious issues that it believed disqualified the system as a GPS contender. We can summarize these concerns briefly.

The 621B system initially consisted of several (three or four) sets of synchronous satellites, most of which were inclined (~30 deg) to the equatorial plane and had moderate eccentricity (~0.22) providing an elliptical orbit. There were two basic types of arrangements, the rotating X and the rotating Y. Both of these had a geostationary satellite in the center (of the X or Y set of satellites), and the inclined, eccentric satellites slowly rotated in near-circular orbits around the center satellite providing rotating-X or rotating-Y satellite

ground tracks. This novel arrangement provided a desirable geometric arrangement for position determination but did not provide coverage of the polar
regions. When NAVSEG's Management Office and Development Group had
first evaluated 621B's performance, it was found deficient in this and some
other respects. One of these concerns was that the 621B initially was planning
to synchronize its onboard clocks with frequent or near-continuous uplink
signals. This was dropped after a short time in favor of the use of multiply
redundant atomic standard clocks in the spacecraft.

The Office of the Joint Chiefs of Staff had established that worldwide
coverage, including the polar regions, was a firm requirement for a DoD
navigation satellite system. For this reason the Air Force's initial 621B system
configurations were unacceptable. To remedy this deficiency, Aerospace
Corporation engineers developed a different arrangement, the "egg beater,"
with a much greater satellite inclination (~60 deg) for the rotating satellites.
This allowed the system to provide full Earth coverage, but it still had a
serious concern.

The 621B was a synchronous constellation, however, each rotating X or
Y component of the constellation remained over a fixed region of the Earth.
This raised the question of whether or not the international community
would accept this coverage aspect of the system. It was unlikely that the
Soviet Union, for example, would accept a U.S. system designed for military purposes that was in continuous orbit over its principal land mass. By
1973, there was precedent for orbiting satellites that passed over all geographic regions of the Earth, but no precedent for regional coverage by
military satellite configurations. For this and other reasons, the Management
Office Development Group and the NAVSEG found the 621B system
unacceptable.

At least a dozen individuals who were Air Force SAMSO and Aerospace
participants in the NavSat Management Office Development Group activities
were well aware of the 621B system's deficiencies at the time of its submission to the DSARC on August 7, 1973. The development group had worked
over an extended period to overcome system problems and concerns, such as
those in 621B, and had developed several system candidates that had none of
the 621B system's shortcomings. These systems had been well coordinated
with the military departments.

The president of the Aerospace Corporation, Ivan Getting was a strong
advocate of navigation satellite systems. He may well have wanted the project
that we understand he took great pride in, the 621B navigation satellite
system, to become the final GPS. There were many aspects of the 621B system that were excellent in their design and indicated knowledgeable and
sound engineering capabilities. In any case, DSARC-2 was scheduled for
December 17, 1973.

DSARC-2: AIR FORCE'S SECOND NAVIGATION SATELLITE SYSTEM SUBMISSION

At DSARC-1 Malcolm Currie, the DDR&E, had asked the Air Force to submit an alternate system that would meet the requirements of all the military departments. Col. Parkinson has written, "Over the Labor Day weekend of 1973, he assembled about a dozen members of the JPO" and then "directed the development of a new design that employed all available satellite navigation system concepts and technology" [20]. The NAVSEG members found that SAMSO's "new" system configuration was essentially identical to the system candidate that NAVSEG had completed development over a year earlier. We can briefly explain the respects in which the Air Force revised system and the NAVSEG GPS were similar.

Both the new SAMSO configuration and the "old" NAVSEG system configuration employed medium-altitude circular orbit satellite constellations, space-based atomic standard clocks for time stability, PRN digital modulation, L-band carrier signals for position determination by ranging, and for ionospheric delay correction, L-band Doppler measurements for velocity determination, as well as a control network of ground monitoring sites and transmitter sites for the uplink of data messages to the spacecraft.

The NAVSEG configuration had been developed, published, circulated, and discussed for about a year. The Air Force presented its similar configuration identified as a SAMSO system to DSARC-2 in December 1973. Probably most of those in the Labor Day weekend group were well aware of the NAVSEG initial GPS design work and related activities. Articles and comments on the DoD navigation satellite, work activities, and system design progress had appeared regularly in *Aviation Week*, other publications, and at technical meetings.

I hasten to add that I have great respect for Col. Parkinson, an energetic and conscientious Air Force officer who faced substantial new tasks and responsibilities at this time. Earlier, in 1971 and 1972, both the Navy and the Air Force had complained unofficially but frequently that the Joint Services (NAVSEG) navigation satellite system selection process was moving too slowly. The Navy and the Air Force each believed that its own system—Timation for the Navy and 621B for the Air Force—was superior and should be selected without further delay. However, by late 1972, NAVSEG had completed a fully developed system configuration and was prepared to submit its DCP for the Defence Systems Acquisition Review Council's consideration. By the end of 1972, the Four Service NAVSEG had in fact reached its assigned objective, and had developed a navigation satellite system that was superior to both the Navy Timation and the Air Force 621B. This was consistent with the direction by the DDR&E originally to the NAVSEG and more recently to the Air Force.

It is not surprising that the DSARC accepted the Air Force's second system submission in December 1973 and designated the Air Force as the executive agent for the system. The Air Force agreed to provide substantial funding to develop and establish a GPS Joint Program Office (JPO) at SAMSO for the final development. They also committed to internal and contractual activities that led to the implementation of GPS. These decisions completed the fourth and final phase of the plans for initiating the GPS program.

This final approval step in the early development of GPS led to the subsequent implementation of the system. These early developments, which set the stage for the operational evolution of GPS, proved to be of immense importance not only for navigation, its original goal, but also for a host of other important, world-changing purposes. In these new roles, GPS has made possible such advances and applications as precision timing, improved accuracy in survey and geodesy, power grid coordination, cell phone location, improved air traffic control, aircraft precision landing, vehicle navigation and tracking, and highly precise and useful measurement applications of a wide and growing variety.

CONCLUSION

Today, 35 years after the surprising December 1973 GPS program denouement, we can perhaps add a positive perspective. By early 1973, after NAVSEG's success in completing their final approval development concept paper, the implementation of the four service GPS program and system design that they had developed required the strong commitment, support, and funding of one of the military departments. The program needed a lead service—an executive agent. The Air Force was the logical choice even though they had earlier, when asked by the NAVSEG, refused to commit to the program. It may have been serendipitous that a bizarre competitive impetus caused the Air Force to reassess its position and take on the GPS program. I believe this was a wise and foresighted decision. Since 1974, in nearly all respects, the Air Force has stood by its commitment and done a sound, professional job in the final joint development, deployment, operation, and maintenance of an invaluable dual-use national and international asset, the global positioning system.

REFERENCES

[1] McDonald, K. D., Contributions to Recommendations for Technical Improvements and Enhancements, "The Global Positioning System – A Shared National Asset," *National Research Council Committee Report on the Future of the Global Positioning System*, Commission on Engineering and Technical Systems, National Academy Press, Washington D.C., May 1995.

[2] McDonald, K. D., "Navigation Satellites: Technology, Status and Issues," invited paper presented as a keynote at the *IEEE Fourth International Symposium on Spread Spectrum Techniques and Applications*, ISSTA '96, Electoral Palace, Mainz, Germany, September 22–25, 1996.

[3] McDonald, K. D., "Status, Capabilities and the Future of GPS: Its Role in a GNSS," presented at the 1996 *Differential Navigation Satellite Systems Conference* (DSNS '96) in St. Petersburg, Russia and published in the conference proceedings, May 20–24, 1996.

[4] McDonald, K. D., "An Analysis of Technical and Policy Issues Affecting the Future of GPS," presented and published in the *Proceedings of the Fourth International Conference on Differential Navigation Satellite Systems (DSNS)*, Bergen, Norway, April 24–28, 1998.

[5] McDonald, K. D., "Requirements, Implementation Candidates and Capabilities of a Second and Third Civil Frequency (L2c, L3c) and Their Impact on Future GPS Performance and Use," GPS Interagency Advisory Council, Inst. of Navigation, and The National Geodetic Survey, Jan. 1998.

[6] McDonald, K. D., "Issues in the Use of GPS: Potential Conflicts for Civil Users," invited paper presented at the *General James H. Doolittle Symposium on the Global Positioning System: Civil and Military Uses*, held under the auspices of MIT, Cambridge, MA, Feb. 1995.

[7] McDonald, K. D., "Technology, Implementation and Policy for a Dual Use GNSS," *Proceedings of the IAIN 9th World Congress*, The International Association of Institutes of Navigation, Amsterdam, The Netherlands, November 18–21, 1997.

[8] Harford, J. J., "Korolev's Triple Play: Sputniks 1, 2, and 3," adapted from Harford's *Korolev: How One Man Masterminded the Soviet Drive to Beat America to the Moon*, Wiley, New York, 1997.

[9] Guier, W. H., and Weiffenbach, G. C., "Genesis of Satellite Navigation," *Johns Hopkins Applied Physics Laboratory Technical Digest*, Vol. 18, No. 2, 1997, pp. 178–181.

[10] Danchik, R. J., "An Overview of Transit Development," *Johns Hopkins Applied Physics Laboratory Technical Digest*, Vol. 19, No.1, 1998, pp. 18–26.

[11] Guier, W. H., and Weiffenbach, G. C., "Theoretical Analysis of Doppler Radio Signals from Earth Satellites," *Nature*, Vol. 181, 1958, pp. 1525–1526.

[12] McDonald, K. D., Nickell, R. S., and Palmer, L. C., "Tactical Navigation Satellite System Study (U)," Technical Report ECOM 02360-3, Third Quarterly Report, U.S. Army Electronics Command, Fort Monmouth, NJ, Contract DA28-043 AMC-02360(E), Communications and Systems, Inc., Falls Church, VA, April 1968.

[13] Wyatt, T., "The Gestation of Transit as Perceived by One Participant," *Johns Hopkins Applied Physics Laboratory Technical Digest*, Vol. 2, No.1, 1981, pp. 32–38.

[14] "The Federal Radionavigation Plan (FRP)," U.S. Depts. of Defense and Transportation, published biennially, available on the web: http://www.navcen.uscg.gov/PUBS/frp2005.

[15] Millman, G. H., and Anderson, R. E., "Ionospheric Phase Fluctuations of Satellite Transmissions," *Journal of Geophysical Research*, Vol. 73, No.13, 1968, p. 4434.

[16] Anderson, R. E., Brisken, A. F., Frey, R. L., and Lewis, J. R., "Ranging and Position Fixing Experiments Using ATS Satellites," Goddard Space Flight Center, Contract NAS-5 11634, Final Rept. on Phase 3, Greenbelt, MD, 19 March 1971 to 1 Dec. 1972.

[17] McDonald, K. D., Rubin, G. R., and Cohen, A. L., "A Mathematical Model for the Performance Analysis of Navigation Satellite Systems," Communications and Systems, Inc., CSI Technical Note, Falls Church, VA, U.S. Army ECOM, Ft. Monmouth, NJ, Sept. 1968.

[18] McDonald, K. D., Nickell, R. S., and Palmer, L. C. "Tactical Navigation Satellite System Study," Final Report, Technical Report ECOM 02360-F, 15 June 1966 to 15 June 1967, Report No, 4, U.S. Army Electronics Command, Fort Monmouth, NJ, Contract DA28-043 AMC02360(E), Communications and Systems, Inc, Falls Church, VA, Feb. 1969.

[19] Strom, S. R., "Charting a Course Toward Global Navigation," *Crosslink Magazine*, Aerospace Corp., El Segundo, CA.

[20] Parkinson, B.W., "Introduction and Heritage of NAVSTAR, the Global Positioning System," *Global Positioning System: Theory and Applications*, Vol. I, edited by B. W. Parkinson and J. J. Spilker, with P. Axelrad and P. Enge, Progress in Astronautics and Aeronautics, Vol. 163, AIAA, Reston, VA, 1996, Chap. I, p. 8.

XM Satellite Radio—The First Ten Years: A Trifecta of Technology, Product, and Market Opportunity

John F. Dealy*

As a participant in or witness to most of the entrepreneurial endeavors in the satellite communications field over the years, I am often struck by the confluence of timing, luck, and visionary leadership so instrumental in the creation of a successful satellite-based enterprise. XM's market leadership and rapid rise to more than 7.6 million subscribers in the satellite digital audio radio system (SDARS) business exemplifies this confluence—along with the hard and intelligent work of a small satellite/radio technology team, a much larger programming cadre building the content for subscribers and an across-the-board marketing thrust involving hundreds of millions of dollars each year.

ELEMENTS OF THE XM OPPORTUNITY

In this chapter, I will describe the essential elements in XM's growth from a raw license applicant in early 1997 to a major media creation/delivery company serving more than 7.6 million subscribers across approximately a 10-year span. By year end 2006, XM was producing revenue of more than $800M a year and achieving positive operating cash flow in the fourth quarter. How did a company in the capital-intensive, long-lead time satellite business raise the money, design the technology, implement the infrastructure, create and distribute the radios, develop adequate marketing momentum, and service more than seven million customers in such a short time frame? In my view, professional execution of three central business philosophies made this happen:

1) XM's core leadership team believed that we had to perform exceptionally in all aspects of the business to provide excellent nationwide service as soon as it was initiated in November 2001.

Copyright © 2008 by John F. Dealy. Published by the American Institute of Aeronautics and Astronautics, Inc., with permission.

*President, The Dealy Strategy Group, Washington, D.C.; Senior Adviser to the CEO of XM Satellite Radio from 1997 to 2008.

2) The team had the conviction (and ability) to raise the necessary funds (in good financial markets and bad) to put in place a robust infrastructure (new generation high-power satellites, extensive terrestrial repeater network and associated broadcast operations equipment/software) prior to the time we entered service.

3) There was full-scale development (and employment) of a programming group to create 100 channels of outstanding music, sports, and information content—including investment in state-of-the-art broadcast studios, extensive library of recorded music, and a full complement of program directors and on-air personalities.

Virtually every satellite communications startup with an entrepreneurial background has sought to move into the industry with adequate facilities to get started, with end-user hardware and marketing activities contracted to others, with inadequate marketing funds to drive penetration—and each has stumbled (in some cases, fatally) because of the inadequacy of their end-user product, their pricing, or their marketing capabilities. XM's leadership from day one was determined to avoid these pitfalls, even though it meant we had to raise more money earlier than those following the more customary path.

When we looked at the potential market for satellite radio in 1997 and the fundraising requirements to achieve our infrastructure and marketplace strategy, we realized the essential need for a strong OEM partner who could facilitate the market penetration by installing a high volume of XM radios in new cars and trucks. We also saw the need to begin construction on custom-designed XM satellites if we were to be credible in raising more than $1B before we generated the first dollar of revenue.

Because of the experience of XM's leaders in prior satellite-related businesses, we understood the fundamental shortcomings of satellite systems as viable investments. First of all, satellites take a minimum of three-and-one-half years from conception to operation in orbit; second, most new satellite ventures go into orbit with very few customers, take years to achieve a reasonable fill on the satellites, and then have a very short high revenue period before the satellite has to be replaced. The business strategy, therefore, in a successful satellite-related company, is to find a service/market/customer segment large enough to generate substantial revenues in a short time frame and loyal enough to maintain and increase that revenue stream across the initial and subsequent satellite lives.

XM's leaders also understood those market areas where satellites have a distinct (and in some cases almost unique) advantage. DirecTV and EchoStar, for example, employing high-power DBS broadcast satellites, have an advantage in reaching rural areas not easily served by cable—their advantage is eroded as cable companies expand into rural areas and major telcos deploy fiber-optic systems directly to the homes. DirecTV and EchoStar, however, cannot effectively reach mobile users traveling in automobiles and trucks.

The digital audio radio system (DARS) opportunity, as seen by XM leaders, fully utilized the best characteristics of satellites: broadcast coverage blanketing the United States and the ability to transmit seamlessly to millions of automobiles and other vehicles on the move across America.

The third piece of the equation (beyond ubiquity and ability to serve the mobile user) fully appreciated by the XM leaders was the economic draw of media content. Because thousands of terrestrial AM and FM radio stations already existed across America, an obvious question facing XM as it discussed funding needs with investors was why should or would large numbers of people pay for satellite radio when they can get terrestrial radio for free? XM's answer is simple (although the execution is quite complex): Terrestrial radio is distance-limited—once you have driven a certain number of miles, you have to seek a new local station in order to receive a radio signal; terrestrial radio is also very limited in the variety and depth of the music it provides. With DARS, you have the ability to receive the signal in your car all across the country and to listen to 100 (and now 170) channels of diverse music, sports, and information. You can explore and find what you like, rather than being tied to the few formats the large, terrestrial radio companies program in each local area.

This trifecta of ubiquity, service to mobile users, and broad range of media content is the opportunity XM identified and set out to capture. When the "commercial-free" music strategy employed by XM is coupled with a modest monthly subscription price of $12.95, you have a compelling value proposition to the consumer—one that has been accepted to date by millions of XM subscribers.

GETTING STARTED

Organizing a business like XM is a combination of foresight and opportunistic decisiveness. Since the late 1980s, entrepreneurs had envisioned a satellite-based DARS serving fixed installations and ultimately mobile users. There were, however, no radio communication frequencies allocated for this service, and the technology of both satellites and end-user receivers (radios) was inadequate to provide a seamless, ubiquitous service at reasonable subscription cost and attractiveness of user hardware (features and price).

By the mid-1990s, some frequencies had been allocated on a global basis, and the U.S. Federal Communications Commission (FCC) had opened a proceeding with four applicants seeking DARS licenses. Shortly after the passage of the Telecommunications Reform Act of 1996, the FCC announced an auction in early 1997 of two DARS licenses (each involving 12.5 MHz of spectrum), but limited the auction to the four existing applicants for DARS service—each of whom had filed its application years earlier on the premise that licenses would be awarded in a traditional manner and not by auction.

The FCC made the case even more difficult by setting a very short time frame (approximately 45 days) for the auction to occur. None of the four applicants had very deep pockets, and each was seeking funds for the auction.

At that time, the only company actively building a satellite DARS system was WorldSpace, headquartered in Washington, D.C., but planning its initial systems for the Middle East and Africa. I was senior advisor to the chairman of WorldSpace (Noah Samara) and discussed this emerging U.S. opportunity with him as soon as the FCC announced it was conducting an auction. Samara quickly became interested in the situation and set out to raise the necessary money for the auction; we then identified the most probable candidate of those in the FCC queue with whom to do a deal. In the interest of time, we decided to focus on one company, American Mobile Radio Corporation (AMRC), a subsidiary of American Mobile Satellite Corporation (AMSC). We knew AMSC had no excess funds and that AMRC had no resources. Within a few days, an investment arrangement was struck with WorldSpace providing virtually all of the money for the upcoming auction. At the auction, AMRC, using WorldSpace capital, obtained one of the two licenses (CD Radio, now Sirius, obtained the other); AMRC/WorldSpace paid almost $90M for its license.

I have recounted this starting point because it illustrates two important elements in creating businesses: one, you have to have realistic people ready to deal at the right time, and, two, you have to find that individual/organization who can assess a situation quickly, make fundamental decisions, and raise large sums of money. Gary Parsons, chairman of AMSC at the time and then chairman of XM, is a realistic businessman who knew he could not raise the necessary funds for the auction and was ready to deal; Noah Samara, chairman of WorldSpace to this day, quickly understood the essential value of a U.S. DARS license, agreed to a reasonable transaction with AMSC, and was able to raise $100 million from private sources within two weeks. This combination is not found too often, and this is the reason many very good business opportunities cannot be pursued effectively.

STRUCTURING THE BUSINESS

With the funds provided by WorldSpace, XM (or AMRC as it was then known) had the financial resources to acquire the license, initiate skeletal operations, develop a technology/operations plan, and begin construction of its satellite infrastructure. Some major decisions were made in this early period. First, it was decided to recruit an executive from the media/entertainment/subscription services world to drive the overall planning and execution of the SDARS business. Second, it was decided to deploy most of the remaining funds from WorldSpace into the design, development, and initial construction of XM satellites. Third, due to financial market considerations and the very early stage of this capital-intensive business, it was decided to seek strategic partners and private equity investment to take XM to the stage where it could be a viable public company. (Its competitor, CD Radio, was already public and was struggling in the financial marketplace.)

By early 1998, XM had selected a vendor for its satellites following an industry-wide solicitation and intense competition between Matra Marconi of France and Hughes Space and Communications of El Segundo, California (teamed with Alcatel of France). Ultimately, Hughes/Alcatel prevailed for three interrelated reasons: Alcatel had the successful heritage of building the payloads for WorldSpace's pioneering SDARS L-band satellites; Hughes had recently developed a very high-power satellite platform, the 702 Class, and sold it to some leading companies in the industry; and the Hughes Electronics (the parent of Hughes Space and a subsidiary of General Motors) senior management were willing to receive a significant downpayment and then fund satellite construction at their own risk for a considerable period of time after the downpayment funds were exhausted. This combination of next-generation high-power satellites, payload heritage, and willingness to carry interim financing gave XM the best chance to provide a robust satellite service in S-band while having the time to complete a substantial private offering.

Shortly after the procurement of the XM satellites, XM filled its major management positions to drive the business—in particular, Hugh Panero as president and CEO, Lee Abrams as vice president of content/programming, Steve Cook as senior vice president of marketing, and Stell Patsiokas as senior vice president of technology (designer of XM network and end-user devices). This team has driven market penetration, content selection, and end-user experience throughout the company's significant growth from nationwide service inception in November 2001.

In 1998–1999, XM made the critical decisions to locate its overall facilities in Northeast Washington, D.C., and its Technology Center (the Innovation Center) in Boca Raton, Florida. By choosing Northeast Washington, XM was able to obtain an excellent structure (a former printing plant) and a central location in the greater Washington, D.C., area. Over the next two years, XM refurbished this company headquarters (and operations facility) into a world-class combination of 82 programming/broadcast studios, sophisticated (and extensive) automated databases/processing for music production and broadcast, satellite uplink facility, and state-of-the-art performance studios (which, over time, have attracted many of the leading musicians in the country to perform before live audiences and do interviews). XM, although located in Washington (as opposed to the media capitals of New York and Los Angeles), has become a "destination" attraction for the music performer.

The conscious strategy of XM's top management was that each element of a media creation/delivery business must be addressed with equal intensity—the resources must be obtained and committed, and XM must have the ability to influence and control each element (sales, content creation and delivery, network, including satellites, end-user devices, and features thereof)—and each element must be firmly in place before XM began nationwide service. In addition, XM's strategy was not to begin service in one area of the country and then gradually roll it out across the nation. XM senior management had been involved in or observed other startup telecommunications/media companies begin with skeletal systems, limited content, unattractive end-user devices, limited area deployment, and no real commitment of resources to marketing—and then seen that the service provided was a failure from day one. Such businesses usually wound up in bankruptcy or were discontinued (often after billions had been spent) within a few years after inception.

With their extensive experience in telecommunications/satellite-related businesses, XM executives were determined that XM's "first impression" in the marketplace be a powerful, positive experience for the consumer. The combination of Gary Parsons (satellite/financing background), Hugh Panero (developing, marketing, and servicing media subscription businesses), Lee Abrams (broad and deep background in music delivery through radio), Steve Cook (marketing subscription services to consumers), and Stell Patsiokas (development of low-cost, attractive wireless end units) working as a team ensured that appropriate emphasis (and resources) were applied to each of the critical business areas both before operations commenced and as the business grew to multimillions of subscribers.

IT ALL TAKES MONEY

Good plans, solid management team, sound technology/designs, and shrewd negotiation of major procurements are necessary conditions to a successful satellite/media business. With the capital intensity of a satellite-based business and the long time frame required to design and produce the satellites before service can commence, these elements alone cannot create a major business. That requires the gathering of enormous amounts of "patient" capital, money committed with no payoff for a long period of time. To raise that type of capital before a company has really built a revenue-generating business requires more than plans and presentations.

XM strategists believed two elements were essential to raising the patient capital (once the funds provided by WorldSpace were exhausted): satellites/infrastructure under construction so a definite time frame for entry into service could be ensured and a strategic distribution partner whose commitment and presence would convince financing sources (and ultimately the public financial markets) that the XM concepts would become a major viable business. By mid-1998, XM had the satellites under construction (as already described), and for the next year it struggled to obtain additional sources of private financing and to secure that major distribution partner. By mid-1999, substantial deferred payments on the Hughes satellites were coming due, extensive work had been done with private equity sources, but no financing had been secured, and XM had zeroed in on General Motors as the most likely candidate to become its major distribution partner. In mid-summer 1999, all of the pieces came together in a $250M private placement with private equity funds, Clear Channel Communications, and General Motors (including its subsidiary DirecTV). One analyst described this as a financing trifecta—smart private equity, strategic player in the terrestrial radio business (thus giving credence to XM as a media company of the future) and the world's largest automobile company (who committed long term to install XM in millions of new cars and trucks).

In the history of a business, particularly an entrepreneurially funded capital-intensive startup, there are usually two or three moments of crisis that determine the character and the future of the enterprise; the first of these for XM was the summer of 1999, when the company was running out of cash. Hughes had stopped work on the satellites due to lack of payment, and the private financing package continued to be delayed. XM survived that crisis, obtained the financing package already noted, and within a few months had a successful IPO, raising additional hundreds of millions. The stock market was strong in 1999 and into 2000, and XM's stock price rose from a $12 per share IPO into the $40 range; the IPO was followed by additional public and private placements over the next few years.

PROGRESS TOWARD SERVICE ROLLOUT—1999 TO NOVEMBER 2001

During this two-year period, XM built its facility in Northeast Washington, D.C. under the leadership of Jack Wormington (formerly of Hughes and subsequently of Boeing), developed its programming staff, assembled the music content, negotiated deals for third-party content (including news and information services), completed the design of its end-user equipment chipsets, and contracted for the first Aftermarket production radios from leading electronics manufacturers such as Sony, Pioneer, and Alpine. GM and XM worked together to provide high-quality XM radios (assembled by Delphi—at that time a General Motors subsidiary) for new production automobiles to be delivered beginning late 2001.

As the year 2000 came to a close, XM realized that the gating item for a successful business rollout would be the completion of its integrated satellite/terrestrial repeater nationwide network. In the DARS auction in 1997, the FCC licensed XM and its competitor to utilize the 12.5 MHz allocated to each both for satellite service and for ancillary terrestrial transmission. This was a breakthrough FCC ruling and is key to the satellite radio companies providing ubiquitous service with high availability (99%+) to mobile users (e.g., automobile drivers and passengers) in rural, suburban, and urban environments. Under the XM system, when a subscriber is in an open-air rural environment, it receives "interleaved" digital streams from each satellite; the radio keeps part of the stream in a cache and utilizes the rest; when the automobile passes under an overpass and its view from both satellites is blocked, the cached content is seamlessly utilized until the car emerges from the overpass and the signal switches back to the non-cached version. Signal availability to the customer on the open road across America is ensured through spatial diversity (two widely spaced satellites both broadcasting to the radio) and time diversity (real-time plus cached signal).

In the urban (and some parts of the suburban) environment, terrain, heavy foliage, and buildings can effectively block both satellites for time durations exceeding the cached content (approximately four seconds of content); to achieve high signal availability in those conditions, XM deployed an extensive terrestrial repeater network (approximately 800 ground repeaters) in each major city and related suburban area. The precise location of those repeaters, the achievement of synchronized repeater timing, and the avoidance of repeater-to-repeater interference (because the XM network is a single-frequency network) required very precise engineering, the effectiveness of which could only be determined after a number of repeaters were deployed and tested. By the end of 2000, early city testing had indicated unacceptable interference among repeaters; in early 2001, the repeater network was redesigned and construction accelerated. At the same time, XM had launched its satellite constellation (XM-1 "Rock" and XM-2 "Roll") and was beginning satellite testing. By midsummer 2001, XM task forces were working with numerous construction contractors to finish the ground repeater deployment and road test the combined satellite/ground repeater signal availability in major cities. It became a race to completion with the scheduled nationwide deployment in November 2001; in fact, critical repeaters in New York City were being turned on and tested the day before formal opening of service ceremonies in New York.

XM SERVICE—EARLY AND CONTINUING CUSTOMER ACCEPTANCE

XM has struggled with many issues throughout its young life: from the repeater build out to radio inventories to progressive degradation of the solar arrays on its first-generation satellites to the escalating cost of programming to the difficulties of branding a nationwide service at reasonable cost to development of a quality customer service experience. To its great satisfaction, however, two core elements of the business have been highly successful from the first rollout: first, signal availability has met or exceeded our expectations and those of our customers; second, customer satisfaction with our content has been very high. Put another way, the quality and variety of our programming has been exceptional, and the signal delivery platform robust. Another key ingredient throughout our rise to more than 7.6 million subscribers has been the quality features and consistently decreasing prices of our end-user radios (a tribute to the cutting edge design efforts of our Florida Innovation Center team and its ability to develop sources of radio production at reasonable cost on a worldwide scale).

With the three-legged stool of excellent content, high signal availability, and attractively priced radios, XM has a solid foundation for long-term revenue growth and profitability. The business is very competitive, not only against the other satellite digital radio provider, but also against entrenched media companies (terrestrial radio operators) and new entrants providing a broad range of digital media experiences—both live and recorded—to an ever more sophisticated and demanding world of consumers.

SATELLITE STORY

At the time XM was conceiving its satellite system—1997 time frame—the choice of technologies was essentially limited to medium high-power satellites (approximately 8–10 kW payload power). This level of power made it technically difficult to communicate ubiquitously with small antennas on car rooftops from a position in geostationary orbit. One could design a system in low to medium orbit; however, this would require a large fleet of satellites to provide continuous service over the continental United States. You could also design a system using elliptical orbits, much like our SDARS competitor did once it found it could not close the gap with the medium high-powered satellites it originally procured. An elliptical orbit approach would require a minimum of three satellites and would not have some of the spatial and time diversity features a two-satellite geosynchronous orbit constellation can achieve. In 1997 and early 1998, XM conducted a competition among the major satellite providers and narrowed its choices to a Hughes 601HP or a Matra Eurobus 2000 Class spacecraft. Each offering was marginally capable of providing the necessary throughput to transmit the XM channels effectively to small antennas on car rooftops in open air; what was not clear was the robustness of these signals to penetrate foliage, particularly during spring wet seasons.

As an alternative, Hughes had also proposed the new 702 High Power bus with a 16-kW end-of-life power capability; this was a configuration yet unproved in orbit but already sold to some major satellite providers whose spacecraft would fly prior to the XM planned launch at the end of 2000. (XM-1 and 2 were actually launched in the spring of 2001.) Because it was clear to the XM technical leadership that our mission was power constrained, we finally decided to select the Hughes 702 bus. With regard to the payload, the only real experience in the industry on an SDARS payload was the Alcatel payload built for WorldSpace on a medium-power Matra bus. With XM's encouragement, Hughes and Alcatel collaborated on their XM bid—the Hughes 702 bus and an Alcatel SDARS payload. When the XM procurement contract was entered into in March 1998, it specified a less than maximum power 702 bus, but contained an option for the standard or classic 702 High Power bus. Considerations of cost led XM to select the lower-powered version of the 702. After contract execution, I continued discussions with Hughes senior management concerning the higher-power version. At the same time, our technical team became more convinced the added power would ensure a more robust service platform for transmission of our XM channels. Accordingly, in June 1998, following extensive and sometimes heated negotiations, XM selected the optional higher-powered 702 classic, and that is the basic power level for all XM satellites to date.

All of the 702 initial design satellites (six were built) employed solar wing "concentrators," that is, mirror-finish panels attached to each solar array to reflect the sun's rays and increase power available to the spacecraft. XM's satellites were the fifth and sixth 702s built; the first 702s appeared initially to function well with their solar concentrators. As XM was in the final stages of rolling out its service in late 2001, a team from Boeing (which had acquired Hughes in late 2000) came to XM and advised me and Derek de Bastos, our vice president of satellite engineering, that there was an unanticipated progressive degradation of solar-array power resulting from the use of concentrators on the initial 702s. This began a five-year effort to monitor this issue, determine whether the satellites would be total losses over time, fund replacement satellites with the same power levels but without concentrators, monitor their construction, and get them launched and into service before XM-1 and 2 power levels dropped to the point where they could no longer provide ubiquitous service with high signal availability.

The capital required for two replacement satellites (including launch and insurance) totaled nearly $500 million, thus putting an enormous additional strain on the finances of an entrepreneurial company. Through almost constant negotiation, aggressive pursuit of insurance claims against carriers who were hit with major claims on all six initial 702s, and innovative fundraising at critical junctures (primarily orchestrated by Gary Parsons, XM's chairman), XM managed to have XM-3 built and launched just in time to avoid service degradation from the declining XM-1 and 2 satellites. In an innovative approach to maintaining full power, XM then relocated one of the two satellites and operated XM-1 and 2 collocated in one orbital slot—each capable of providing power to one-half the XM channels for a temporary period of time. XM completed the procurement for XM-4 and Boeing won in a tight competition with Loral; XM-4 was launched in October 2006 and went into service on 15 December.

So XM today has two new satellites in orbit, each with more than 15 years of expected life and with power equal to or better than the original power levels on XM-1 and 2. Part of XM's ability to get these satellites produced was the willingness of Boeing to extend payment deferrals during construction so that XM would have the time to raise the incremental funds necessary for these unanticipated expenditures. One reason the vendors are willing to do this is XM's excellent record of paying its vendors when amounts are due. Remember that during all of this time, XM was significantly cash negative from operations (as it built its consumer base) and was experiencing major perturbations in the financial marketplace for capital-intensive entrepreneurial companies. This "integrity" of XM in meeting its obligations has enabled me to negotiate numerous temporary payment deferral arrangements with satellite, launch-vehicle, and terrestrial equipment suppliers.

CHOOSING A LAUNCH VEHICLE—SEA LAUNCH

To create the business of XM from an entrepreneurial start and to achieve the technical performance required of our satellite/terrestrial infrastructure, we had to take a number of significant risks. The art of managing the situation was to make sure the risks were prudent against the results that had to be obtained and the probability of failure. At the same time we selected a new satellite platform (but from a very respected supplier, Hughes), we explored the launch-vehicle world to find the right launch service provider. Sea Launch, a partnership involving Russian, Ukrainian, Norwegian, and U.S. suppliers, was a very young company launching satellites from the middle of the Pacific Ocean on a floating launch platform (a self-propelled vessel sailing from Long Beach, California, to a position on the equator south of Hawaii). The advantage of Sea Launch was two-fold: a launch from the equator required XM to burn significantly less fuel to get to geostationary orbit (thus increasing satellite useful life), and the Sea Launch price was equal to the best in the industry. The disadvantages of Sea Launch were the number of entities who had to collaborate effectively, the lack of detailed information concerning Russian and Ukrainian design/production, and Sea Launch's early history of launch problems. The cost advantage of selecting Sea Launch was somewhat offset by the higher insurance premiums XM had to pay for launch and in-orbit insurance.

Balancing all of these factors, XM selected Sea Launch as its launch services provider but contracted with Hughes/Boeing for a delivery-in-orbit program. Each Sea Launch campaign has been an adventure with last-minute aborts, sea conditions that prevented launches until the last possible try before returning the ship to Long Beach, technical issues unresolved until shortly before the launch command is given, etc. Yet, all four XM launches have been on Sea Launch, and each has been remarkably successful—placing the XM spacecraft in nearly perfect geotransfer orbit positions—thereby increasing the in-orbit life potential of each satellite. Three of the launches were delivery-in-orbit (DIO) by Boeing, who was primarily responsible for the interface with Sea Launch (of which Boeing is a substantial part owner); the fourth was delivery on ground to XM with XM under direct contract with Sea Launch for the launch mission. By now, we could clearly see the pluses and minuses of each approach. In the DIO approach, XM is somewhat removed from direct access to Sea Launch information but has the launch services resources of Boeing more engaged in ensuring a safe mission; Boeing as part owner of Sea Launch and prime contractor also has more leverage in getting Sea Launch to conduct tests, explore issues in depth, replace parts, etc. Where XM is the contracting party with Sea Launch, we have more direct access to Sea Launch information, which leads to earlier and better program issues/problem identification—as it did on the XM-4 mission—but less access to Boeing experts and Boeing judgments than in the DIO scenario. This led to some tense moments on XM-4 where Boeing refused to provide its internal judgments and rationale on launch-vehicle issues, asserting it had no contractual relations with XM or Sea Launch on the launch vehicle, even though Boeing has nearly $30M in in-orbit incentives dependent on satellite performance, most of which are not payable in the event of a launch failure! Such, we have found over many years, are the ways of very large companies whose actions are often legal defense rather than customer or business reality driven.

NEGOTIATING WITH GIANTS

Whether it be satellite procurements, car distribution deals, radio production, sales in the aftermarket, content arrangements, or record label licensing, XM found itself dealing with many of the largest industrial and media enterprises in the United States. These companies all have agendas of their own, and their relationship with XM—while often critical to us—is important but not essential to their success. In addition, the larger the enterprise, the greater the chance that managers will regularly be replaced and decisions made at levels barely aware of the XM relationship. Successful management of these David and Goliath industrial relationships is an art form unto itself. Some of the key ingredients in making these interfaces productive are to ensure there is continuing visibility at the top of the giant enterprise concerning who is XM and the potential of the relationship, injecting competition into the equation wherever possible, understanding the box middle management of giant enterprises is in, and helping them to do things compatible with their constraints and objectives. Knowing the interest of the very senior management in the XM relationship is critical; if they believe the relationship is insignificant or not in their company's strategic direction, all of the good work done at the middle management level will be for naught.

XM is an exciting place with hundreds of programmers and on-air personalities creating a new media experience each day, all produced and delivered at our company headquarters in Washington, D.C. Bringing senior executives of our major business partners into Eckington Place opens their eyes to this excitement and unleashes a different part of their personalities—their interest in music, their love for sports, their work on a high school or college radio station; a personal connection, which can be critical to the resolution of some future impasse between the companies, is established.

ROLE OF THE RADIOS

Although XM requires an extensive and expensive infrastructure to deliver its media content to the subscriber, most of that infrastructure is invisible to the potential and current subscriber; he or she interfaces with a device in the car to get the music. The story of the XM radio devices is a fascinating business saga exemplifying two themes: necessity is the mother of invention, and technology leadership is a fragile race you can never stop running. At the inception of the XM buildup, we knew we needed attractive radio products for new and used vehicles; they had to be reasonably priced and sized, with user features competitive with state-of-the-art consumer electronics products. This area was more easily addressed on the OEM side, where the auto companies had standing relationships with world-class car radio manufacturers, particularly GM's relationship with Delphi, the world's largest manufacturer of vehicle radios. Because substantial and early market penetration of the OEM segment was dependent on GM installing the radios in the factory (not at dealers) and integrating the XM feature into the normal entertainment console (AM/FM/tape/CD), the task was a technical one of antenna development, design and pricing of chips, avoiding interference, and working closely with the car manufacturer's radio/electrical system teams.

XM's Innovation Center was established in 1998 under the leadership of Stell Patsiokas who brought with him a team from Motorola with exceptional experience in radio and chip design/integration; Patsiokas is also a natural businessman who understood the need to build credibility with the car manufacturer's technical staff and with the senior management of Delphi. The work of this team (substantially augmented over the years) has been critical to the major success the company has enjoyed with GM and later Honda (XM's second strategic OEM partner; both GM and Honda made substantial equity investments in the company at critical junctures). Building on this record, XM has been able to develop long-term relationships with other major car companies such as Nissan and Toyota, as well as fast-growing entrants into the U.S. market such as Hyundai.

Challenging as the OEM radio market has been, it pales compared to the issues in the aftermarket, where the consumer has to figure out what radio to buy, how to install it in his/her car, how to make it work with the existing radio system in the car, how to provide power to the XM radio, how to receive the XM signal through a separate antenna, and related issues—all the while wrestling with the upfront cost of the new XM radio and the monthly cost of a subscription (originally $9.95 and now $12.95) for something (radio) that traditionally had been "free," that is, the radio came with the car and terrestrial radio was advertising-supported. XM's original approach to the existing car market (which is 15 times the size of the new car market) was to go to established radio manufacturers with world-class brands, for example, Sony, Pioneer, Alpine, and positions in the aftermarket retailers, for example, Best Buy, and have the retailer build the XM radios with chips/technology developed by XM. To address the customer barrier to entry situation, XM worked with Sony to develop a Plug'n'Play radio not requiring professional installation and capable of being moved from the car to the home, a radio that could be played in either place through the use of small cradles or base kits. Problems with this approach were soon apparent: the radios were expensive at retail (more than $300 when the necessary kits were included), design changes were slow to happen and not always optimized to what XM perceived to be the customer desire, and adequate inventory at retail was dependent on large company decisions based on historical sales rather than aggressive future growth expectations.

Within a year after service launch in November 2001, it was evident to XM management that people in large numbers would pay for satellite radio and that the flow of attractively priced radios had to be accelerated in the aftermarket. Patsiokas and his team set out to make this happen—their solution, the SkyFi Plug'n'Play radio line—designed by XM, built by contract manufacturers (Flextronics) in Asia, branded and marketed by Delphi, its first entry into aftermarket radios, and priced to meet mass market demand. Over time, this product line evolved into wireless FM modulation capability (substantially easing the customer installation issue), memory in the radios (to enable recording of favorite songs for later replay), price points below $100, more information on the songs being played, display of stock quotes, and other user-friendly features, as well as stand-alone, low-cost "boom boxes" to facilitate home and outdoor use. All driven by the XM Innovation Center, SKYFi evolved into the Roady line of smaller, even more economical radios. With the flexible manufacturing arrangements established by XM and the success experienced by Delphi, production and inventory flow increased dramatically, fueling the acquisition of millions of aftermarket (now expanded to home and portable use) subscribers over the next three to four years.

The necessity of improving price points, user-friendly features, and ease of installation in the aftermarket thus drove the invention of new radios, new business relationships. and a greater role of XM in the design and manufacture of the end-user devices. The second point I made earlier (that essentially you can never stop improving your technology even when you think you have the lead) is exemplified by XM's development of traffic, weather, and navigation platforms, which has facilitated its major relationship with Honda and its evolving partnerships with other OEM car manufacturers. For many years, XM's Innovation Center has been ahead of the marketing department in the sense that the Innovation Center has been pushing forward with chip development before the marketing department had defined the desired product and with product applications, such as navigation/traffic, before the marketing department had plans or resources to exploit the market potential of these new applications. Some of these Innovation Center initiatives enabled XM to establish a leadership position in the delivery of real-time weather data to the private aircraft market; others enabled XM to meet the expectations of Honda, which has established the early leadership position in integrated vehicle navigation/entertainment systems. The result—XM/Honda NavTraffic systems are standard equipment on all current Acura RL model automobiles—and XM has a powerful deployed example to attract other car companies who realize integrated entertainment/navigation/traffic systems are a key factor in the marketplace success of certain segments of the competition for new automobile purchasers.

CONTENT IS KING

We have discussed extensively the infrastructure, financing, and radio development/production essential to XM market penetration and early success; for long-term viability and growth, however, the customer must be continually excited by the delivered media product and believe he or she is receiving full value for the monthly subscription being paid. Each new subscriber has historically been enthused by the range of programming provided by XM and the "discovery" of specific content unavailable through traditional radio formats. How to keep this content "fresh" and stimulating to the customer each day is the challenge of a radio subscription service. Some media, by its nature, if professionally presented, has the freshness and stimulation built-in like exclusive sports programming (such as every baseball game and virtually every ice hockey game), breaking news events as found on CNN or the BBC, talk shows, and traffic and weather stations. Most music, however, needs the art of presentation to avoid the Muzak or staleness of repetitious content and set it aside from listening to your preselected CDs. XM employs numerous stratagems to provide a distinctive presentation of its music content—from live DJs interfacing with subscribers, to interstitial material concerning the artists and their work, to historical vignettes from the period of the music—for example, audio clips from the 1940s on the 40s on Four channel through numerous XM-created programs such as *Artist Confidential* where the artist is interviewed at XM in front of a live audience, explaining and playing his/her music and sharing poignant life stories; musical legends such as Bob Dylan have their own exclusive presence on XM.

XM's ultimate objective is to be the music source or the "home" for music lovers—with simply more and better content for those really interested in music (or in some particular genre thereof—such as folk, classical, or jazz). XM is also seeking to provide sufficiently attractive and exclusive content to those with specific life interests, such as baseball, hockey, women's issues, Latin culture, and to make them real "fans" of XM, thus ensuring long tenure as a subscriber. Much like the technology issues we discussed earlier, content is a continually evolving world, seeking to present what is attractive to the changing tastes of subscribers and bind them to the service.

As content and its demands evolved, so did XM's management of content. Lee Abrams remained a central figure, but XM has also hired a young Executive Vice President of Programming, Eric Logan, with a background of country music and management of programming for extensive radio networks.

MARKETING STRATEGY

How did XM generate 7.6 million subscribers in little more than five years from commercial service rollout? It starts with an excellent product delivered through a robust satellite/terrestrial infrastructure, but it ends with effective marketing strategies in both the OEM and aftermarket distribution channels. The first challenge was awareness, and XM led the industry in branding XM as the future transmission band: first AM, then FM, now XM; the second was obtaining shelf space at the major retailers, hence the early reliance on Sony, Pioneer, and Alpine with their established position at retail; then the extensive effort to ramp factory installation of XM in new automobiles being produced by GM and Honda. XM's original marketing team led by Steve Cook had extensive experience in retail distribution, consumer electronics marketing, and the sale of subscription services; everyone learned on the job the art of working with major automobile companies (helped by the presence on XM's board of GM and Honda executives committed to making the relationship as productive as possible).

XM was also stimulated from day one by the presence of an aggressive competitor in the market, CD Radio (now Sirius), whose sometimes intelligent and sometimes irrational marketing thrusts kept the pressure on XM to innovate in every aspect of its business. At year end 2006, Sirius remains where it has been from the inception of the industry—one full year behind XM in subscriber count (6.0 million to XM's 7.6 million) and more than one year behind in the efficient delivery of new subscribers (the cost of obtaining a new subscriber or CPGA).

XM NOW AND TOMORROW

The current XM of 7.6 million subscribers was built essentially under the leadership of the founding team and initial senior executive hires; our competitor is on its third management team. During 2006, XM made its first major moves to augment the executive team—bringing in a chief operating officer, chief marketing officer, new head of customer service, head of integrated product development, and numerous additional marketing and sales executives; its existing head marketing executive was reassigned to focus on the greatly expanded OEM partnerships developed over the years and critical to our next growth spurt. The new executives bring fresh perspectives and different experiences to the growth and profitability challenges of the SDARS business; at the same time, they can integrate with XM's outstanding cadre of existing talent and together fuel a renewed energy to capture additional consumers and expand the addressable market for XM's content.

On 19 February 2007, XM and Sirius announced an agreement to merge the companies; the merger approval process will take many months (requiring both antitrust review and FCC approval). The combined company would create an enterprise serving more than 13 million subscribers with greater capability to provide programming choices for consumers and to combat the various forces arrayed against this still very young business. The focus of competition at that point would gravitate away from tactical maneuvers versus each other toward strategic thrusts across many market sectors making up the audio/data entertainment industry. To fulfill part of their FCC license requirements, Sirius and XM have been codeveloping an "interoperable" radio capable of receiving both services' signals simultaneously. A merger will clearly accelerate the introduction of this technology to the marketplace with its benefit of 250–300 channels of audio/data entertainment accessible through one subscription and one device.

When the XM founding team began its quest to bring America (particularly the American driving public) an exciting new medium to relieve the tedium of congested commutes and long drives, they could not foresee each of the technological, regulatory, economic, and competitive challenges they would have to overcome to build a viable business. What they did foresee is that there would be challenges and they would have to anticipate (as well as react) if they were to create a major new business. As already described, the founding team, the organization they built, and the dedicated talent they hired have accomplished that fundamental objective. Ahead is the ability to move to the very different level of a mature, cash-flow-positive enterprise with the economic clout to survive and thrive. How to accomplish this next step without diminishing the ardor, creativity, enthusiasm, and determination of the entrepreneurial XM organization to date is the challenge of the recently restructured XM management and the combined XM–Sirius company.

In July 2008, after nearly one-and-one-half years of regulatory submissions, government reviews, and concerted opposition from terrestrial radio, the XM-Sirius merger finally closed. The combined company has more than 19 million subscribers at year-end 2008 and approximately $2B in annual revenue. I (and my Dealy Strategy Group team) left XM shortly after the merger, as did XM's CEO to whom I was Senior Advisor. The merged entity is named SiriusXM Radio and is currently struggling to refinance certain debt obligations in the face of a very difficult economic climate.

Chapter 14

INTELSAT VI (F-3) REBOOST MISSION: AN EPOCHAL EVENT IN THE HISTORY OF THE COMMERCIAL SPACE INDUSTRY

LEONARD R. DEST*

INTRODUCTION

The history of the commercial space industry is marked by the periodic convergence of the three main pillars of the successful expansion of commercial space activities in the 20th century: a supportive governmental space agency, a powerful aerospace company, and a confident operational organization.

In May 1992 the three titans of the space industry at that time—NASA; the space division of legendary Hughes Aircraft Company, Hughes Space and Communications Company and the first recognized international commercial satellite operator; the 121-member global cooperative named the International Telecommunications Satellite Organization, Intelsat—joined forces to accomplish a space mission that transcended the mere recovery and repair of a large stranded telecommunications satellite. The mission captured the imagination of citizens of the world as man tackled an unruly machine of his creation in the hostile environment of space. And while the world was captivated by daily live television coverage of astronauts battling a huge satellite that would not be captured, there is also a marvelous untold story that occurred behind the scenes of the battle of man's intellectual knowledge of engineering and physics pitted against the dynamics of machines in space and the wisdom of executives and managers that supported the engineers.

The facts are relatively straightforward. On 14 March 1990 as a result of a wiring malfunction on a Commercial Titan launch vehicle, the Intelsat VI (F-3) communications satellite remained attached to the launch-vehicle's upper stage and failed to achieve its intended geostationary transfer orbit

*President and Chief Executive Officer, RD AMROSS, West Palm Beach, Florida; former Vice President, Hughes Space and Communications International, Los Angeles, CA.

(GTO), which would have permitted the satellite's onboard propulsion system to place the satellite into its final geostationary orbit (GSO). Intelsat mission controllers utilized the spacecraft's command system to separate the satellite from the lower-stage perigee kick motor (PKM) and positioned the satellite in a safe low Earth orbit (LEO). Intelsat then contracted with Hughes and NASA to rendezvous the Space Shuttle Endeavour with the Intelsat satellite, to have astronauts capture the satellite, to attach the satellite to a new PKM, and finally to reboost the satellite into the intended GTO trajectory. The NASA shuttle mission began with the inaugural launch of the Space Shuttle Endeavour on 7 May 1992 23:45 GMT from Kennedy Space Center at Cape Canaveral, Florida. After multiple attempts the Intelsat VI (F-3) was captured by three astronauts, attached, and re-boosted on 13 May 1992. The successful 8.89-day mission ended with a successful Space Shuttle Endeavour landing on 16 May 1992 at 21:03:15 GMT at Edwards Air Force Base in California. The STS-49 mission was the first U.S. orbital flight to feature four extravehicular activities (EVAs) and the first flight to involve three crew members working simultaneously outside of the spacecraft.

The Intelsat VI (F-3) was successfully placed on orbit and continues (as of December 2008) to be successfully operated by Intelsat as IS-603 at 340° E Longitude over the Atlantic Ocean to provide services between and among countries in North America, South America, Europe, and Africa.

The story behind the facts is complicated, intriguing, and in the end an overwhelming positive reflection of the engineering and operational expertise existing in the space industry in the early 1990s, as well as a positive acknowledgment to a generation of senior managers and executives who grew up with the space industry in the period from 1960s to the1990s and who were willing to accept risk and celebrate the benefits of success.

It is equally noteworthy that subsequent stranded satellites since Intelsat VI (F-3) in 1990 were most likely maneuvered into orbital reentry and destruction, not because the technical challenges were deemed to be insurmountable or because engineers did not desire to save the satellites, but rather because the satellite industry executives and risk managers of the new century were more likely to have taken the advice from their lawyers and insurance brokers to minimize liability or because these managers simply failed to accept the financial benefits of success, let alone the risk of failure, with their executives, owners, and shareholders.

BACKGROUND

The history of commercial communications satellites begins with Hughes, INTESLAT, and NASA. Early Bird, the world's first commercial communications satellite (see Chapter 2), was built for the Communications Satellite

Corporation (COMSAT), the organizational forerunner of Intelsat by the Space and Communications Group of Hughes Aircraft Company, later Hughes Space and Communications Company, and now Boeing Satellite Systems. Intelsat launched Early Bird on 6 April 1965, using a NASA Delta launcher. For the next 20 years, most Intelsat launches were conducted by NASA. Intelsat also became an early user of the Ariane launcher developed by the European Space Agency (ESA) and the French national space agency CNES. As NASA gained confidence in the space shuttle in the early 1980s, the U.S. government mandated that U.S. commercial launches be conducted solely by the space shuttle.

In 1982 Hughes Aircraft Company's Space and Communications Group (known in 1990 as Hughes Space and Communications Company) built the Intelsat VI series of spacecraft. The Intelsat VI series encompassed five satellites used to provide domestic and transoceanic voice, television, and data services to Earth stations in 180 countries. Each Intelsat VI spacecraft, with 48 active transponders, 38 at C-band and 10 at Ku-band, had a capacity for three TV channels and 120,000 simultaneous two-way telephone circuits, using digital circuit multiplication equipment. Intelsat VI was the largest and the last series of the Hughes dual-spin-stabilized satellites. Hughes had taken the concept of spin-stabilized spacecraft from the early single spinner of the Syncom and Early Bird designs forward by first introducing the concept of dual-spin-stabilized spacecraft with the Intelsat IV series design and later in the Intelsat IVA and COMSTAR designs of the early 1970s. Hughes increased the power capability of the satellites in the late 1970s by developing an extendable deployable solar array for the HS-376 series of satellite. Intelsat VI incorporated all of these design features, as well as all the state-of-art technologies of the early 1980s. The Intelsat VI was launched as a simple single-spin-stabilized spacecraft as had been all Hughes spacecraft prior to the introduction of the Hughes body-stabilized HS-601 design.

As a result of the space shuttle launch requirement, when Intelsat began procuring launch services for the Intelsat VI generation of large satellites (approximately 4500 kg at time of launch) in 1982, Intelsat specified the Intelsat VI spacecraft to be compatible with launching on either the Ariane 4 launch vehicle or the NASA space shuttle. Hughes proceeded with designing the spacecraft to be compatible with both launch systems. For Ariane launches, the Ariane 4 used the upper stage of the launcher to spin the spacecraft and place the satellite directly into a geostationary transfer orbit. For shuttle launches, the spacecraft included a separate perigee stage to provide the necessary propulsive energy to place the Intelsat VI into GTO using a solid-propellant rocket motor referred to as a PKM. The combination of the satellite and perigee stage was to be mounted horizontally in the shuttle cargo bay, and using a system of latches and springs, the combined payload would be rotated and separated from the shuttle cargo bay via a "frisbee" type of ejection that would provide adequate spin stabilization. Hughes had used a similar type of shuttle separation system for the U.S. Navy LEASAT satellites. Because of spacecraft development problems, no Intelsat VI had been launched by NASA on the space shuttle by January 1986.

On 28 January 1986, during initial ascent, the NASA Space Shuttle Challenger suffered a catastrophic failure resulting in the deaths of the seven-member crew of astronauts and mission specialist. One of the mission specialists that died was a Hughes Aircraft Company engineer, Greg Jarvis, who worked on many programs at Hughes and was well known to Intelsat managers and engineers.

With the failure of the shuttle, satellite operators and NASA faced an enormous problem of how to launch commercial satellites that had been manifested for launch on the NASA shuttle. In mid-1986, the U.S. government and NASA concurred with the commercial space transportation act to permit existing launch vehicle manufacturers to market and sell "commercial launch services." The two most prominent launcher companies, McDonnell Douglas with the Delta launcher and General Dynamics with the Atlas-Centaur launcher, entered the commercial marketplace. However, because of the large mass of the Intelsat VI satellites, no Atlas or Delta launcher configuration was capable of lifting the mass into space. Intelsat was faced with launching all five Intelsat VI spacecraft solely on the Ariane launch vehicle and not utilizing the already manufactured PKMs.

After some delay, in the summer of 1987 Martin Marietta, the manufacturer of the Titan launch vehicle that was primarily used by the U.S. Air Force and for a few NASA missions, decided to develop a commercial Titan launch service offering using the U.S. Air Force Titan 34D launcher. Hughes, which was also actively seeking to have an alternative heavy lift launcher to launch many Hughes satellites planned to be launched on the shuttle, was an immediate supporter of the Martin Marietta venture, agreeing to ensure compatibility to launch two Hughes satellites on the first two commercial Titan launches. On 1 January 1990, the first commercial Titan successfully launched the SkyNet 4A satellite for the U.K.'s Ministry of Defense and the JCSAT 2 commercial communications satellite manufactured by Hughes for the Japanese JSAT customer. Hughes used a perigee stage for the JCSAT satellite that was very similar to that to be used for the next mission, which was the Intelsat VI (F-3) satellite.

INTELSAT VI (F-3) MISSION AND FAILURE RESPONSE

On 14 March 1990, the commercial Titan lifted off at 11:52 GMT from the Cape Canaveral U.S. Air Force Launch Complex LC40 from launch pad LC40. Although the propulsion systems and guidance and control of the stages of the Titan performed properly, the payload composed of the perigee stage and Intelsat VI (F-3) satellite failed to separate because of a wiring error in the stage separation electronics, stranding the combined payload in LEO.

In a purely technical sense, a quick response by the combined Intelsat and Hughes launch mission team saved the Intelsat VI (F-3) satellite by separating the satellite from the perigee stage by using a contingency ground command. The perigee stage with the PKM remained attached to the Titan second stage and reentered the Earth's atmosphere within one day. Within a few days after properly orienting the satellite to maintain an acceptable power and thermal control balances, Intelsat and Hughes began to determine how best to save the spacecraft for a future recovery or repair mission in which a PKM would need to be attached to the satellite either in orbit or back on Earth to get the satellite to the desired GTO and eventually on to the final geosynchronous orbit. To safeguard the Intelsat VI (F-3) satellite's solar-array panels, the orbit of the satellite was raised to a safe and medium-term survivable orbit of 299 by 309 n miles using the onboard propulsion system.

From a much broader perspective, although the decision to separate the Intelsat VI (F-3) satellite from the Titan launcher represented a significant opportunity for a recovery or repair mission, it also represented considerable liability and financial exposure to Intelsat due to insurance coverage ambiguity and recovery costs.

During the launch sequence of events, immediately upon the determination that although the launcher telemetry indicated that the payload separation command had been sent, the satellite was not monitored by the first ground tracking station to be in the predicted orbit; Intelsat in consultation with Hughes began to execute a pre-established contingency plan that had been prepared as part of the mission planning. By following the contingency plan, the technical staff concluded that the payload might still be attached to the Titan upper stage. When the orbital analysts were able to generate a "no separation" orbit and point the ground antennas at that orbit, the Intelsat VI (F-3) satellite's telemetry was observed. With spacecraft telemetry and the ability to measure the current orbital parameters and predict the future orbit, the mission team was able to advise senior managers and executives of Intelsat, Hughes, and Martin Marietta of options available to consider for implementation. These ranged from "do nothing," to defer any decision as long as technically possible, to immediately proceed with a satellite separation and then consider alternatives with the satellite in LEO.

At this point the senior managers and executives of Intelsat including the Director General Dean Burch, Deputy Director General, Engineering & Operations John Hampton, Director of Engineering Pierre Madon, and senior mangers, including D. K. Sachdev, Simon Bennett, and Bill English, had the ultimate "high risk and low or high reward scenario" placed in front of them for consideration.

From a liability perspective and an insurance perspective, the failure while not then definitively determined, seemed to have been a result of a failure or error of either the Martin Marietta Titan launch vehicle or of the Hughes manufactured perigee stage. Although Martin Marietta and Hughes had been successful 75 days earlier with the JCSAT 2 mission, clearly some failure or error had occurred on this mission. By making no decision, the payload and second stage of the Titan would reenter Earth's atmosphere with 24 hours, and any liability as a result of debris hitting the Earth would have been covered by either the U.S. government's third-party liability insurance that exists for all U.S. launches, Intelsat's insurance, and/or insurance and liability financial reserves in place by two very large U.S. aerospace companies; Hughes Aircraft Company and Martin Marietta.

But by Intelsat taking any proactive action such as separating the satellite from the other hardware, Intelsat would radically change the risk profile. First, if the Intelsat VI (F-3) still reentered the Earth's atmosphere, Intelsat would now have sole responsibility and liability. Second, the separation in itself would change the orbital parameters of the Titan stage and remaining perigee stage combination and possibly expose Intelsat to the liability of any damages from the debris for this hardware re-entering the Earth.

On the positive side was the potential of either recovering the Intelsat VI (F-3) satellite or performing an in-orbit repair. Both had been accomplished, in fact by Hughes. Two Hughes smaller HS-376 satellites (Palapa B-2 and the Westar VI) on a common NASA shuttle mission had each experienced the same PKM ignition problem and were left stranded in LEO. NASA and Hughes conducted the NASA shuttle mission STS 51A in April 1985 in which astronauts captured and berthed both satellites in the shuttle orbiter's cargo bay and returned the satellites to Earth. Both spacecraft were refurbished and later successfully launched into orbit. In August 1985 NASA conducted the shuttle STS 51I mission in which an astronaut was able to successfully repair a Hughes U.S. Navy LEASAT satellite stranded in LEO when a sequencer failed to operate properly. Based upon both Hughes and NASA's experience working together and successfully saving what appeared to be lost missions, Intelsat management saw a great possibility of somehow saving the Intelsat VI (F-3) satellite, which was a critical replenishment satellite for the Intelsat international telecommunications network. The remaining unknown question was how much a repair or recovery mission would cost Intelsat, as the viability of proceeding with a possible saving of the Intelsat (F-3) satellite in the end would need to be financially sound.

Intelsat VI launch missions were conducted in the Intelsat Launch Control Center located in the middle of the Intelsat Headquarters in Washington, D.C. In addition to the launch mission team composed of Intelsat technical staff, COMSAT technical personnel, and Hughes engineering staff, the entire Intelsat management team was in close proximity. Intelsat was fortunate to have a seasoned technical management, including Pierre Madon, Simon Bennett, D. K. Sachdev, and the late Alan McCaskill with decades of experience. In addition, Hughes always had senior managers and executives at the customers' control center.

Therefore it was not difficult to convene an executive management meeting in the so-called "prayer room" directly behind the mission director's position in the control center. The Mission Director Erland Magnusson and Deputy Mission Director Leonard Dest manned the director's console receiving periodic reports from spacecraft, orbital, and vehicle dynamics specialists. Magnusson was the relay to the executive conference, updating the attendees on all relevant information. When a reentry calculation predicted reentry within the hour, the mission team commenced to configure the spacecraft to be separated from the perigee stage as per the contingency plan. The command sequence was performed up to the critical commands to initiate the separation. While executives of Intelsat continued to lean toward saving the satellite, with lawyers and insurance risk managers beginning the mandatory briefings on liability and risk, the mood of Intelsat management began to waver, and the possibility of permitting the payload to reenter seemed more likely.

There are conflicting recollections of what happened next, but the commonly held facts were that Magnusson entered the executive conference room, provided the latest information including the prediction of pending reentry within one hour, and that the mission team had configured the spacecraft to be commanded to separate from the perigee stage and only needed the green light to implement the separation. In a rapid sequence, which remains in question, Magnusson advised Dest to proceed and virtually instantaneously the commands were executed, and the Intelsat VI (F-3) was separated from the perigee stage and Titan launcher. Spacecraft telemetry permitted the team to spin the spacecraft for dynamic attitude control stability, perform attitude orientation maneuvers, and begin to establish a safe power and thermal condition for the satellite.

Intelsat management never second guessed the decision, although there are some who believe that the mission management and engineers heard what they wanted to hear and that no decision was ever issued to proceed to separate the satellite. Nonetheless, Intelsat now was in the position to determine the future of the Intelsat VI (F-3) satellite and about to commence a journey of technical, management, and individual challenges and growth never previously experienced by the organization.

MAINTAINING A STRANDED SATELLITE AND PLANNING FOR RECOVERY

For an operational organization, which did limited internal research and development activities, Intelsat now had a huge research and design (R&D)

project on its hands. The space environment a large satellite is exposed to at 200–300 n miles is considerably different than the environment at 23,000 n miles where Intelsat controlled and operated its fleet of satellites. Intelsat had no first-hand experience, and no other commercial operator had the experience. Further Intelsat was not organized to undertake a full repair or rescue program, in addition to the ongoing programs and operations.

Intelsat management immediately made a series of decisions that established an organizational structure and means to work with future team members in a pragmatic and efficient manner. The management of the project was located at Intelsat headquarters to facilitate interaction with management as well as U.S. government agencies, in particular, NASA. Lakhbir (Lakh) Virdee, a senior member of the spacecraft and operations organizations, was given overall program management responsibility. Virdee was able to establish a matrix team of Intelsat specialists, external specialists, and contractors to evaluate how to proceed and then implement a program.

Intelsat immediately began a survey of the industry on how to proceed. Numerous organizations with interest in the success of the Intelsat VI (F-3) satellite, as well as the typical "ambulance chasers" unfortunately always affiliated with a disaster, were quickly approaching Intelsat with schemes to repair or recover the satellite.

Of immediate assistance were specialists from NASA and ESA who had experience operating satellites in LEOs. Although gravitational effects, aerodynamic drag, and other orbital disturbances were fairly well understood by Intelsat from launch-mission planning modeling of the short-term operations in LEO, other environmental issues were lesser well understood.

It was quickly discovered that at altitudes in the 200+ n mile regime satellites encounter the very low-density residual atmosphere composed primarily of oxygen in an atomic state. [On ground, this is encountered predominantly in the molecular (O_2) state, but at the top of the atmosphere solar UV breaks the molecular bonds.] A satellite moves through the atomic oxygen at a velocity of about 7.5 km/s. Although the density of atomic oxygen is relatively low, the flux is high.

The large flux of atomic oxygen, which is in a highly reactive state, can produce serious erosion of surfaces through oxidation. Some surfaces respond differently by changing dramatically their surface structure and therefore properties, which are important for spacecraft thermal control.

The flux of atomic oxygen depends on the atomic-oxygen density, the relative spacecraft velocity, and the orientation of spacecraft surfaces. The spacecraft velocity in such an orbit is about 7.5 km/s relative to the atmosphere. To a first approximation, the recession rate is proportional to the fluence (time-integrated flux). This rate is material dependent. Kapton, used for thermal blankets and other forms of insulation, recedes at about 3 μm per 10**20 atoms/cm², whereas a surface in LEO can easily accumulate 10**21 atoms/cm² in a matter of months. The worst-case fluence is to a ram surface and is greatest at solar maximum when the atmosphere expands.

Clearly, predictions made with solar activity estimates at the two-sigma confidence level lead to pessimistic values, and results will be doubly pessimistic if both atomic-oxygen densities and orbital decay are predicted on the basis of this worst-case solar activity.

Because atomic-oxygen density varies strongly with altitude, strong variations are also expected in atomic-oxygen effects. Anti-sun-pointing surfaces in circular, low-inclination, LEOs accumulate more atomic oxygen than sun-pointing ones because peak atomic-oxygen density is after noon local time.

Intelsat, under the leadership of Andrew Dunnet, began a worldwide survey of how atomic oxygen would damage the Intelsat VI (F-3) satellite until it was repaired or recovered. The space and scientific community came together immediately to begin detailed analyses and tests to determine the viability of the spacecraft to survive. Immediately upon being contacted, experts from NASA, ESA, Los Alamos National Laboratory, and other research institutes agreed to contribute to expert analyses and testing.

It was determined that the Intelsat VI (F-3) spacecraft was designed for service in geosynchronous orbit and contained several materials that were potentially susceptible to attack by atomic oxygen. Analysis showed that direct exposure of the silver interconnects in the satellite photovoltaic array to atomic oxygen in LEO was the key materials issue because the silver is exposed directly to the atomic-oxygen ram flux. Available data on atomic-oxygen degradation of silver were limited and showed high variance, and so solar-array configurations of the Intelsat VI type and individual interconnects were tested in ground-based facilities. The results of ground testing reported atomic-oxygen degradation testing of Intelsat VI type silver foil interconnects both as virgin material and in a configured solar-cell element indicated that more than 80% of the original thickness of silver in the Intelsat VI solar array interconnects should remain if a repair or rescue mission would be completed by 1992.

NASA, which at the same time was evaluating a possible space shuttle rescue mission of the Intelsat VI (F-3), agreed in preparation for a possible rescue that solar arrays, similar to those on the satellite, would be exposed to the conditions of low orbit to determine if they were in any way altered by the atomic oxygen present. The returned arrays would then be closely examined to judge if Intelsat's arrays and its other systems would be seriously damaged by those effects, thereby posing retrieval risks. In October 1990 the Space Shuttle Discovery STS-41 mission included the Intelsat Solar Array Coupon (ISAC) flight experiment. Several materials for which little or no flight data existed were also tested for atomic-oxygen reactivity. Dry lubricants, elastomers, and polymeric and inorganic materials were exposed to an oxygen atom fluence of $1.1 \times 10**20$ atoms cm**2. Many of the samples were selected to support the International Space Station Freedom design and decision making. Upon review of the results of the STS-41 mission, it was concluded that there was a very high probability that the Intelsat VI (F-3) would be in satisfactory condition if the spacecraft were either repaired or rescued by 1992.

With the spacecraft in a relatively safe orbit, with routine attitude control implemented to maintain power and thermal control as well as to minimize atomic-oxygen deterioration to the solar arrays, the planning of a repair or recovery mission was the primary objective to be achieved.

Intelsat, after reviewing various offers and proposals, decided to proceed with the most promising approaches that were developed by Hughes Aircraft Company, which had designed and manufactured the Intelsat VI series as well as which had implemented prior recovery and repair missions working with NASA Johnson Space Center of Houston, Texas. Hughes had a well-deserved reputation of engineering excellence, confidence (some believed it to be arrogance), systems engineering, and programmatic expertise that was the envy of the commercial space industry and in many cases the overall space industry.

Hughes management recognized both the risks and rewards of undertaking any type of effort to safely place the Intelsat VI (F-3) on orbit. Hughes, which had significant technical problems developing the sophisticated Intelsat VI series of satellites and which has lost significant money in the design phase and continued to lose money building the flight spacecraft, had not unexpectedly had a very adversarial relationship with Intelsat primarily because of the launch delays and lack of additional Intelsat funding. Rebuilding the relationship with the management of Intelsat was important to Hughes management. As such, in rapid sequence Hughes management made the decision that if Hughes received full payment for the delivered and launched Intelsat VI (F-3) satellite with no fault placed on Hughes, if Intelsat would pay a reasonable price to develop a repair/recovery mission and if both Hughes and Intelsat could persuade NASA to perform such a mission, Hughes would support the concept.

When the Intelsat VI (F-3)/perigee stage payload failure investigation was completed at Martin Marietta and it was determined that the Titan command wiring used on the Intelsat VI (F-3) mission was wrong and that the onboard sequencer sent the payload separation command to the phantom second spacecraft as it did on the prior flight's dual spacecraft mission, the fault of the Intelsat VI (F-3) separation failure was clearly at Martin Marietta and not at Hughes. Intelsat and its insurers and Martin Marietta now had the issue of claims and reimbursements, freeing Intelsat to make all final payments to Hughes and absolving Hughes of any fault in the failure.

Intelsat had performed a financial analysis and determined a value it was prepared to pay to perform a repair or rescue mission. Negotiations commenced with Hughes to agree to a price to pay for systems engineering, program management, and mission planning in a coordinated effort with NASA.

Hughes assigned a senior program manager and the manger of the LEASAT repair mission, Charles (Chuck) P. Rubin, to be the Intelsat (F-3) program manager. Rubin had the enthusiastic support of the Hughes technical organizations and management, as well as multiple personal contacts into the NASA organization at multiple levels and in particular into the NASA astronaut corps.

NASA in 1990 was a very different NASA than when Hughes had performed the WESTAR/PALAPA and LEASAT recovery and repair missions in 1985. NASA had gone through the excruciating internal review of every process associated with the space shuttle after the Challenger failure. NASA no longer had a mandate to launch commercial space payloads. NASA through the astronauts had established a new regime of safety over mission management's objectives of meeting costs and schedules. The NASA Johnson Space Center (JSC) had new management, now aligned with the NASA Headquarters approaches. But, to its credit, NASA was still a place where the challenges of space were revered, and there were people at every level who believed the successful repair or recovery of a stranded satellite as being the type of task that was part of the DNA of NASA.

HUGHES, INTELSAT, AND NASA PLAN FOR RECOVERY MISSION

Hughes, Intelsat, and NASA held a series of meetings at multiple levels both at NASA Headquarters and NASA JSC. Virdee from Intelsat and Rubin from Hughes led the discussions. These two program mangers determined that a forceful and low key approach was critical to persuading the various NASA decision makers of the value of such of mission to future missions of NASA space shuttle with the International Space Station (ISS). Astronauts needed such practical work tasks to prove processes and procedures and develop tools and techniques. Mission planners needed to conduct shuttle orbiter rendezvous with in-orbit spacecraft. Mission managers needed the real-time interactions of working with outside organizations that were not 100% under the control of NASA. Here was the perfect pathfinder for each of these real requirements facing NASA in the spring of 1990 as it pursued a decade of launching the ISS.

As with most processes, and in particular processes with the government agencies, progress came in sudden unpredictable leaps forward followed by long periods of inactivity. Slowly but surely a consensus was developed within NASA, endorsed by Hughes and Intelsat that a recovery-and-return mission was not practical because of both the mass of the Intelsat VI spacecraft and a combination of safety and lack of provisions to secure such a large mass in the cargo bay for return to Earth. Therefore, by default, a repair mission was the only mission possible. But, unlike the LEASAT mission, no electrical work around repair could make the Intelsat VI (F-3) satellite get on orbit. The repair had to be a repair mission to attach the Intelsat VI (F-3) satellite to a perigee stage that was in the shuttle orbiter cargo bay and then eject the combined payload into a preliminary orbit prior to the PKM ignition and propulsive maneuver to place the satellite into the desired GTO.

Three technical and programmatic problems faced the combined Hughes, Intelsat, and NASA program teams: 1) how to recover the satellite, 2) how to attach the satellite to the perigee stage, and 3) how to organize a shuttle mission to have a perigee stage in the cargo bay of the space shuttle. The last issue was a very large issue. In the spring/summer of 1990, NASA had a full space shuttle schedule for the next few years as NASA had only recently resumed shuttle operations after the Challenger failure investigation and corrective actions were implemented. There were simply no open missions to launch the Intelsat perigee stage. NASA mission management was seeing possible opportunities three to five years in the future, dates that were neither compatible with Intelsat's fleet deployment plans nor compatible with the survivability of the Intelsat VI (F-3) satellite in LEO.

Two additional serious management bottlenecks existed before Intelsat could formally proceed. The first was NASA's initial unwillingness to undertake the rescue. The Intelsat director with responsibility for launch and satellite operations, D. K. Sachdev, and other senior managers made several trips to NASA Headquarters in Washington, D.C., to meet the Director, Space Shuttle Operations, Robert (Bob) l. Crippen. With great difficulty, Sachdev persuaded Crippen to undertake the mission. The first NASA response was that they would charge full cargo bay fee (above $200M) for a launch whenever a mission opportunity opened. This NASA proposal would have never been approved by the Intelsat management and the Intelsat board. After more lobbying by Intelsat, Crippen managed to consider the rescue a training mission and reduced the NASA price to $90M. The Intelsat board supported the Intelsat executive management plan and gave approval for spending nearly $200M, including the cost of the cradle to be designed by Hughes and other costs.

Perhaps it was a "push" solution or perhaps it was a "pull" solution, but NASA proposed that the maiden launch of the new Space Shuttle Endeavour, which had been planned as earlier missions to be purely an instrumented initial mission with no particular mission plan, could be upgraded to be a full training mission with mission objectives and that the Intelsat perigee stage could be launched as a cargo bay test payload. The Space Shuttle Endeavour's mission was targeted for the first half of 1992, well within the targeted period to meet both Intelsat's in-orbit deployment plan as well within the safe period for the spacecraft's survivability in LEO. NASA had made a major commitment to meet Intelsat's requirements, as well as an approach within the budget that Intelsat had allocated for NASA.

With NASA commitment, Intelsat in July 1990 decided to proceed with a full recovery and repair mission, renamed the Intelsat Reboost Mission. Hughes was awarded a $43M contract for special hardware, a new PKM, and mission support, and NASA received a contract for a space shuttle mission.

It was now up to the engineers at Intelsat, Hughes, and NASA to design the hardware, tools, and mission plan to accomplish a space shuttle mission to somehow rendezvous and capture a 4500-kg spinning satellite in LEO, attach it to a perigee stage mounted in the shuttle's cargo bay, and then two years later eject the combined payload into a comparable orbit as the one that the Titan had not properly place it during the launch, which would be no small feat for even this talented team. Intelsat as the end customer took a leading role in the program and worked closely at every step with Hughes and NASA in planning the reboost mission.

Hughes took primary responsibility for the hardware design, systems engineering, and spacecraft maneuvering portion of the rendezvous with the Space Shuttle Endeavour. Hughes established a complete program management team with Chuck Rubin as program manager. Hughes developed a solid systems engineering team. Hughes selected Don Patterson a veteran of many Hughes and NASA missions to coordinate activities between the two organizations. In 1991 Len Dest moved from Intelsat to Hughes and served as deputy program manager and designated deputy mission manger under Intelsat's Erland Magnusson and alongside the Intelsat program manager, Lakh Virdee. Keith Volkert of COMSAT served as Intelsat's residence systems engineer at Hughes.

Hughes was responsible for and built the special hardware for the extravehicular activity and provided a specially designed cradle located in the aft payload bay to support the PKM and associated equipment (Fig. 14.1).

The design approach was driven by simplicity, safety, and success. The concept was straightforward. The satellite was to be recovered by the astronauts using a capture bar that would attached to the satellite's existing separation ring and would also serve as a guide tool to locate the satellite to a satellite to motor adapter (SMA) via a separation ejection mechanism (SEM). The solid-propellant PKM was mounted in a cradle that was attached to the shuttle cargo bay.

The new PKM, built by United Technologies Corporation Chemical Systems Division, was an Orbus 21S solid rocket motor weighing 10,430 kilograms (23,000 pounds). It is called a PKM because the motor boosts the satellite from perigee (or lowest point in its elliptical orbit).

Hughes also designed and built a docking adapter assembly. By design, the astronauts would latch the adapter's manually operated mounting clamps in space. The adapter was to be released by pyrotechnically actuated bolt cutters by command from the ground after the motor is fired.

NASA astronauts visited the Hughes facilities to assist the engineers in the design of the interface equipment to ensure that reliability, safety, and

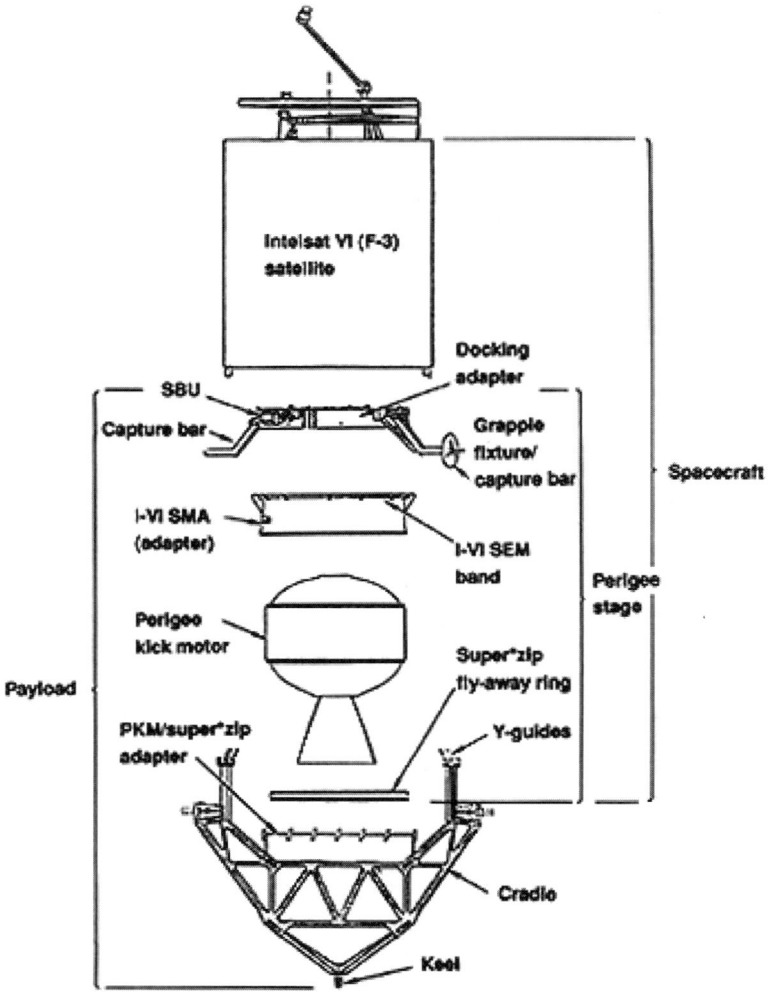

Fig. 14.1   Layout for Intelsat VI F-3 reboost payload.

operational functionality were included in the final flight equipment. Training was conducted underwater to simulate the effects of zero gravity, using NASA's weightlessness environmental training facility (WETF) in Houston, Texas. In providing mission support, Hughes engineers worked 25 feet underwater side by side with the mission astronauts who worked in space to perform the reboost mission (Fig. 14.2). As can be seen in this figure, NASA designed a duplicate of the aft portion of the Intelsat VI spacecraft to be used for training at the NASA JSC facility. NASA also designed a capture bar to be used by the astronauts to get proper hold of the spinning Intelsat VI space-craft as it would approach the cargo bay of the shuttle (Fig. 14.3).

Fig. 14.2 Engineers work underwater alongside reboost mission astronauts. (See also color figure section at the back of the book.)

In early November 1991 the Intelsat VI (F-3) reboost hardware was completed and delivered to the NASA Kennedy Space Center at Cape Canaveral, Florida, for launch site operations that began in late January 1992.

Fig. 14.3 The capture bar is designed by NASA. (See also color figure section at the back of the book.)

The rendezvous of the stranded Intelsat VI F-3 and the Space Shuttle Endeavour would involve intricate coordination of maneuvers of the two spacecraft (Fig. 14.4). By the mission plan immediately after shuttle launch, engineers in the Intelsat Launch Control Center in Washington, D.C., and at stations around the world would begin sending a complex set of commands to the satellite to lower it from its 300-mile orbit. This was the first time that two spacecraft maneuvered simultaneously to achieve a rendezvous. To execute these independent mission orbital operations required a very high level of coordination between the Intelsat and Hughes spacecraft orbital and dynamic analysts and the NASA flight director (Al Pennington) and NASA flight dynamics officers (FDOs) (Mark Haynes, lead FDO). To facilitate this coordination, early in the mission planning phase NASA assigned an individual,

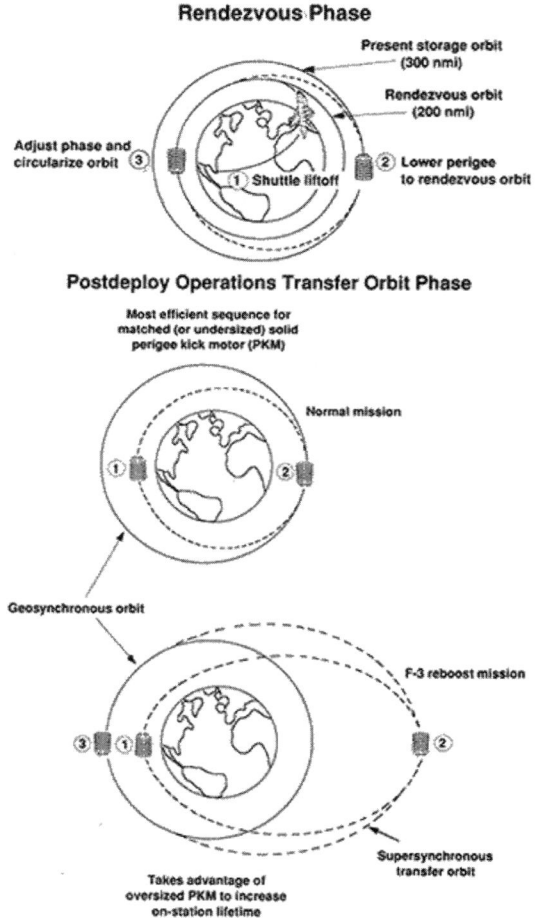

Fig. 14.4   Diagrams show coordination of the spacecraft.

Sally Davis, who had both flight operations and flight dynamics experience. Sally Davis was integrated into the Intelsat and Hughes mission team at the Launch Control Center in Washington, D.C., with both technical and management responsibility to and from the NASA JSC Mission Control Center via the flight director and FDOs. The Intelsat reboost mission was important for Davis as she was preparing to perform the same function with the Russians for the ISS mission activities and later became one of the first Russian interface officers (RIO), who served as the primary interface between the U.S. and Russian control teams. She later became a flight director for ISS missions.

The coordination between the NASA Flight Director Pennington, the Intelsat Mission Director Magnusson, the Intelsat Orbital Director Bill Kinney, the NASA Lead FDO Haynes, the Spacecraft Attitude Control Director Loren Slafer of Hughes, and the Intelsat Mission Planner Jerry Salvatore of Hughes with the NASA Coordinator Davis was a key to mission success.

Both Intelsat and NASA were schooled in the principles of launch rehearsals and contingency planning activity. Numerous "live" rehearsals were conducted for the baseline mission plan as well as for multitudes of contingencies. Repeated practices of the mission plan and overall coordination were continued until all parties were satisfied that the reboost mission team was at near perfection.

Intelsat VI F-3 and Endeavour were planned to rendezvous within a "control box" volume that was defined by NASA approximately five hours after Space Shuttle Endeavour's takeoff. The control box extended over six degrees of arc in a roughly 200-n-mile circular orbit with an inclination of 28.35 degrees. Intelsat VI F-3 had to be in the control box within 46 hours after shuttle liftoff, and Endeavour completed the rendezvous 24 hours later.

About six hours before Endeavour's approach, the satellite was to be spun down via ground commands to its onboard thrusters and attitude control electronics. The spin speed was to be reduced to approximately 0.65 revolutions per minute. After the spin down was completed, the satellite was to be commanded into a safe configuration for rendezvous.

The Space Shuttle Endeavour, which was to be commanded by astronaut Dan Brandenstein, was to approach the satellite on flight day 4. Once the satellite and orbiter rendezvoused, the Endeavour astronauts were to commence an external vehicular activity (EVA) outside the shuttle to capture the satellite, bring it into the shuttle bay, and attach the new PKM.

Approximately 90 minutes before the satellite was to be captured, two astronauts were to enter the payload bay. One astronaut was positioned on the manipulator foot restraint, which was attached to the remote manipulator system (RMS). After the RMS positioned the astronaut near the aft end of the satellite, he was to attach the capture bar to the satellite. Any satellite rotation was to be halted by the astronaut using a "steering wheel" in the center of the capture bar. The astronauts then used the capture bar to dock the satellite to the shuttle. After the satellite was brought into the cargo bay, the two astronauts were to position it diagonally on opposite sides of the cradle. One attached an extension to the opposite end of the capture bar to help position the satellite within special alignment guides on the cradle and the docking adapter. These guides were to place the satellite within range of the docking adapter clamps. The astronauts then were to latch the clamps.

Once the new motor was attached, the crew was scheduled to activate the power and the timer switches on each of the two staging and boost electronic units. The astronauts were expected to complete the tasks in 4.5 to 6 hours.

The crew members then moved into the Space Shuttle Endeavour's airlock prior to spacecraft deployment, which was to occur at a time determined by Intelsat. The Space Shuttle Endeavour crew in the cabin would then fire the Super *Zip to release the cargo bay's hold on the satellite.

Intelsat controllers would establish a command link with the satellite 35 minutes later, after the Intelsat VI (F-3) had reached a safe distance from the shuttle. The satellite's control electronics would then be commanded on, its spin rate increased, and it was to be placed in the proper attitude and prepared for motor ignition. The PKM was planned to be fired using a variable time delay that was commanded by the Intelsat Launch Control Center.

The PKM propulsive energy would thrust the satellite into a supersynchronous elliptical transfer orbit with an apogee (or highest) altitude of about 45,000 n miles (twice that of synchronous orbit). The higher transfer orbit altitude would allow a more efficient use of the PKM and reduce the amount of liquid propellant required to rotate the satellite's orbital inclination toward Earth's equatorial plane. By this means, the useful orbital lifetime of the satellite had been extended approximately 1.2 years to an estimated 10.8 years.

Over a period of several days, the controllers would send a series of commands to the satellite to take it to geostationary orbit. Once geostationary orbit was achieved, the controllers would send another series of commands to deploy the satellite's solar drums and communications antennas.

That was the plan.

What happened was something completely different.

INTELSAT VI (F-3) REBOOST MISSION

The most complex satellite rescue mission ever attempted as of the date of the mission was undertaken with the Intelsat VI F-3 Reboost Mission in May 1992. It involved the first dual-active rendezvous between a satellite and a space shuttle and the first in-orbit attachment of a solid-propellant rocket motor.

STS-49, the first flight of the Space Shuttle Endeavour, lifted off from NASA Kennedy Space Center's launch pad 39B on 7 May 1992 at 23:45 GMT. The primary objective of the STS-49 mission was the capture and redeployment of the 4500-kg Intelsat VI (F-3) satellite.

SPACE SHUTTLE STS-49 ASTRONAUT CREW

The STS-49 crew was composed of Daniel C. Brandenstein, commander; Kevin P. Chilton, pilot; Pierre J. Thuot, mission specialist 1; Kathryn C. Thornton, mission specialist 2; Richard J. Hieb, mission specialist 3; Thomas D. Akers, mission specialist 4; and Bruce E. Melnick, mission specialist 5 (Fig. 14.5).

The capture of the Intelsat VI (F-3) satellite would require three EVAs rather than the one as planned. During the first EVA, astronauts Pierre J. Thuot and Richard J. Hieb were unable to attach the capture bar to the satellite from a position on the shuttle RMS. A second unscheduled EVA was an identical unsuccessful attempt on the following day. Finally, an unscheduled but successful third EVA resulted in the hand capture by Pierre J. Thuot and fellow crewmen Richard J. Hieb, and Thomas D. Akers successfully snared the Intelsat VI satellite as commander Daniel C. Brandenstein delicately maneuvered the Space Shuttle Endeavour to within a few feet of the 4500-kg slowly rotating satellite. An assembly of station by EVA methods (ASEM) structure was erected in the cargo bay by the crew to serve as a platform to aid in the hand capture and subsequent attachment of the capture bar. The three astronauts hand-grabbed the errant satellite, pulled it manually into the cargo bay, and attached the satellite to the perigee stage before its release. Figures 14.6–14.12 summarize some of the key events of the dramatic events that followed.

While the video, first shown only on the NASA closed-circuit television but by days two and three on broadcast television throughout North America, Europe, and elsewhere worldwide, was most dramatic in its viewing of astronaut Pierre Thuot attempting time after time to capture the Intelsat VI (F-3) satellite and eventually the three astronauts literally manhandling the satellite, there was a comparable drama in the NASA Mission Control Center in Houston and the Intelsat Launch Control Center in Washington, D.C. Although the orbital and attitude maneuvers of the Space Shuttle Endeavour were highly complex and precisely executed by the STS-49 Commander Dan Brandenstein and Pilot Kevin Chilton, the maneuvers of the Intelsat VI satellite were significantly more challenging. The large 4500-kg satellite was simply not designed to operate in LEO and not designed to operate when the

Akers   Brandenstein   Chilton   Hieb   Melnick   Thornton   Thuot

Fig. 14.5   The STS-49 crew. (See also color figure section at the back of the book.)

Fig. 14.6   Intelsat VI (F-3) satellite is shown prior to rendezvous with Space Shuttle Endeavour. (See also color figure section at the back of the book.)

Fig. 14.7   Mission Specialist Thuot positions the capture bar at the aft end of Intelsat VI F-3.

Fig. 14.8   EVA Mission Specialist Thuot standing on the RMS end effector with the Intelsat capture bar. (See also color figure section at the back of the book.)

Fig. 14.9   STS-49 crew captures Intelsat VI above OV-105's payload bay during EVA. (See also color figure section at the back of the book.)

Fig. 14.10   Three mission specialists perform a manual capture. (See also color figure section at the back of the book.)

satellite was not spinning like a top at 50 revolutions per minute. As a virtually nonspinning mass, with limited dynamic control and limited propulsive control, maintaining any form of stability was critical. And, as a top, the physical mass properties of the satellite were such that the vehicle had a natural tendency to want to rotate about its principle axes, the so-called "flat

Fig. 14.11   Intelsat VI F-3 separation and ejection from the STS cargo bay. (See also color figure section at the back of the book.)

Fig. 14.12   Intelsat VI F-3 drifts in space after servicing. (See also color figure section at the back of the book.)

spin" state. In fact, after a number of the unsuccessful attempts by astronaut Thuot, Intelsat VI (F-3) was on the verge of going into a flat spin. Engineers who had only an intellectual concept of what a "flat spin" would be were watching live video of it on their monitors. The dynamics knowledge and execution of corrective actions via spacecraft commanding by the Intelsat attitude control team headed by Hughes' Loren Slafer were beyond comprehension. Time and time again control of the satellite was achieved, attitude reorientation was performed, and the process was repeated again—without success. The efforts of Slafer, Kinney, and Salvatore to maintain satellite dynamic control, attitude control, and orbital control in close proximity to the astronauts and the Space Shuttle Endeavour were a testament to engineering and physics intellectual knowledge and the confidence of experts.

And it was also a testament to the wisdom of senior mangers at NASA, Hughes, and Intelsat that they continued to proceed with this dangerous and risky mission through to a successful conclusion. There were abundant opportunities for the NASA astronaut crew, the Intelsat and Hughes engineers, and the managers to give up and accept defeat. But it was the fortitude of convictions, the determination to succeed, and the willingness to accept risk regardless of the reward that propelled a small group of space and satellite veterans to accomplish one of the great successes in the history of space—the successful capture and reboost of a satellite that would serve worldwide communications for over a decade.

The following records were set during the STS-49 mission: first EVA involving three astronauts; first and second longest EVA to date: 8 hours and 29 minutes and 7 hours and 45 minutes; first shuttle mission to feature four EVAs; EVA time for a single shuttle mission: 25 hours and 27 minutes, or 59:23 person-hours; and first shuttle mission requiring three rendezvous with an orbiting spacecraft attaching a live rocket motor to an orbiting satellite.

CONCLUSION

NASA would go forward to pursue the future via the International Space Station and suffer the catastrophic loss of Space Shuttle Columbia and the resulting second assessment of the space shuttle program resulting in the plan to end space shuttle flights in 2010.

Martin Marietta would merge with Lockheed in 1995 to become the world's largest defense contractor, Lockheed Martin. The Titan launch vehicle would never serve the commercial marketplace, but rather become the primary heavy lift booster for the U.S. Air Force until it was retired in 2006. The Titan was replaced by the U.S. Air Force evolved expendable launch vehicle (EELV), the Lockheed Martin Atlas V using a Russian RD-180 booster engine, and the Boeing Delta 4 launcher. In late 2006, Lockheed Martin and Boeing merged their EELV launch vehicle engineering, manufacturing, and operations into a new company, United Launch Alliance (ULA). As such, both the Atlas and Delta launchers, which were at one time the dominate launch vehicles, are effectively removed from the commercial marketplace, 20 years after being reintroduced in 1986.

Hughes would be successful in commercial space for another five years before being sold in pieces to Boeing and Raytheon, with commercial satellites taking a smaller and smaller part of the resulting space activities.

Intelsat would transform into a private company and then merge with its primary competitor, PanAmSat, to again become the world's largest commercial satellite operator.

The members of the NASA, Hughes, and Intelsat Reboost Mission team would populate the space and high-technology industries for the next decade and beyond. But for virtually everyone associated with the Reboost Mission, that period of time and what they and their peers accomplished will remain imprinted in their brains and souls as one of the greatest events in their professional career.

BIBLIOGRAPHY

Dest, L., Bouchez, J-P., Serafini, V., Schavietello, M., and Volkert, K., "Intelsat VI–Spacecraft Bus Design," *COMSAT Technical Review—Special Issue on Intelsat VI*, Spring 1991.

Dick, S.J., and Garber, S., *A Chronology of Defining Events in NASA History, 1958–1998*, Feb. 2005.

Donatelli, P.A., Purohit, G.P., and Debolt, R.D., "Intelsat VI Reboost Mission Perigee Kick Motor," AIAA Paper 93-2516, June 1993.

"ESA Space Environments & Effects—Atomic Oxygen Effects & Analysis," ESA 2003.

"Intelsat VI F-3—Major Points in the Mission," Boeing Integrated Defense Systems, Los Angles, CA, SCG 922059/6000/4-92, April 1992.

Koontz, S. L., Cross, J. B., Hoffbauer, M. A., and Kirkendahl, T. D., "Atomic Oxygen Degradation of Intelsat VI-Type Solar Array Interconnects: Laboratory Investigations," NASA Technical Memorandum 102175, March 1991.

Koontz, S.L., King, G., Dunnet, A., Kirkendahl, T., Linton, R., and Vaughn, J., "Intelsat Solar Array Coupon Atomic Oxygen Flight Experiment," *Journal of Spacecraft and Rockets*, Vol. 31, No. 3, 1994, pp. 475–481.

NASA Report, Media Resource Kit, available online at http://www.jsc.nasa.gov/history/shuttle_pk/mrk/FLIGHT_047-STS-049_MRK.pdf.

NASA Report, Press Kit, available online at http://www.jsc.nasa.gov/history/shuttle_pk/pk/Flight_047_STS-049_Press_Kit.pdf.

Rees, D., "COSPAR International Reference Atmosphere: 1986, Part I: Thermosphere Models," *Advances in Space Research*, Vol. 8, No. 5–6, 1988.

Virdee, L., Dest, L., Rubin, C., Davis, S., Haynes, M., Seaman, C., Huning, S. W., Roberts, W., and Wardlaw, R. "Intelsat VI (F-3) Reboost Mission," AIAA Paper 92-1951, March 1992.

Chapter 15

EARTH STATIONS: FROM VERY BIG TO VERY SMALL

MARK DANKBERG*

As the satellite systems have moved closer to the ultimate user, the importance and design vectors for the ground segment have changed dramatically. This is equally true for both commercial and government applications. One relatively young company that has assumed an impressive share of this industrial transition is ViaSat. The founder and head of this company tells the whole story of the evolution of the ground segment for satellite systems.

Satellites and ground stations have been used to transmit phone calls, television, and other information for only about four decades. During that time, the ground terminals have undergone an amazing transformation. The first ones were enormous structures requiring specialized engineers to install and operate and could only communicate a handful of TV channels or telephone calls to national telephone companies or broadcasters. Today, a 60-cm home television dish in many countries can deliver literally hundreds of high-quality TV channels, including some with high definition and surround sound. Tens of millions of people around the world receive satellite "free to air" or with monthly fees comparable to cable systems. Those terminals are so inexpensive and simple they can be carried home in a box and might even be self-installed. Similarly, in North America it is possible to get 128 by 512 kbps two-way home broadband Internet services via satellite through antennas as small as 65 cm, with monthly fees as low as $49. And, of course broadband Internet carries myriad forms of information including streaming video and voice-over-IP (VoIP) telephone calls.

An equally dramatic transformation has occurred with mobile satellite terminals. It would have been hard to even relate the concept of "mobility" with the first satellite Earth stations. Yet today, there are handheld satellite terminals providing voice and low-speed data services at prices not much higher than terrestrial cellular services were just a few years ago. Handheld versions of digital satellite radios appeared on the market in 2005, offering hundreds of channels of high-quality music, sports, news, and other audio entertainment. A few handheld satellite video channels are available as well. Meanwhile, digital satellite radios are factory options in dozens of automobile models and in some markets include video, too. Satellite voice and data terminals have been common on commercial flights and business jets, as well as thousands of maritime vessels for many years. More recent developments have increased data rates to multimegabit speeds and reduced air time prices to be competitive with other forms of Wi-Fi hotspots.

The transformation of satellite communications in defense and military applications has also been profound. In the 1970s and 1980s military systems pioneered the creation of transportable, airborne, and shipboard satellite terminals. In most cases commercial market forces have driven terminal sizes, costs, and capabilities to be better than their military counterparts. So, in many cases military organizations have focused their own "organic" capabilities on security and robustness while also becoming a major consumer of commercial services around the globe. The amount of satellite bandwidth used by military forces has increased by orders of magnitude in the last decade.

Overall, this remarkable transformation of satellite ground terminals has been driven by multiple factors including ground technology, space technology, capital markets, regulatory policies, international treaty organizations, standards bodies, military strategies and tactics, and the twists and turns of dynamic, competitive, telecommunications, and media markets. The results have been orders of magnitude benefits in size, weight, Earth station costs, service quality and quantities, and operating fees.

But, ultimately, it seems that the single most important factor in driving ground terminal improvements is *market demand.* Although technology investments, standards, capital investments, and all of the other factors have certainly influenced the outcome, the history of satellite services shows that tapping into underlying user market needs is what drives progress. That turns out to be true for different market segments, different satellite systems, and different geographic regions.

It is a difficult challenge to try to organize the story behind these improvements in any kind of linear way; chronological order probably is not the most meaningful approach. So, instead, we will consider a number of the major technology factors that are critical because they *enabled* progress when the supply of satellite services and products became aligned with demand.

Then a number of specific satellite communications market segments are discussed to illustrate economic, market, and policy factors that have been the most significant. The examples are intended to show how these factors have combined to yield some of the most interesting and compelling advances in satellite systems and applications.

TECHNOLOGY FACTORS

In the end, technology is *the* enabling factor in driving down size, weight, complexity, and costs while also driving up performance, functionality, and mobility. But, it is also important to note that it is not always *satellite communications* technology that is the most decisive. Satellite systems have in many ways exploited mainstream technologies associated with other domains, often benefiting more than corresponding terrestrial transmission systems.

Because technology advancements represent the foundation of progress across all of the systems, we will consider that first. Then, we can use that as a reference point in considering the different market applications.

SATELLITE-CENTRIC TECHNOLOGIES

There have been many, many advances in satellite communications technologies in the past four decades. The focus in this chapter is on factors that have contributed to improvements in Earth stations, but that perspective must also take into account the way that spacecraft and payload technologies have paved the way for smaller, simpler, and cheaper ground equipment.

SPACE TECHNOLOGIES.   Fundamentally, satellite communications is about closing microwave communications links over distances ranging from hundreds of miles [for low Earth orbit (LEO) satellites] to tens of thousands of miles [geostationary orbit (GEO) and highly elliptical orbit (HEO)]. Most directly, the satellite end of the link benefits from high-power amplifiers, low noise receivers and frequency converters, and high-gain antennas. Indirectly, any technology that makes more mass, volume, surface area, and dc power available to those payload elements that actually deliver effective isotropic radiated power (EIRP) and gain to noise temperature (G/T) to the communication links also helps. In one sense, almost the entire repertoire of satellite technology has been brought to bear including structures and materials, propulsion systems, thermal management, solar arrays, and others outside the scope of this chapter. For instance, attitude control and stabilization are important because they allow high-gain antennas to be aimed accurately and consistently at the desired spots on the ground, minimize fuel consumption, and make more resources available to the payload.

The combination of satellite traveling wave tube amplifiers (TWTAs) and high antenna gains now easily enables downlink power flux densities that would exceed regulatory limits. (This has been true for quite a number of years.) This means that improvements in satellite terminals have for some time been driven not by sheer downlink RF power and signal-to-noise ratio considerations, but by using that power in innovative ways to meet interesting market opportunities. For two-way systems that need low-cost, simple ground stations, the uplink budget is often a driving factor. In this case, probably the single biggest technical factor onboard the satellite is the uplink receive antenna gain, which is primarily driven by the aperture size. From that perspective some of the most important satellite technologies include multiple feed antennas, antenna packaging and deployment mechanics, and platform stabilization.

Although some very challenging space technologies have been demonstrated and implemented, they do not always align with specific market demands. For instance, onboard processing (OBP) including satellite modulation/ demodulation, coding, and signal routing combined with TDMA, spatial "beam hopping," was one of the most innovative and challenging aspects of the 1980s-era NASA Advanced Communication Technology Satellite (ACTS). Although the technologies were intended to improve utility and simplify Earth stations, and they were shown to perform as intended, they have not yet been used by any of the commercial systems that have subsequently driven Earth station economics. For instance, the first two Spaceway satellites, which do include OBP and routing, have initially been deployed for bent-pipe video transmission—turning off and bypassing all of the onboard digital functions. A third Spaceway satellite was launched in 2007 and is just now entering data service and will test the market value of OBP. On the other hand, the multibeam spot beam Ka-band RF technology, which was also part of the ACTS system, has turned out to have multiple applications in both data and video services. The U.S. military's MILSTAR, Advanced EHF (A-EHF), and planned TSAT systems prominently feature digital onboard processing technologies including onboard modulation, demodulation, switching/routing, beam-hopping, beam steering, and cross-banding. To the extent that "single hop" satellite links directly connecting small terminal to small terminal is important (for instance, in the military or enterprise markets), digital OBP will likely remain an important enabling technology.

A closely related technology is onboard digital band-pass filtering/processing. This has primarily been applied to mobile satellite systems such as Thuraya and Inmarsat-4, which operate in congested mobile frequencies. Although not involving explicit onboard modulation and demodulation, the filtering does use digital signal processing components designed and refined to operate in space. The underlying technology is a close cousin of the form of OBP demonstrated in ACTS.

Finally, another technically challenging space technology is electronic beam-forming. Multibeam satellites have already shown their value in simplifying and reducing Earth station costs and/or enhancing capacity or performance for mobile, video, and broadband data services. Electronic beam forming has been implemented in the mobile band (for instance, Thuraya and ICO) and at Ka-band (for instance, Spaceway). Electronic beam forming also involves substantial amounts of onboard digital processing, relying on many of the digital signal processing technologies as in full digital OBP and the band-pass processing in advanced mobile systems. Recently, a number of very advanced mobile satellite systems have begun construction using ground-based beam forming (GBBF). This approach retains the use of electronically steered phased arrays onboard the satellite, but moves all of the beam-forming calculations to the ground segment via an innovative system of feeder links and calibration techniques. GBBF promises substantial advances in frequency reuse, antenna gain, even more flexible real-time adaptive ground coverage patterns, adaptive interference cancellation, and other potential benefits that could decrease user terminal size and cost while improving performance. ICO's GEO S-band mobile satellite service (MSS) with GBBF was launched in early 2008, but has not yet entered commercial service.

GROUND TECHNOLOGIES. It is difficult to consider "ground" technologies separately from "system" factors. In this context, system refers to how the communications link (and/or network) is designed to accomplish its mission. System design factors probably account for the vast majority of link and network improvements that have enabled Earth stations to be smaller, simpler, and cheaper. Often, a system design approach exposes some specific aspect of terminal design that can be optimized and simplified in the context of that system.

SYSTEM TECHNOLOGIES

ASYMMETRIC ARCHITECTURES. Sometimes taken for granted, the advent of asymmetric network architectures was enormously important in opening the door to smaller, cheaper Earth stations. Most of the earliest satellite ground terminals were designed for symmetric point-to-point connections. For instance, long-distance telephone circuits were by nature symmetric links. Station A had to transmit and receive equals volumes of information to Station B. Often the same Earth stations used to transmit transoceanic television signals were used to receive broadcasts from the other end of the link.

The concept of asymmetric hub and spoke architectures was a fundamental change. Basically hub and spoke concepts build on two insights:

1) There are many applications where the flow of information is highly (or completely) asymmetric. Obvious examples are video and audio entertainment, but there are many others including data transaction processing and broadband Internet access. Most of the highest value satellite applications are inherently asymmetric.

2) Complexity in asymmetric applications can be transferred away from the more numerous subscriber (spoke) Earth stations and into the small number of hubs and satellites. It usually makes economic sense to allow the satellites and hub Earth stations to be more complex when that added complexity reduces the size or cost of the spokes (because the ratio of spokes to hubs and

satellites can range from hundreds to literally millions to one). This basic insight has underpinned advances ranging from the first business very small aperture terminal (VSAT) terminals, to cable head-end receive-only terminals, to direct broadcast satellite (DBS), to satellite radio.

In some cases, system designers have had to go to extreme lengths to accomplish the desired asymmetries because technical factors were inconsistent with prior uses of satellite systems. For instance, DBS satellite depended on opening new frequency bands in order to obtain its intended benefits.

DIGITAL COMMUNICATIONS TECHNOLOGY ADVANCES. The other critical factor that drove system performance and enabled enormous simplifications in Earth terminals has been advances in digital communications technology. Key factors here include 1) the use of digital modulation and source coding techniques vs analog, 2) steady advances in channel coding that reduced the required E_b/N_o (and associated C/N_o or S/N), 3) improvements in modem design that provided both theoretical gains (e.g., more effective signal sets or demodulation concepts) and lower implementation losses, and 4) source coding (compression) techniques that required fewer bits to represent audio, image, or digital data within acceptable distortion limits.

As an example, the difference in SNR needed to transmit a single voice conversation between early analog FM with discriminator detection vs using a 4-kbps vocoder with a modern iterative decoder can exceed 100:1.

Many of these techniques are reaching known theoretical limits. For instance, modern modems are very close to Shannon's information theoretic limits for capacity in AWGN channels. Voice codecs appear to be close to information theoretic limits when taking into account subjective factors such as speaker recognition, tonal inflections, and ability to tolerate ambient acoustic noise. There do appear still to be opportunities in video source coding, especially taking into account current interest in both very high-definition (HD) and low-resolution (mobile) markets. Also, there has been renewed interest in "rate distortion theory," which effectively combines both source and channel coding to optimize link performance.

NETWORKING TECHNOLOGIES

It can be difficult to directly associate networking technologies with terminal size and cost reductions, but the effects are there nonetheless. In this context, networking technologies are considered to include multiplexing, multiple access techniques, switching and routing, and security. In some cases, these technologies effectively reduce the number of bits that Earth stations must process to perform their mission, which can directly or indirectly reduce link budgets and the associated implementation costs. Multiple access techniques can also significantly reduce satellite bandwidth requirements to achieve a specified quality of service, dramatically improving economics, and driving market demands that reduce manufacturing costs. In other cases network features enable capabilities or business models that were critical to motivate the creation of the class of Earth station products at all.

For instance, one of the most important enabling technologies is "conditional access," a security concept that enables "pay-for-use" business models that are the foundation of most satellite video and audio broadcast systems. Satellite TV drove the use of digital conditional access for mass markets, which enabled operators to charge for TV reception and more consistently reduce "piracy."

Another significant networking and multiplexing technology was the creation of the DVB-S (digital video broadcast - satellite) standard for digital television transmission. The DVB-S standard, combined with a rapidly growing home satellite TV market, led to highly integrated, low-cost integrated circuits that substantially reduced the size and cost of set top box receivers.

SATELLITE GROUND TERMINAL TECHNOLOGIES

Certainly there have been major strides in cost, integration, and performance of the actual ground components and subsystems. Some of the most obvious advances over the years have included 1) greater integration of digital components, especially modems, channel and source codecs, multiplexers, and switches/routers; 2) higher level of integration of microwave RF components, especially low-noise block down-converters (LNB), and for two-way systems, block-up converters (BUCs); and 3) higher power and higher frequency satellites enabled inexpensive, small antennas that could be made of stamped metal or formed from composites.

Although such technologies are undoubtedly associated with smaller, less expensive terminals, it is not clear whether the availability of such technology actually *caused* that transformation or whether they resulted because of transforming events in the markets for satellite terminals. This somewhat provocative perspective of causes and effects will be considered in the following section on the market factors that accompanied the evolution of simple, low-cost ground terminals.

MARKET FACTORS

It is fascinating to consider satellite communications in the overall information and media technology marketplace. There has been a truly symbiotic relationship. Satellite communications has absolutely and fundamentally changed business and consumer markets, but also paradoxically has depended on unrelated advances in those markets for its own growth opportunities.

One of the most basic questions that the satellite industry confronts is separating the causes and effects of its own successes and failures. Satellites offer a powerful, unique communications capability, and operators are constantly seeking new applications that drive demands for service (especially transponder leases). There is a perspective that says "if only ground terminals were this small, or that cheap, then we could tap into these new growth markets." Certainly, there is lots of empirical evidence that satellite terminals have gotten smaller and cheaper, and there certainly are more in use now than there were 20 years ago. But, have lower size and costs caused new markets? Or, have new markets caused smaller size and lower costs?

Let us consider some critical new satellite terminal markets, some highly successful, and others spectacular failures, to examine that question.

TELEPHONY SYSTEMS

Although telephony applications are currently waning, this was one of the very first productive markets for satellite communications. Satellites made practical in the 1970s and early 1980s what otherwise was almost impossible: direct point-to-point long-distance connections between cities that were thousands of miles apart and separated by continents or oceans. Costs and design and operational complexities were so daunting—as were regulatory issues—that a dedicated intergovernmental organization (Intelsat) was needed to make the satellite solution possible. Intelsat, and its U.S. affiliate, Comsat, along with its technology arm, Comsat Labs, were responsible for two very important leaps in establishing satellite terminals and paving the way for many, many future systems.

One big accomplishment was in creating the original Intelsat Standard A Earth stations (Fig. 15.1). Although the Standard A was not small or simple and was *very* expensive, it did *standardize* and reduce to engineering practice

Fig. 15.1 Intelsat Standard A Earth station (courtesy of Globecomm, Inc.). (See also color figure section at the back of the book.)

what was otherwise a very customized, "science project" effort—the design of a useful satellite Earth station that could reliably communicate with a satellite and a similar Earth station at the other end of the link. Because pair-wise long-distance communication among distant points was so expensive at the time, the Standard A was an economical solution, and it represented a major step forward in risk reduction and convenience compared to the prior state of the art.

Subsequently, given the underlying demand for long-distance communication, terrestrial costs began declining steadily, and the Standard A became impractical for many point-to-point applications. At this point, the emphasis evolved from *being able to set up links at all* towards the need to drive capital and operational costs down to more favorably compete with emerging terrestrial alternatives. Intelsat responded with systems designed to support smaller and smaller (and less expensive) Earth stations, such as Standards B and C, and then added digital interoperable service standards such as international business service/international data rate (IBS/IDR). Although all of these systems might appear expensive compared to some of the consumer services now available, they were very important evolutions in the satellite communications markets of the 1980s and early 1990s. These systems helped spur many of the space and ground technologies, including the evolution from analog to digital, that were employed so productively in the examples to follow.

One of the important consequences of the standardization among the various Earth station terminals and services was the creation of entire classes of ground system component and subsystem technology companies. Earth terminals, especially the smaller ones, went from being designed, built, and integrated as a single *whole* to being constructed of collections of standardized components such as high-powered amplifiers (HPAs), up and down converters, frequency references, modems, multiplexers, redundancy switches, low-noise amplifiers or low-noise block downconverters (LNAs or LNBs), power supplies, and many others. The standardized Earth station market spawned many of the industries and companies that were essential to the technology innovations needed for future markets. Declining costs also showed that satellite markets had some elasticity. As costs and complexity declined, the underlying addressable markets expanded, improving production learning curves and creating economies of scale. Intelsat Earth stations were often the very first satellite terminals every deployed in many, many countries and paved the way for regulatory and policy conventions and treaties that were absolutely essential to virtually all modern satellite applications.

Ultimately technology and cost reductions that were pioneered by Intelsat in the international, cross-border, marketplace reached price and performance levels that enabled domestic, mesh (any-to-any, peer-to-peer switched satellite services), which was an important satellite communications market in the 1990s. The cost and size reductions that occurred in the point-to-point telephony market were quite impressive, in context. Earth station sizes shrank by about 1 order of magnitude and costs by 2 orders of magnitude. Ultimately though, terrestrial transmission and switching costs, driven by fiber optics, declined by even more orders of magnitude, and the market for point-to-point satellite telephony has been declining for the last decade.

CABLE-TELEVISION DISTRIBUTION

The U.S. cable-television industry helped create much of the framework for the advancement of the satellite television market. Initially the cable-television industry had a single value proposition: it would use coaxial cable to supply otherwise free VHF and UHF over-the-air television broadcasts to rural markets that were beyond the reach of network and independent broadcast towers. But, then, three discrete changes in the market unleashed a sequence of events that first drove the proliferation of cable applications of satellite and ultimately demonstrated the market for direct broadcast satellite to the home.

The first leap is attributed to Ted Turner, who grasped the potential for satellite to transform an unknown Atlanta independent TV station (WTBS) into a national media outlet. In 1976, using a broadcast dish from Scientific Atlanta, Turner made WTBS available nationwide on the Satcom-1 satellite, albeit to only about four different local cable distributors. It was not the first application of satellite for program distribution; HBO began distributing its pay TV service via satellite in 1975. But, nevertheless it was the first channel in what would eventually become known as "basic cable." Many more such channels followed, such as CNN, ESPN, MTV, and a host of others. The cable distribution system essentially created the market for the video contribution uplink terminal, creating the same sort of standardization and economies of scale that were so essential in the voice and data markets.

The second leap was really at the receiving end and had a symbiotic relationship with the uplink terminals. The existence of the new satellite-only broadcasters created a dramatically different value proposition for the fledgling cable industry. Instead of merely just retransmitting otherwise free broadcast stations, a cable system that added a satellite receive-only dish suddenly could become a source of programming and content that was not available anyplace else. It was exclusive to the cable industry. It created an amazing inversion in the cable industry. Now, instead of focusing on rural markets outside the reach of terrestrial broadcast, cable systems flourished in urban markets that were hungry for ever greater diversity in program content, had the disposable income to pay, and sported the demographics that commanded advertising dollars. The market for receive-only satellite terminals created something of a miniboom and again led to standards and industry participants that paved the way to the future. But, it was the third step that probably had the greatest impact.

By the early 1980s there was a lot of domestic TV programming being distributed by satellite. This included the major network program feeds, contribution
uplinks from sporting and special events around the country and around the
world, and, of course, all the new basic cable channels, as well as a few premium
pay-per-view services. It became evident that anyone with some know-how, a
C-band dish, and a motorized antenna steering mechanism could watch virtually
anything available. Entrepreneurs quickly went to work, driving down costs and
improving convenience with standardized kits of components, and published
programming guides (complete with orbital slot, transponder frequency, and
polarization!). If the cable receive-only terminal was a mini-boom, the backyard
C-band business was a bonanza, with over a million terminals deployed in the
United States alone (as compared to thousands for the cable industry). Prices
plummeted by at least an order of magnitude, and convenience improved substantially. C-band dish systems went from being the domain of engineers to an
off-the-shelf item at discount chains such as Costco.

Ironically, it was the very success of the backyard C-band business that led
to its demise. The key insight was to contemplate that if over a million people
would go to that much trouble to put up an 8-foot antenna and struggle with
complex program guides (not to mention ever-increasing antipiracy efforts)
*then imagine how big the market would be if we made this really cheap, easy,
and convenient.* Although that insight might seem obvious in retrospect,
many leading players in the satellite and entertainment businesses of the
period balked at the prospect of launching a purpose-built direct-to-home
space system. Still the cable industry and the backyard dish business was
probably an essential predecessor of the DBS business. It demonstrated
demand, fostered consumer-priced technology, educated consumers, pointed
the way to viable distribution channels, helped promote the creation of the
basic cable channels that formed the foundation of DBS content, and helped
sustain the companies that participated in creating DBS components.

ENTERPRISE VSAT

The VSAT industry has been one of the first and leading contributors to the
size and cost transformation of two-way satellite terminals. Corporate private
network communications represent the largest portion of the VSAT market.
Applications include point-of-sale transaction processing at retail chains, and
gas stations, as well as automated teller machines (ATMs) and lottery terminals.

Early VSAT terminals were much larger and more expensive than today's
versions. In the 1980s a VSAT could involve a 2.4- or 3.7-meter (or even larger)
antenna and could easily cost $20,000 (compared to current versions that often
use 1.2 meter or smaller dishes and cost closer to $1000.) Yet, some companies,
Wal-Mart being among the most notable, determined that superior information
technology would yield commanding and enduring competitive advantages for

their core business and that VSAT was the only practical way to connect their many geographically diverse operations in a consistent and reliable manner. It is not always easy to distinguish cause from effect, but there is a good argument that VSAT terminals became smaller and cheaper because advances in enterprise information technology created value out of a form of communications that only satellite could provide reliably at that time. Once competing businesses in those (and other) vertical industry groups associated specific competitive advantages with the information technology applications enabled by VSAT, they slowly and steadily adopted VSAT networks of their own. For instance, it is very rare in the United States now to find a gas station that does *not* use VSAT for pay-at-the-pump credit card authorizations. That steady growth in VSAT demand drew multiple entrants to the industry, increased sources of supply for components of all types, and drove costs and prices down—in the range of 20% per year for about the past two decades!

DIRECT BROADCAST SATELLITE TV

Let us quickly review the environment that led to DBS systems in the United States. Cable TV initially drove satellite usage to uplink and distribute programming like ESPN, CNN, and MTV. Channel choices blossomed from a handful in even the largest media markets to dozens almost everywhere. Soon entrepreneurs realized individuals could buy their own "backyard" C-band terminals, which spawned the direct-to-home industry with well over one million users. But terminals were still in the $2000 range, and operation was very cumbersome. Ultimately, savvy corporations and entrepreneurs realized that if so many people were willing to go to such great lengths to have satellite TV then there *must* be a huge market for a service that was actually small, convenient, and competitively priced. But, there was a big problem. There was no physical way to make home satellite terminals small enough, convenient enough, and cheap enough to tap into that demand *using existing satellites.* What was needed was a new type of DBS satellite with higher frequencies, more EIRP, and wider orbital spacing than any existing satellite could offer. What was needed was a *purpose-built* DBS satellite—a daunting proposition. If the market did not materialize, the DBS satellites would not be sufficiently useful for more conventional applications to recover the investments needed to create it.

Nevertheless, DBS TV was introduced to the United States in 1994 by DirecTV (Fig. 15.2) and quickly became the fastest selling consumer electronics item ever up to that time – even surpassing CD players. Amazingly, DBS carried almost every available programming type *except the local broadcast signals, which had so shortly before been the very definition of television entertainment.* DirecTV did carry national network channels, but no local programming—and still tapped into enormous market demand. Although DBS was originally aimed at rural customers who could not receive cable or broadcast

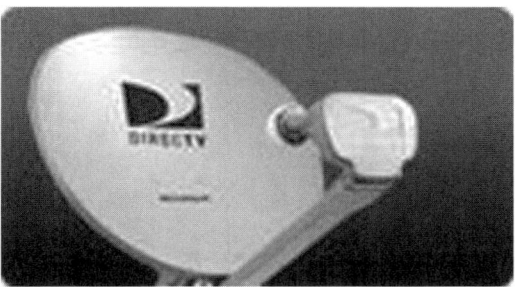

Fig. 15.2   A DirecTV user terminal.

TV, it quickly spread to suburban environments where many customers preferred it over terrestrial cable. DBS was priced a little lower than cable, but also had another competitive advantage in that it used digital transmission. For a variety of economic reasons, cable kept their "basic tier" programming on analog, which resulted in noticeably inferior quality, and then imposed a premium fee for "digital cable." The cable industry argued that satellite TV was inferior because it lacked local channels, entitling them to a premium for their own digital service. Satellite operators recognized that lack of local channels was a constraint on their market and responded by launching more satellites and progressively adding local channels to more than half the market areas in the country. Ultimately, the analog tier decision has likely been an important factor in allowing satellite TV to reach over 25% of U.S. households and causing the first ever declines in terrestrial cable market share.

Again, it is a little tricky to sort out cause from effect. Did economies of scale and low costs for satellite terminals lead to market penetration and greater growth? Or, did a confluence of market factors, competitive decisions, and underutilized terrestrial technologies create so much market demand that production volumes soared, driving costs down? There is a good argument that the latter is closer to the truth in the United States. Cable TV had transformed the definition of pay TV service, established demand for and created programming beyond the broadcast networks, created low-cost TV receive-only (TVRO) terminals, pioneered conditional access, and demonstrated enormous pent-up demand (Fig. 15.3). That created the opportunity for well-conceived, effectively promoted DBS services to exploit the market, and ride the learning curve to lower and lower costs and higher levels of integration.

MOBILE SATELLITE SERVICES

In contrast, consider the handheld mobile satellite services sector. The two LEO systems that did come to market, Iridium and Globalstar, were each enormous technical accomplishments—representing over $10B in

Fig. 15.3   A DISH user terminal.

investments combined. At least $1B more went into other LEO or medium Earth orbit (MEO) systems that never made it to market. Success in the MSS market, whether LEO, MEO, or GEO (outside of Inmarsat), has been mixed at best. Yet, the list of technical advancements pioneered by one or both of the LEO systems that made it to market is very impressive:

1) Dual satellite and terrestrial phones, although not as small as current cellular phones, were quite comparable to the first analog AMPS cell phones that had sparked the mobility market just a few years prior.

2) Each service offered mobile satellite phones you could carry in a briefcase or purse and buy for about $1000 or less, price points that were unprecedented in the MSS industry.

3) They pioneered "assembly-line" produced LEO satellites. The two systems combined were able to launch and manage over 100 spacecraft, quickly vaulting them to global leadership in commercial satellite fleets. Although the systems as a whole were expensive, they achieved breakthrough results measured in terms of cost per satellite in orbit.

4) Their two-way satellite terminals were the simplest to operate ever, not much more complex than dialing a cell phone. Iridium had true global coverage.

5) Iridium represented the broadest application ever of onboard processing, Ka-band feeder links, and intersatellite links.

Despite all of the investments, technical accomplishments, advancements in miniaturization, convenience, ease of use, and service pricing, the LEO/MEO sector was an economic disaster. There were over $11B in losses

Fig. 15.4   The GlobalStar phones.

and write-offs. The MSS sector probably discouraged capital markets from new satellite investments for several years. The technology investments, for the most part, did reach the size, weight, performance, and costs targets intended (Fig. 15.4). But, the market analyses that had projected demand for products and services at those performance, convenience, and price points, in hindsight, were highly flawed. There were two very fundamental problems:

1) The original concept was to portray the satellite services as *just like cellular* but without the coverage limits of conventional cellular radio towers. But, the just like cellular value proposition raised expectations that satellites could not possibly meet. For, instance, satellite phones would not work indoors—a big, big disappointment when positioned as just like cellular.

2) Terrestrial cellular systems made huge strides during the time it took to finance, build, and launch the LEO systems. Driven by enormous market demand, coverage increased, handsets got smaller and less expensive, air time got cheaper, and convenience improved dramatically. Ironically, the LEO systems had intended to tap into a small portion of cellular demand, but the fact that demand was *much greater* than they had anticipated resulted in less market penetration and much, much less revenue than forecast. Cellular market growth and improvements meant that satellite became much less competitive on dimensions that were absolutely fundamental to the definition of the mobile market: convenience (e.g., handset size and coverage) and price.

In retrospect, the LEO systems did a lot of things right and one very big thing wrong. They identified a big growth market. As with DBS, they recognized that merely trying to use existing satellites would not really address the market, so they designed *purpose-built* space systems. They made enormous

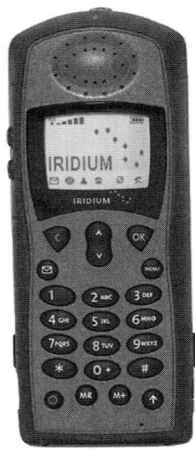

Fig. 15.5   The Iridium phone.

investments in technology and came to market with dramatic improvements in size, weight, cost, and convenience (Fig. 15.5). But, unlike DBS they chose the wrong dimensions on which to compete. DBS provided better quality, more channels, and comparable pricing with cable—enough to overcome the initial lack of local broadcast. But, compared to cellular, LEO service was much less onvenient, more expensive, and besides, what kind of cell phone only works outdoors? It missed the mark on dimensions that were fundamental to the just like cellular value proposition. Dramatic reductions in size, weight, and terminal price did not break the MSS market out of its niche.

SATELLITE BROADBAND

The case of satellite broadband is also illuminating in comparing the impacts of the market environment versus technology investments in fostering new satellite communications applications. Chronologically, investments in satellite broadband tracked slightly behind, but mostly parallel to those of the LEO/MEO mobile systems. Satellite broadband projects, mostly aimed at opening and exploiting Ka-band frequencies, emerged and gathered momentum in the 1990s and boomed until the Internet/telecom investment "bubbles" burst. The most complex, ambitious satellite broadband projects including Astrolink, Celestri, Cyberstar, Skybridge, Spaceway, and Teledesic resulted in cumulative investment write-offs in excess of $5B. Of course, that amount pales in comparison to comparable cumulative terrestrial broadband communications services investment losses in fiber (e.g., Global Crossing, Enron), cable broadband (e.g., @Home), DSL (e.g., Northpoint, Covad), and fixed wireless (e.g., Winstar, Teligent). Two of the most modest broadband satellite concepts were actually the first (and so far only) ones to reach the market—Wildblue Communications (sharing a satellite with Telesat Canada, which offers an equivalent service) in North America and IPStar of Thailand in Asia Pacific. There were also a number of attempts to target the broadband market using existing Ku-band FSS satellites including services branded as Starband and DirecWay (previously called DirecPC, and recently renamed HughesNet). The Ku FSS systems offer an interesting comparison reminiscent of direct-to-home backyard C-band TV services compared to the purpose-built DBS space systems.

The Internet boom featured rapid growth in the number of Internet users, rapid growth in the number of paid subscribers (for instance, with AOL), dramatic increases in Internet backbone traffic, rapid adoption of Internet protocol based systems in enterprise private networks, and, of course, eye-popping market capitalizations for companies that could participate in the on-line, broadband revolution. Interestingly, although there was little doubt about growth of backbone traffic (handled by fiber-optic links), there seemed to be significant doubt about demand for consumer broadband "last mile" connections in the United States (which was the initial target market for several of the satellite systems).

Coincidentally (or not), one result in the United States has been a large surplus of unused "dark" fiber and significant geographic areas that have no terrestrial broadband access available. In retrospect, much of the common views of the period were generally correct (substantial and enduring long-term growth in bandwidth utilization and number of on-line users and associated commerce) but lacking in key details (overestimating near-term rates of backbone traffic growth and underestimating demand for affordable broadband access lines).

One of the most confusing (and ultimately economically devastating) aspects of the broadband revolution in that time was accurately identifying the underlying value propositions. It was not clear whether delivering broadband services meant merely high peak connection speed, managed telecommunications services, managed content, aggregated content, or any of a large number of on-line service business models. It was not clear to what extent there was value in retail services, wholesale services, hub/spoke Internet access connections, or direct point-to-point links. It was not clear what the subscriber traffic models would be, which congestion statistics would be important, or how consumers or enterprises would value different types of offerings. It was not clear whether on-line services were popular because they connected users directly to the Internet (such as Earthlink) or because they provided content in a more accessible form than the Internet did (e.g., AOL).

The answers to many of these questions are probably still not clear today. But, there is a simple way to consider broadband access that seems consistent with much of the financial outcomes. It has become increasingly evident that broadband means *low-cost bandwidth* as much as it means *high-speed bandwidth.* A comparison behind the transmission economics of a telephone call versus delivery of an iTunes video makes the point.

1) Assume a phone call has a retail value of 5 cents per minute and is digitized using an 8-kbps compression algorithm. In this case there is little cost of content (the "content" is the conversation), so that most of the economic value is due to transmission. Simple arithmetic yields 5 cents retail value for 8 kbps/sec = 1 kbyte/sec = 60 kbytes/minute, or just under $1 per megabyte.

2) Compare that to a 60-minute iTunes video with a retail price of $1.99. Note that none of that revenue is actually transferred to the subscriber's broadband service provider itself. But, it does provide some indication of the value of a bit of broadband content. Here, the value of the content is much greater than the value attributable to transmission. Also, note that a lot of Internet broadband content is "free" (no direct cost to the consumer). If the video is encoded at 512 kbps and lasts one hour, the file download is about 60 megabytes. If the transmission value were as high as 10% of the retail price (it is likely much lower), then the imputed retail transmission value would be in the range of about 0.3 cents/megabyte. That is, this exercise yields a retail value of transmitting a bit that is roughly three orders of magnitude lower than for voice.

3) Another way to value broadband access is to consider the retail value of home Internet access versus the amount of capacity consumed. Just using orders of magnitude, retail subscription pricing is in the range of $10 to $100 per month (typically about $25 to about $50), and bandwidth consumption is in the average range of about 1 to 10 gigabytes per month. That would yield a range of about 0.25 cents to about 5 cents of retail revenue per megabyte of consumption (in the same order of magnitude range as the video clip transmission example).

The point of these examples is that the retail value of consumer broadband data transmission is probably two to three orders of magnitude lower than the value of voice transmission. This also makes intuitive sense because fiber optics reduced the cost of terrestrial data transmission by four orders of magnitude or more, and hybrid fiber coax cable networks and DSL also achieved transmission cost reductions of two to three orders of magnitude.

It follows that for satellite service providers to compete in an enduring (versus simply opportunistic) manner in broadband transmission, then satellites would also need to achieve at least one to two orders of magnitude reduction in transmission costs. (Given the relatively high margins that profitable satellite operators already have, plus the opportunity to charge premium prices in hard to reach locations, let us grant an order of magnitude leeway for purposes of illustration next.) There are clearly many factors that must be considered in constructing a complete broadband business model, but for a satellite system, it is hard to ignore the satellite itself. A simple and obvious metric would be the capacity throughput of the satellite (measured in gigabits/sec aggregate), compared to its capital cost in space (satellite procurement, launch, plus insurance). Consider two cases: 1) a "pre-broadband" high-performance C/Ku-band FSS satellite versus 2) a special-purpose-built Ka-band broadband satellite (such as from one of the broadband space systems just identified). Analysis will also be "order of magnitude."

1) Assume an FSS satellite with about 40 total transponders, each carrying about 30 Mbit/s, costing about $200M in orbit, including basic ground infrastructure. It has about 1.2 Gbps of capacity.

2) Consider an onboard processed Ka-band system project that costs $3B of startup investment (system design, technology development, satellite construction and test, launch and insurance, and basic ground infrastructure) that would yield two in-orbit satellites, each yielding 10 Gbit/s of throughput (assuming all of the beams could be fully utilized at maximum capacity). If all of the subsequent satellites cost about $200M in orbit, you could say it cost $3B of investment to gain one order of magnitude (10 Gbytes per satellite versus 1 for the FSS satellite).

3) Now consider the bent-pipe transponded systems. The Wildblue / Telesat satellites yielded about 10 Gbps for two Ka-band payloads in orbit (including system development costs) for about $500M. IPSTAR yields somewhere in the range of about 20 Gbps for one Ku-band spot beam payload. These systems also provide about one order-of-magnitude improvement relative to FSS, but for a much, much lower investment.

4) A bent-pipe satellite with capacity on the order of 100 Gbps would provide two orders of magnitude cost reduction in transmission cost. Such a satellite seems technically and economically feasible and ought to compete reasonably well versus alternative terrestrial *access* systems (though not necessarily in backbone). In early 2008 ViaSat contracted with Space Systems/Loral for construction of a Ka-band spot beam satellite that would yield in excess of 100 Gbps of throughput when combined with a purpose-built corresponding direct-to-home ground network. Simultaneously Eutelsat began construction of a similar satellite from Astrium anticipated to deliver about 70 Gbps of total throughput with the same ground segment.

The point is that if *low-cost* bits represent the fundamental value propositions for broadband, then it explains why FSS systems and expensive onboard processed Ka-band systems have been so unsuccessful. There are other important related factors. The core value proposition of Ku FSS has been its ability to "reach" far flung users of low volumes of point-to-point bandwidth (e.g., VSAT or telecom trunking) or the ability to broadcast bits to many users at once (e.g.,video distribution). Demand for those services remains high even with per bit pricing that is much, much higher than that for broadband access. There is no market incentive to reduce the cost of Ku FSS bandwidth by satellite operators. There is no evidence that demand for existing, *proven* applications using Ku FSS capacity would expand substantially if bandwidth pricing were sharply lower. In fact, given the prices that SES Astra and Eutelsat command in Europe, there is a strong argument that broadcast services are relatively insensitive to the cost of transponder leases. And, there is great risk that experimenting with Ku FSS pricing to test data market pricing elasticity would simply undercut existing markets.

Spot beam satellites, by contrast, can dramatically reduce per bit costs (through frequency reuse and the higher EIRP gain of narrow beams) without undercutting the "reach" value proposition of Ku FSS. (Because the beams are so narrow, they are not really good at connecting distant sites point to point or at national or regional broadcast.) Spot beam satellites really represent a different "business" for satellite operators—more than they represent a different technology. The customer base is different, the economic value proposition is fundamentally different, and the selling channels are much different. Although Ka-band has some technical advantages versus Ku-band (e.g., smaller reflectors at the same orbital slot spacing), it is more likely that Ka-band is associated with broadband because the underlying business is so different (and consequently unproven) that there is no economic incentive to displace profitable broadbeam Ku FSS satellites with the unknown value of the spot beam satellites it would take to appeal to broadband customers.

Even the Ka-band OBP systems are not really about "cheaper" bits, but "better" bits. (Better could include switched backbone bits, valued comparably to frame relay, instead of access bits, comparable to DSL.) Services that deliver "faster" bits, but with insufficient volume to accomplish the intended purpose (e.g., delivering the complete iTunes video at the advertised rate), have been shown to be unsuccessful when promoted as broadband (which implies speed *and* volume, both).

If these interpretations are correct, then satellite broadband is in the nascent market phase as opposed to having missed a market window. Low-cost bandwidth would tap into substantial market demand, driving volumes of terminals and demand for more efficient satellites. That would lead to smaller, less expensive terminals in a virtuous cycle comparable to what happened with DTH TV once space systems that were "good enough" were in place (Fig. 15.6). Given the relative adoption rates of Wildblue service in the United States compared to the much more mature Ku fixed satellite systems (FSS), which even benefited from better brand recognition because of comarketing with DirecTV, there is evidence that satellite broadband might enjoy mass market success. In fact from the direct-to-home satellite broadband market in the United States has grown from about 150,000 subscribers in 2005 to about 800,000 subscribers in early 2008.

The broadband market has yet to fully play out, but indications are that this is a story about how space systems are designed to appeal to new growth

Fig. 15.6   The WildBlue antenna.

markets with lessons learned quite comparable to those of the DTH TV and LEO MSS experiences.

DIGITAL SATELLITE RADIO

Digital satellite radio in the United States has achieved impressive market penetration—with growth even faster than that for DTH TV. And, there have been corresponding achievements in terminal size, costs, prices, and user convenience. In fact, when digital satellite radio capability is offered as original equipment in a new car, the end user can perceive the equipment as completely "free." The receiver comes built-in to the car stereo, even if the owner never subscribes. It takes up effectively no space in the vehicle, and operation is almost completely transparent and intuitive. It is hard to get much smaller, cheaper, or convenient than that! Furthermore, digital satellite radio addresses much more than the automobile-bound user—an expansion of its market appeal in a manner reminiscent of DTH TV. There is clearly demand for personal receivers users can carry around (Fig. 15.7) and low-cost home "repeaters" make service available even indoors. There is a version of the service for businesses, and programming is also delivered over the Internet to satellite radio subscribers. Services have expanded beyond radio to include TV channels, and there are some data applications.

There are some interesting parallels between the satellite radio and LEO MSS markets from a technical perspective—but the economic outcomes have been enormously different. The two US systems, Sirius and XM, each made

Fig. 15.7   Satellite radios.

billion dollar up-front investments to develop and productize space and ground terminal technology. It's interesting to note that the MSS systems only had financing problems *after* they were brought to service (for the first to market systems: Iridium and Globalstar). Satellite radio had financing issues *before* they came to market. Once in service, the main business issue for satellite radio has been the high cost of obtaining compelling content (especially radio personalities and professional sports programming).

The investment thesis for the LEO MSS systems was that mobile satellite service somehow *enhanced* (or at least could piggyback on) the mobile cellular value proposition. But, of course, it didn't because it lacked very fundamental parts of that target value proposition, such as convenience. Conversely, arguments ran that satellite radio only *competed* with alternative terrestrial media, especially CD players and free-to-air local radio. There was much focus on the *audio quality* value proposition for satellite radio, which terrestrial was also working on with *In Band–On Channel* digital broadcast in the AM and FM bands in the United States and S-band digital radio elsewhere. But, it turned out the real value proposition of satellite radio has been *programming variety and selection.* Satellite radio dwarfs terrestrial selection by an order of magnitude, even in media rich markets. And, what is fascinating is that much of its programming content is delivered with audio quality that is noticeably *worse* than even AM radio, let alone FM or CD. For instance, sports programming on satellite radio, such as Major League Baseball or National Football League football, has obviously perceptible source coding artifacts. Clearly satellite radio broadcasters have eagerly traded down quality to increase quantity and variety of programming.

But, the *variety* value proposition has tapped into market demand for radio listeners. As of this writing, XM and Sirius are in excess of 15 million subscribers, combined. The high demand for radio receivers has quickly driven terminal size, cost, features, and functionality well beyond what was initially offered. In fact, the demand for satellite radio services has been so significant that a number of new terrestrial services are angling to compete. However, it will certainly be difficult for those terrestrial services to bring to bear the amount of spectrum and national coverage that satellite has. It will be interesting to see how those market factors influence the competition between the satellite and terrestrial services. XM and Sirius have merged to reduce programming costs and increase operational efficiencies.

SUMMARY

This chapter has covered a broad range of satellite technologies and applications. It should be clear that even the most modern and recent of systems have built on the achievements of those that have come before. There is a vast portfolio of space and ground technologies needed to enable these systems. Those technologies can be combined in myriad ways to achieve startlingly different economic effects. Often it is confusing to try to sort out the causes from the effects. But, ultimately, this author believes that tapping into user demand and carefully (or serendipitously) finding the right dimensions of value to emphasize, at just the right point in the market penetration cycle, are what have driven communications satellite terminals from very big to very small.

Chapter 16

DARPA's SPACE HISTORY

OWEN BROWN,* FRED KENNEDY,† AND WADE PULLIAM*

Five decades ago, it was just a simple "beep-beep" that surprised the free world and especially the United States. That beeping came from a crude transmitter inside a 84-kg satellite circling the globe in low Earth orbit (LEO). The radio signal, first heard from the sky on 4 October 1957 from Sputnik I, did not just signal the beginning of the space race between the United States and the Soviet Union, but also began a series of events that brought about the formation of DARPA, or the Defense Advanced Research Projects Agency. This agency is best known for its part in developing the Internet and stealth technology, but it also has a proud history in space technology, playing a key part in the development and demonstration of the space infrastructure, both military and civilian, that we know today. (DARPA was known as ARPA, the Advanced Research Projects Agency, from its founding in 1958 until March of 1972 when its name was officially changed to DARPA with the addition of Defense. The agency's name reverted to ARPA between 1993 and 1996, at which time it again became known as DARPA. This chapter will use the name of the agency correct for the time frame of the events being described.) Although the United States was beaten to the punch in launching the first artificial satellite, ARPA had a hand in first demonstrating many of the practical applications of satellite technology now prevalent, including communications, weather forecasting, early warning, reconnaissance, and geo-location. Additionally, ARPA provided the early leadership and funding for the launch systems that would later be used by NASA to win the space race by landing a man on the moon.

This material is declared a work of the U.S. Government and is not subject to copyright protection in the United States.

The views, opinions, and/or findings contained in this article/presentation are those of the author/presenter and should not be interpreted as representing the official views or policies, either expressed or implied, of the Defense Advanced Research Projects Agency or the Department of Defense.

*Program Manager, Tactical Technology Office, Defense Advanced Research Projects Agency, Arlington, Virginia.

†Lieutenant Colonel, United States Air Force, Carlise, Pennsylvania; former Program Manager, Tactical Technology Office, Defense Advanced Research Projects Agency, Arlington, Virginia.

Although DARPA has expanded its mission to one of preventing techno-logical surprise in domains other than space, it has remained active in space technology development and demonstration to this day. In this chapter, the history of DARPA's beginnings will be told. DARPA's space programs over the last 50 years ago will be summarized. As stated, many of these space programs were precursors to disruptive space capabilities familiar to us all.

THE LAUNCH HEARD AROUND THE WORLD

The Defense establishment must therefore plan for a better integration of its defense resources, particularly with respect to the newer weapons now building and under development. These obviously require full coor-dination in their development, production, and use. Good organization can help assure this coordination. In recognition of the need for single control in some of out more advanced development projects, the Secretary of Defense has already decided to concentrate into one organization all the anti-missile and satellite technology undertaken within the Department of Defense.

President Dwight Eisenhower, 1958 State of the Union Address

The launch of Sputnik I in October of 1957 should not have been a complete shock to the United States, or the West as a whole, but once the small satellite was placed into orbit, the sense of surprise was undeniable (see Fig. 16.1). The Soviets successfully tested an intercontinental ballistic missile (ICBM) in the summer of 1957 and had been announcing for months

Fig. 16.1 Soviet launch of the world's first artificial satellite, Sputnik I, provided the impetus for changes in U.S. scientific, educational, and military policy, including the formation of ARPA (downloaded from http://nssdc.gsfc.nasa.gov/planetary/image/ sputnik_asm.jpg). (See also color figure section at the back of the book.)

that they planned to launch an artificial satellite. In fact, the Deputy Director of the CIA Herbert Scoville stated in an open meeting on 4 October, only hours before the actual launch, that the "Soviets could launch a satellite this month, this week, or even today" [1]. The military threat was clear; soon the Soviets would have the capability to deliver nuclear weapons by missile to the continental United States using the R-7 launch vehicle that had orbited Sputnik. However, what may have been worse was the blow the launch had to the American psyche. The belief in "American ingenuity" and the technological and educational superiority of the West was put to the test.

Publicly, Eisenhower expressed confidence that the U.S. program was not that far behind. Privately, Eisenhower placed part of the blame for America's lagging space program on inter-service rivalries. Each service was pursuing a separate space program, which risked a lack of focused effort and a potential waste of resources. Just in the area of rockets, each service was developing its own system: the Army's Jupiter-C, the Air Force's Atlas, and the Navy's Vanguard. Additionally, each service had satellite programs that it was pursuing. In short, each service was in effect ignoring some of its established military responsibilities by funding programs in the area of space in order to capture the new mission.

When Sputnik I was quickly followed in November by Sputnik II, a 500-kg satellite, the public and congressional apprehension became fervor, as it now became even clearer that the Soviets had the capability to deliver nuclear weapons by ballistic missile, which further created consternation within the Eisenhower administration.

This concern was further heightened by the launchpad explosion of the first U.S. satellite launch attempt, Vanguard, on 6 December (see Fig. 16.2). Branded "Flopnik" by the press, the failure of Vanguard and the success of the Sputnik launches seemed to indicate that not only were the Soviets ahead, but also that the United States might not be able to catch up. [The decision to proceed with the Naval Research Laboratory Vanguard program in 1955 resulted in the cancellation of a satellite launch by the Army Ballistic Missile Agency (ABMA), under the direction of Wernher von Braun, a system that in retrospect had a much higher probability of success given its direct lineage from the German V-2 program. In fact, it was the proposed Jupiter-C rocket that was ultimately used to place the first U.S. satellite into orbit, Explorer 1 [2].] The political pressure on the administration, especially from Senator Lyndon Johnson, became overwhelming. Over the next year, large changes in public policy were enacted to respond to the perceived problems: the teaching of science and mathematics was promoted by the National Defense Education Act, the National Advisory Committee on Aeronautics was reorganized and dubbed the National Aeronautics and Space Administration (NASA) and the job of sorting out the military's space program was given to a newly created organization, ARPA.

Fig. 16.2   The failure of the Vanguard launch by the U.S. Navy, what became known in the press as Flopnik, created more concern within the United States that the nation was behind in the space race (downloaded from http://exploration.grc.nasa.gov/education/ rocket/gallery/history/vanbloom.jpg).

The President announced the new agency 9 January 1958 in his State of the Union address, and it was made official on 7 February when Secretary of Defense Neil McElroy issued DoD Directive 5105.15 formally establishing ARPA and gave it the responsibility "for the direction or performance of such advanced projects in the field of research and development as the Secretary of Defense shall, from time to time, designate by individual project or by category." The immediate effect of this directive was the transfer of all military space projects to ARPA.

EXPLORER

It was not that the United States did not have a program to launch a satellite. As part of its commitments to the worldwide cooperative scientific program, called the International Geophysical Year (IGY), the United States had decided in 1955 to pursue a satellite launch. Unlike the Soviets, the United States was pursuing separate tracks for its rocket development: a military rocket program to develop ICBMs and the Vanguard program to develop a launch system for the IGY satellite experiment. (The United States decided to pursue separate tracks for its launch systems for larger national security reasons. During this time, the United States was developing its reconnaissance satellites, but freedom of space overflight was not an established principle. It was hoped that a scientific payload tied to the IGY using a non-ICBM launch system could establish the precedent that overflight of a nation in orbit did not violate national airspace and that precedent could subsequently be used by the reconnaissance satellites. Ironically, the Soviets themselves provided that precedence with Sputnik [3].) The failure of the first Vanguard launch provided an opportunity for Wernher von Braun, and his team at the Army Ballistic Missile Agency (ABMA), to take the shot that they were denied in 1955, when the decision was made to proceed with the Vanguard program, based on the Viking missile [2]. Although this decision denied the United States its shot to be first into space, the AMBA team was a ready backup and was quickly ordered to move ahead with a launch to attempt to orbit a Jet Propulsion Laboratory built satellite, Explorer 1.

On 31 January 1958, Explorer 1 was placed into orbit—officially entering the United States into the space age. From one point of view, the launch of the 14-kg satellite did not even come close to demonstrating the heavy-lift capability that the Soviets launch systems did with Sputnik II. However, what was often overlooked was that instead of the simple beeping of Sputnik 1, Explorer 1 carried an 8-kg scientific payload design by James Van Allen of Iowa State University, which discovered the existence of a belt of charged particles trapped in the Earth's magnetic field. The discovery of the Van Allen belts was not only the most important outcome of the International Geophysical Year, but also it clearly demonstrated the lead that the United States held in electronics and manufacturing technologies. Although the immediate military implication of being able to loft larger warheads understandably centered all of the attention on the throw-weight lead of the Soviet systems, once the United States closed that gap, its lead in other technology areas allowed it to quickly demonstrate and take advantage of space systems to a much higher degree than the Soviets.

ARPA DEMONSTRATES THE CAPABILITIES OF SATELLITES

Provided with the impetus of presidential support and the feeling of national emergency, ARPA quickly began to sort out the U.S. space efforts and establish direction to what had been disjointed or overly ambitious efforts of the military services. The ideas and capabilities for many various space systems had been floated before often with proposals ready to be acted upon. However, these concepts had often languished without any focus or were wasting critical resources through duplication across the military services. ARPA spent the first seven months of its existence, during which it had decision-making authority over the complete U.S. space program (before the formation of NASA), sorting through these proposals and overlapping efforts to impose order to the overall space program and to fund those space concepts that might have the greatest military utility. Yet in just these short few months, ARPA set the United States not only on a path toward winning the space race, but also toward demonstrating many of the space capabilities that are well known today.

SCORE — WORLD'S FIRST COMMUNICATION SATELLITE

> This is the President of the United States speaking. Through the marvels of scientific advance, my voice is coming to you from a satellite traveling in outer space. My message is a simple one: Through this unique means l convey to you and all mankind, America's wish for peace on earth and good will toward men everywhere.
>
> *President Dwight Eisenhower, 19 December 1958*

This presidential message was one of the first examples of new space capability that ARPA was demonstrating during the first few years of its existence. SCORE, or Signal Communications Orbit Relay Equipment, was the world's first communication satellite. As with many of ARPA's early efforts in space, the first proposals for what would become SCORE had been pursued by the Signal Research and Development Laboratory (SRDL) at Fort Monmouth since 1955. (Theoretical studies had been in the open literature from as early as 1952, notwithstanding even early suggestions in science fiction [4].) This proposal was picked up in July 1958 soon after the creation of ARPA.

In what would become a model for the agency of quick demonstrations of military capability, SRDL was given *60 days* to assemble a demonstration communication satellite to be placed in LEO on top of an Air Force Atlas ICBM. The objectives of the SCORE program were twofold: primarily, that the Atlas launch system could place a payload in orbit, and secondarily, that a communication repeater installed onboard could be used to transmit messages worldwide [5]. (The early Atlas systems were an unusual configuration in which two of the three engines would be jettisoned, but a majority of the structure and fuel tanks would remain.) Because the Atlas could only reach LEO and only a limited number of ground stations existed, the SCORE system did not operate like the "bent-pipe" geostationary communication satellites of today. Instead the SRDL payload included both a real-time communication repeater for some over-the-horizon relaying of messages and a store-and-forward system that allowed the demonstration of worldwide message delivery through uploading of coded messages onto an onboard tape recorder and downloading by the receiving station when the satellite came into view. It was this tape-delay system that was used to relay President Eisenhower's message to ground stations across the globe. Although the

project schedule soon slipped to 90 days because of delays in the Atlas production, the full system was ready by December 1958, less than five months
from project go-ahead.

The launch was a complete success. The Atlas placed 4000 kg into LEO
(the SCORE payload weighing 68 kg), clearly demonstrating the capabilities
of the Air Force Atlas ICBM to place payloads into orbit. (The Atlas had
previously been demonstrated as an ICBM in December 1957.) The communication payload also operated successfully, transmitting voice and tele-
type messages in both real-time and tape-delayed mode 78 times during its
12-day lifetime. Through this experiment, the United States had demonstrated
more capability with launch systems and payload capability than the Soviet
Union had, passing the Russians just 14 months after the space race began.

As with the other projects it began, ARPA took an agile management role
in the SCORE system development, with the Air Force being handed the
responsibility for the Atlas, built by General Dynamics, and SRDL for the
payload, built by RCA. This approach of maintaining low overhead was, and
still is, a vital part of the success of the agency, especially during the early
years when so many space systems were being developed and demonstrated.
In what would become standard procedure, the agency concentrated on
developing and fostering innovative concepts, as well as making key project
funding decisions, while leaving the burden of facility and workforce maintenance, system development and production, and operations to industry,
research labs, and the military services. At first this decision was made mostly
for the bureaucratic reason of limiting the objections of the services at having
their space programs transferred to ARPA [1]. In the long run, however, it has
served the agency well in keeping it nimble, capable of changing directions
quickly and able to manage many efforts rather than just a few.

CORONA—WORLD'S FIRST RECONNAISSANCE SATELLITE

> And if nothing else had come out of it [the space program] except the
> knowledge that we've gained from space photography, it would be worth
> 10 times what the whole program has cost. Because tonight we know
> how many missiles the enemy has and, it turns out, our guesses were way
> off. We were doing things we didn't need to do. We were building things
> we didn't need to build. We were harboring fears we didn't need to
> harbor. Because of satellites, I know how many missiles the enemy has.
>
> *President Lyndon Johnson, 1967*

No other space effort, either within or outside of ARPA's history can say to
have been more important to the national needs of the time than CORONA.
During a time when the discussion in U.S. national security circles was of the
"bomber gap" and the "missile gap," it was unclear where the United States
stood vis-à-vis the Soviet Union in the early stages of the Cold War, and the

fear was that the United States was falling behind both militarily and scientifically. Even though it was clear that the United States required a strategic reconnaissance system to gather necessary intelligence from the Soviet Union, the correct system to pursue was not necessarily clear. During the late 1950s, the United States operated the U-2 over Soviet airspace in an effort to gather that information. However, even as it was being built, Kelly Johnson (the U-2's designer) knew that the U-2 was a stopgap measure, realizing that before long the Soviets would find a way to bring the aircraft down. Once the SA-2 was fielded in operationally significant numbers, the U-2's days as a strategic reconnaissance asset against the Soviet Union were numbered, which was demonstrated clearly by the downing of Gary Powers.

During this time period, the United States was considering diversifying its strategic intelligence capability by developing reconnaissance satellites. This concept was first proposed in the Project Feed Back by RAND in 1954 and by 1956 the Air Force had initiated a program dubbed WS-117L. The system was centered on a complex photography system, calling for film to be processed onboard and then the images scanned and beamed down to ground stations. Although the system had positive attributes such as quick delivery of intelligence, a simple concept for data recovery, and a long mission time (as much as a year), these features led to stressing requirements on other components and on the system design itself.

When ARPA was created in 1958, the WS-117L program was stuck in technology development delays, mostly caused by its overly ambitious requirements. Once the WS-117L program was transferred to ARPA management, it became three separate efforts: DISCOVERER, SENTRY, and MIDAS [6]. It was hoped that by splitting and therefore focusing the effort, each would be able to more quickly make progress than if they had as a unitary program.

The SENTRY Project, later called SAMOS, would continue to develop photographic reconnaissance systems utilizing radio transmissions and was immediately turned back over to the Air Force to run. In later years the developments of this program led to the series of satellites that provide strategic reconnaissance for the United States today. MIDAS would develop infrared, wide-area missile warning satellites, an early precursor of the current DSP satellites.

The third program, DISCOVERER, would develop a reconnaissance satellite based on a film return system—what would become CORONA (Fig. 16.3). The concept called for the mating of a Thor booster with an Agena second stage to put a satellite containing a small-grain film camera system in a polar orbit. After just 17–18 orbits, stabilized through the three-axis orientation rocket system on the Agena, the film, protected in a reentry pod, would be ejected from the satellite, reenter the atmosphere, be decelerated, and then be recovered in mid-air by specially equipped aircraft. This system concept, although complex from an operational point of view, was judged more

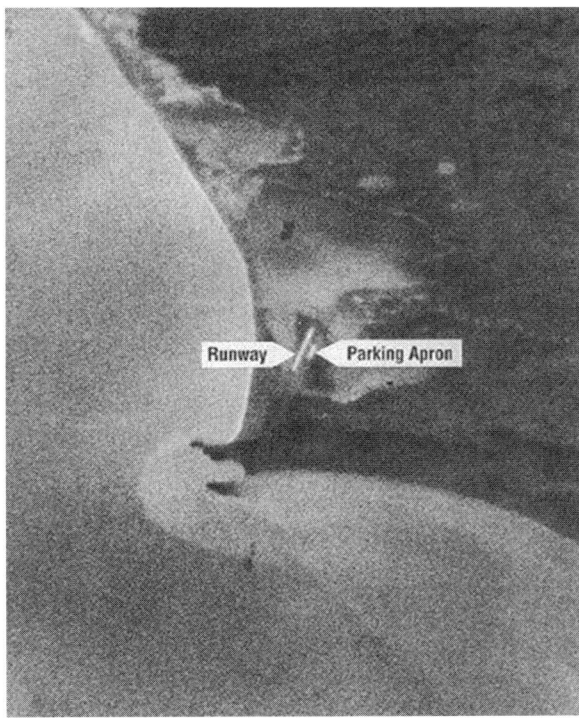

Fig. 16.3 The first picture taken over the Soviet landmass by the world's first reconnaissance satellite, DISCOVERER XIV. The CORONA program was one of ARPA's first efforts in space and produced in 30 months a vital asset for the United States, filling in the gap in strategic intelligence left by the downing of Gary Power's U-2 (downloaded at http://www.nasm.si.edu/exhibitions/GAL114/SpaceRace/sec400/sec420.htm).

technologically feasible at the time than that of SENTRY. Given the high priority the strategic situation imposed on satellite reconnaissance, the CORONA effort was begun and initially funded by ARPA, and then management of its development was handed over to the CIA. (Final operational control of the program was transferred to the Air Force, which then formed the National Reconnaissance Office around it.)

Although not successful until its 14th attempt just 30 months after program inception (and 18 months after the first launch) the CORONA program was a resounding success. DISCOVERER XIV returned over 3600 feet of film, covering 1.5 million square miles of the Soviet Union (approximately 17% of the Soviet landmass)—more than had been collected over the Soviet Union during the entire U-2 program [7]. This first mission dramatically changed the strategic calculus for Washington, showing that instead of the Soviets having scores of ICBMs, they had approximately six [3]. If there were a missile gap, it was in favor of the United States. This information, and that which followed in subsequent missions, allowed the United States to be more forceful in confrontations with Moscow and more circumspect in its own weapons programs, saving valuable resources. CORONA would remain the backbone of U.S. strategic photographic intelligence into the 1970s when it was finally superseded by satellites that could relay imagery electronically to the ground.

TIROS — WORLD'S FIRST WEATHER SATELLITE

The satellite that will turn its attention downward holds great promise for meteorology and the eventual improvement in weather forecasting. Present weather stations on land and sea can keep only about 10 percent of the atmosphere under surveillance. Two or three weather satellites could include a cloud inventory of the whole globe every few hours. From this inventory meteorologists believe they could spot large storms (including hurricanes) in their early stages and chart their direction of movement with much more accuracy. Other instruments on the satellites will measure for the first time how much solar energy is falling on the earth's atmosphere and how much is refracted and reflected back into space by clouds, oceans, the continents, and by the great polar ice fields.

"Introductions to Outer Space," President's Science
Advisory Committee (PSAC), March 1958

Although incredibly successful, the CORONA system had one major drawback; the film return design meant that each mission had a limited lifetime, approximately 27 hours, in which the timing of the photography was preprogrammed so as to maximize the intelligence captured over the Soviet Union. A launch on the wrong day over Russia would provide just very expensive pictures of the tops of clouds. Out of this limitation grew an urgent requirement for a meteorological satellite system to assist in the operation of optical reconnaissance satellites, a need which was filled by the TIROS (Television Infrared Observation Satellite) program [8].

The desire for a weather satellite actually dated back to the 1940s and had been part of the IGY plans [8]. But it was in the shakeout of the Air Force WS-117L program where TIROS was born. The technical means to take television pictures of necessary resolution for strategic reconnaissance and to transmit them back to Earth was still years out when ARPA was handed the responsibility for the WS-117L program at its creation. ARPA staff realized that although the available television camera technology of 1958 provided insufficient resolution for targeting and surveillance missions within the allowed weight limit, it was completely adequate for the meteorological requirement. Therefore, in July 1958, ARPA redirected RCA, which had been working on the television surveillance system for WS-117L, to begin the

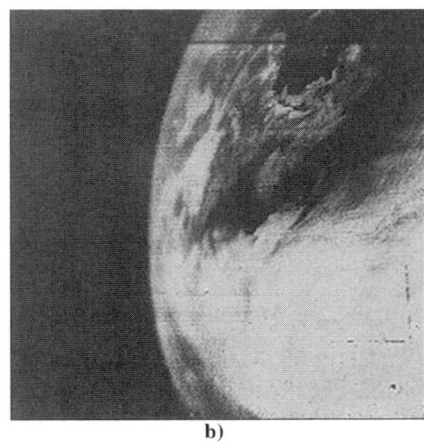

a)   b)

Fig. 16.4   The first weather satellite program, TIROS (a), launched by NASA on 1 April 1960 (b), was started by ARPA. Improved weather predictions made capable by satellite data have vastly improved military planning and saved countless civilian lives through advanced warning of large storms (downloaded from http://history.nasa.gov/SP-168/ p202a.jpg and http://earthobservatory.nasa.gov/Library/RemoteSensingAtmosphere/ Images/first_tiros.jpg). (See also color figure section at the back of the book.)

development and construction of a demonstration weather satellite, TIROS I. (Management of RCA's contract was handed to the Army Signal Corps R&D lab.) The satellite was a spin-stabilized, pill-box design (see Fig. 16.4a) that included two television cameras—one low and one high resolution (550 lines per frame)—and was powered by 9200 solar cells [8], the first major use of solar cells to power a satellite. By the time the TIROS project was handed to NASA in April 1959, the plans and funding for the payload construction, launch, and data analysis were already in place.

Launched a year later on 1 April 1960 on top of an Air Force Thor booster (Fig. 16.4b), TIROS I was an instant success. During its 78-day lifetime, it provided 19,389 pictures of cloud cover that were considered valuable (out of nearly 23,000 taken) and also some pictures of sea ice useful in ice reconnaissance [9]. Some of these first pictures showed features that were identified as large storms, immediately demonstrating the capability to assist in weather forecasting. Although worldwide data without interruptions were not achieved until 1966, TIROS was immediately considered semi-operational [10]. However, the legacy of the TIROS program is much larger than improved operational planning for reconnaissance satellites; it ushered in a new age of weather forecasting, both civilian and military. Not only was day-to-day forecasting improved, but also the ability to track major storms over the ocean was revolutionized, providing coastal communities and ships at sea the warning that they needed to save countless lives. Beside the development of the

Internet (and maybe satellite communications), few other ARPA programs have had such a profound impact on the entire world.

TRANSIT — WORLD'S FIRST NAVIGATION SATELLITE

[T]he discovery of the longitude, the perpetual motion, the universal medicine, and many other great inventions brought to the utmost perfection.

Doctor Lemuel Gulliver when asked to imagine what he might be able to see if he were immortal

Used in *Gulliver's Travels* as synonymous with attempting the impossible, accurately locating position on the Earth, especially on the open sea, has been an age-old problem. The inaccuracies in the indirect method of determining position, typically one or another version of dead-reckoning guidance, led ships of Swift's time to miss their resupply ports, take inefficient routes, and cluster around known shipping lanes, making easy prey for piracy. In hopes of solving the problem, the British Parliament in 1714 passed the Longitude Act, which offered support and incentive awards to inventors who could devise a system for accurately determining longitude for the Royal Navy. (Latitude was determined through the use of a sextant.) To administer these awards, the act set up the Board of Longitude, perhaps the world's first research and development organization. (Although disbursing more than £100,000, the prize was never fully paid off. The problem had been adequately solved for finding locations the size of ports by John Harrison who developed a more accurate, pendulum-free timepieces that could be used to calculated longitude given the calendar date [11].) Through support of the TRANSIT program, the world's first navigation satellite, ARPA was able to bring to final conclusion what that first research and design (R&D) organization could not, accurate geo-location on the open sea.

As with the TIROS satellite system, TRANSIT was born out of requirements of the strategic struggle between the United States and the Soviet Union, but became a dual military-civilian asset (Fig. 16.5). It would become the forerunner of the modern global positioning system (GPS) constellation. In this case, the requirement was to provide accurate position data (within 0.1 nautical mile) to Polaris-armed submarines during their patrols so that launch trajectories could be quickly calculated day or night with enough precision to hit their targets [12]. With the accuracy that TRANSIT facilitated, the ballistic submarine leg of the U.S. nuclear triad could with confidence hit the targets assigned to it. In a destabilized strategic environment underlined by the potential of a knock-out Soviet first strike, the acknowledged capability of the nearly invulnerable submarine force to hit targets through the Soviet Union provided all of the deterrence the United States would need—clearly demonstrating the importance of the TRANSIT program among those programs begun in ARPA's early years. But besides being a critical part of the

Fig. 16.5   The ARPA TRANSIT program was the first operational satellite positioning system that remained in operation until 1996 when the current GPS constellation became operational. Besides the general improvement in geo-location for military units, the TRANSIT satellite system was fundamental to the accuracy of the submarine-launched POLARIS missiles and therefore a vital component in the United States' nuclear deterrent during the Cold War (downloaded from http://www.nasm.si.edu/exhibitions/gps/transit.gif).

U.S. strategic system, TRANSIT provided position data to both military and civilian ships, oil rigs at seas, and even land surveying, including those of the Defense Mapping Agency (now part of the National Geospatial Agency).

Interestingly, the idea for TRANSIT was another unintended consequence of the Sputnik I and its incessant beeping. Researchers at the Johns Hopkins Applied Physics Laboratory, or APL (which was part of the Navy's system of research labs and was involved in the Polaris program), became fascinated with the Sputnik launch, as had most of the world. Their fascination led to efforts to determine its orbit from the Doppler shifts of the beeping. Once successful, it did not take long for them to realize that the problem could be inverted—that a location on Earth could be calculated by knowledge of the orbit of a satellite, which could be updated from ground tracking stations once the satellite was in orbit, and the Doppler shift of a signal from that satellite, measured with a ground receiver/computer system (a different operating principle than the current GPS navigation system). Within just months of the Sputnik launch, APL had been able to fully develop the concept for the proposed satellite navigation system. The benefits of this solution, especially for submarines as compared to other forms of navigation, were overwhelming. The polar orbits of the satellites allowed worldwide coverage of the system. Because the measurements of angles or directions are not required, simple omnidirectional antennas can be used (very useful on a pitching ship and with a submarine that only has to expose a small antenna at appropriate times). As the method uses radio frequencies, the solution is all weather. Now a submarine in the open ocean could in any conditions and with minimal exposure determine its location worldwide within 200 meters. Almost by accident, APL had found the solution to the problem of determining missile launch coordinates on a moving platform.

The TRANSIT program proved to be an example of the benefit of having an R&D organization like ARPA outside of the services, a benefit that the Department of Defense (DoD) has reaped time and time again since the founding of ARPA. In the TRANSIT example, because the DoD responsibility for space programs lay outside of the Navy in 1958, APL brought the proposal to ARPA, where in October 1958 it was quickly approved and funded for all of the pieces of a demonstration system including the construction of a demonstration satellite, tracking/orbit update stations, launch systems, and the development of ground navigation equipment/receivers.

The first successful TRANSIT launch was in April 1960 (TRANSIT 1B), just 18 months after project approval, and demonstrated the feasibility of satellite navigation. Just as other early ARPA programs, TRANSIT had secondary demonstration goals. The Able-Star second stage demonstrated for the first time an engine restart in space. Also, TRANSIT 1B tested the first magnetic torquer device used for satellite attitude control and a solar attitude detector. Future versions of TRANSIT demonstrated for the first time gravity-gradient stabilization, allowing a directional antenna for transmitting its signal, greatly decreasing onboard power requirements. Another side benefit of the TRANSIT program was greatly improved scientific knowledge of the Earth's gravitational field through the large data set provided by the TRANSIT satellites and tracking stations. This geodesy knowledge was the fundamental limiting factor to position calculations as slight changes in the Earth's gravitational field would slightly alter the satellites orbit. By 1965, the gravity model had become sufficiently accurate to reduce the positional accuracy of TRANSIT to less than 120m, the requirement set for POLARIS [8]. By 1968, the TRANSIT system was declared fully operational by the Navy and had already begun to be used for civilian applications. The system remained in active service until 1996 when the current GPS system became fully operational.

VELA HOTEL

Each of the Parties to this Treaty undertakes to prohibit, to prevent, and not to carry out any nuclear weapon test explosion, or any other nuclear explosion … in the atmosphere; beyond its limits, including outer space; or under water, including territorial waters or high seas.

Limited Test Ban Treaty of 1963

During the second ever large thermonuclear test at Bikini Atoll in March of 1954, poor prediction of the fallout pattern from such a large explosion led to the unintentional exposure of hundreds of Marshall Island natives to high, but nonlethal, levels of radiation. The ironically named *Fortunate Dragon*, a Japanese fishing boat, was in the area and also received very high doses of radiation, which proved to be fatal for its captain. This "Bravo" nuclear fallout accident focused international attention on nuclear testing and spurred President Eisenhower to consider proposing a nuclear test ban as a solution to the fallout issue and possible slow down the arms race [1]. By 1957, the issue had been given to the Presidential Science Advisory Committee (PSAC) to consider both the implications for a nuclear testing ban and methods in order to verify such a treaty. The latter effort was, in part [8], undertaken by the High Altitude Detection Panel (the Panofsky panel) of PSAC, which recommended that a satellite system be employed to detect atmospheric or space nuclear tests as part of a verification system for a possible nuclear test ban treaty. [Separate panels were assigned to examine parts of the testing ban, such as underground (the Berkner Panel).]

In a continuation of presidentially directed programs, Eisenhower assigned ARPA in the summer of 1959 the task of developing the technologies necessary for the detection of nuclear tests, what would become Project VELA (Fig. 16.6). (Vela means "watchman" in Spanish.) This program would examine technologies for detection space and atmospheric tests by satellites (VELA HOTEL) or by ground-based system (VELA SIERRA) and for underground tests by large seismic arrays (VELA UNIFORM).

By September, ARPA had already planned for the required launchers for VELA HOTEL and had assigned Los Alamos and Sandia the responsibility for designing the satellite system and execution of the effort to the Air Force. The goal of the program was to develop a satellite system that would be able to detect a minimum 10-kiloton nuclear explosion taking place on the surface of the Earth from as far away as 160 million kilometers. By 1963, the first of the satellites was ready for launch and just in time. In April 1963 the United Kingdom, the United States, and the Soviet Union signed the Limited Test Ban Treaty (banning atmospheric and space testing of nuclear weapons), which went into effect on October 10 of that year. Just seven days later, the first pair (of six) of VELA satellites was launched on top of an Altas-Agena rocket and placed in 115,000-km circular orbit, beyond the outer Van Allen belt and nearly a quarter of the way to the moon, higher than any military satellite had been placed before. Spaced 180 degrees apart from each other and at an altitude that could examine a complete hemisphere, this first pair quickly established a verification method for the new treaty. The second pair of VELA satellites detected the Chinese Lop Nor test in 1964, quickly demonstrating their capability and utility [13].

Fig. 16.6   VELA 5 satellite: a presidentially directed effort, the ARPA VELA HOTEL
satellites, part of the larger VELA program, provided the United States with the means
to detect atmospheric and space nuclear detonations, a key verification capability in support of the Limited Test Ban Treaty (downloaded from http://imagine.gsfc.nasa.gov/
Images/vela5b/vela5b_5sm.gif). (See also color figure section at the back of the book.)

Under the ARPA program, six pairs of VELA satellites were eventually
placed into orbit from 1963 through 1970. The first two pair included neutron
and gamma ray detectors, designed by Los Alamos and Sandia to measure the
characteristics of a nuclear explosion in space. As has been true of many
ARPA programs throughout its history, science and equipment that have first
been developed and tested for military purposes have the unintended consequence of vastly improving or creating a new area of science. In the case of
the VELA HOTEL satellites, the gamma ray detectors that were developed to
detect a nuclear test in space kick started the field of gamma ray astronomy
[1], key in the identification of supernovae and measurements of the early
universe just after the Big Bang. On the third pair of spacecraft, launched in
1965, an optical instrument, called a bhangmeter, was added to provide a
separate modality to identifying and verifying atmospheric tests. The bhangmeter, developed by Los Alamos for this purpose, measured the double flash
signature that is characteristic of an atmospheric nuclear detonation. The

utility of the bhangmeter in the third pair was limited by the satellite's spin stabilization. But by 1967, the fourth pair was gravity-gradient stabilized (Earth oriented), which greatly improved their capability. The last two pair (1970) also included an electromagnetic pulse detector, even further increasing the data sets necessary to reduce false positives [8].

The VELA program was completely transferred to the Air Force in 1970 after the launch of the last pair. Since July 1983, the GPS satellite network has carried the nuclear test detection equipment, including X-ray, bhangmeter, and EMP detectors, and in September 1984, the VELA satellites were shut down [8]. However, during the time that the VELA HOTEL satellites were operational, it is believed they did not miss a single nuclear event that they were in a position to observe, providing the verification capability necessary to support the Limited Test Ban Treaty [14]. Even more importantly, the known capability of the United States to detect atmospheric nuclear tests, and through the VELA UNIFORM program underground tests, has been a disincentive to aspiring nuclear powers to attempt to test nuclear weapons in secrecy.

ARPA's ROLE IN LAUNCH-VEHICLE DEVELOPMENT

In June 1958, the National Security Council invited ARPA to present its plans for developing launch vehicles. ARPA still had complete control over the U.S. space program, but plans had already begun for a civilian agency, NASA, to take over the civilian aspect of the program. However, in that meeting, ARPA personnel set the ground work for the much of the launch architecture that would be used by NASA in reaching the moon and exploring the other planets—the use of clusters of engines for large launch vehicles (used in the Saturn series) and hydrogen-oxygen upper stages (Centaur) [15].

CENTAUR

Centaur was "the" rocket by which NASA would conduct extensive earth orbit missions, lunar investigations, and planetary studies. Aside from military missions assigned to Centaur, which were to be considerable, NASA planned to launch one operational Centaur every month for a period extending well into the 1970s and beyond.

History of Centaur [16]

The advantages of using LH_2/LOX as a rocket propellant had been recognized even by early rocket pioneers. However, working with and designing for cryogenic liquids could not be adequately overcome; producing, handling, and storing LH_2/LOX on the rocket proved to be too difficult for the engineering solutions of the time. During the early 1950s, though, the Atomic Energy Commission (AEC) had produced major advances in large-scale hydrogen generators and storage systems for its work in thermonuclear devices [8]. Additionally, during the mid-1950s, the Air Force, under contract to Lockheed, was pursuing a supersonic surveillance aircraft, SUNTAN, using LH_2 as its fuel. Pratt and Whitney was subcontracted to Lockheed as the propulsion lead and successfully demonstrated that LH_2 could be used in a turbojet engine. The newfound industrial capability to produce, handle, and utilize it in a propulsive device reenergized the interest in LH_2/LOX as a solution to high-energy launch systems.

Before ARPA's creation, Convair had proposed to the Air Force that it develop a LH_2 upper stage for its Altas booster using the same thin-skinned, pressurized structure technology. Pratt and Whitney was also proposing using its experience in LH_2 propulsion to develop an upper stage. ARPA combined the proposals and in August 1958 funded through the Air Force a joint Convair–Pratt effort to develop a LH_2/LOX upper stage for the Atlas booster, the Centaur.

Although the program was quickly transferred to NASA in October 1958, the Centaur was already on a path for success (Fig. 16.7). The program was

Fig. 16.7   The Atlas Centaur launch system. ARPA funded the development of the LH_2/ LOX Centaur upper stage, as a high-performance stage necessary to lift heavy payloads. Combined with the Atlas and Titan boosters, Centaur was the upper stage for launches of probes to Mercury, Venus, Mars, Jupiter, Saturn, Uranus, and Neptune. Centaur continues to serve as the Untied States' most powerful upper stage (downloaded from http:// exploration.grc.nasa.gov/education/rocket/gallery/atlas/AtlasCentaur.jpg). (See also color figure section at the back of the book.)

the first to demonstrate the cooling of the nozzle with cryogenic fuel, the pumping and control of these fuels in zero gravity, and that thin-skinned metal structures could survive cryogenic embitterment. Also, as Centaur was intended as an upper stage, the program demonstrated engine restart and precision navigation, both necessary to achieve accurate orbits. All of these advances would be used in later launch systems, including the Saturn family and the space shuttle. The final version of Centaur has an operational proven I_{sp} of over 444 seconds (the Atlas booster, which uses RP-1 as fuel, achieves approximately 353 seconds), underlining the vision of the ARPA judgment in LH₂/LOX technology and its teaming of Convair and Pratt. When mated to an Atlas booster, the Centaur is capable of placing four tons in LEO, two tons in geostationary orbit (GEO), and one ton to escape velocity [8]. As stated by a Glenn Research Center history, Centaur is simply "America's Workhorse in Space" (data available on-line at http://www.nasa.gov/centers/glenn/about/history/centaur.html).

SATURN

> But why, some say, the moon? Why choose this as our goal? And they may well ask why climb the highest mountain? Why, 35 years ago, fly the Atlantic? Why does Rice play Texas? We choose to go to the moon. We choose to go to the moon in this decade and do the other things, not because they are easy, but because they are hard, because that goal will serve to organize and measure the best of our energies and skills, because that challenge is one that we are willing to accept, one we are unwilling to postpone, and one which we intend to win.
>
> *President John F. Kennedy, Rice University, 1962*

Although President Kennedy provided the vision and drive behind the space program, and the moon shot in particular, to the nation as a whole it was decisions and a time-saving suggestion that were made during that short period that ARPA controlled the U.S. space program that jumpstarted the U.S. booster program to be in a position to meet the challenge. Von Braun's group at ABMA had been working on a concept for a large booster since early 1957, work that received a big push after the launch of Sputnik. Their concept, a major feature of ABMA's National Integrated Missile and Space Development Program, proposed to speed the development of a large booster by ganging a cluster of existing missiles together and using a to-be-developed Rocketdyne E-1 engine burning RP-1/LOX. After being given the space mission at its formation in February 1958, ARPA picked up on the concept and made a critical suggestion, recommending substituting eight existing Thor/Jupiter S-3D engines for the still-to-be-developed E-1, saving $60 million and two years in development time. With this change, ARPA provided funding of $92.5 million in August of 1958 to get the Juno V (soon to be changed to SATURN) project started [8] ARPA continued its funding support through ground testing and the study of launch facilities right up until the booster development was transferred to NASA.

The ARPA/ABMA first stage was successfully launched in October 1961 as part of the SATURN C-1 configuration. NASA used 10 SATURN C-1 launchers, which also included CENTAUR LH$_2$/LOX engines on the second and third stage, to test APOLLO procedures and equipment. The follow-on SATURN 1B, which used the same first stage and CENTAUR engines for the third, was used to test APOLLO systems and engines and demonstrate docking maneuvers right through 1966 when the SATURN V test flight began (Fig. 16.8) [8]. Although the final SATURN V configuration differed from these early models, it was the early support of ARPA for the cluster concept, the suggestion of using existing engines during early development and the technology demonstration of LH$_2$/LOX engines in the CENTUAR program, that accelerated the U.S. space program in its quest to beat President Kennedy's deadline and land a man on the moon by the end of the 1960s.

Fig. 16.8   Saturn 1B launches the Apollo 7 spacecraft. The Saturn family of launch systems was derived from the Juno (Jupiter-C) rocket developed by the Army Ballistic Missile Agency, funded by ARPA, and then transferred to NASA upon its creation (downloaded from http://exploration.grc.nasa.gov/education/rocket/gallery/atlas/Saturn1b.jpg).

A New Mission

ARPA's control of the U.S. civil space program was short lived. Formed in February 1958 to accelerate the U.S. space effort, it would be nearly completely out of the business in the civil effort by November 1959. It was clear even during the creation of ARPA that President Eisenhower and Senate Majority Leader Johnson desired for a civilian agency to pursue the space race. Although it took a little longer to pass the Space Act legislation (the authority already existed for the Secretary of Defense to create a new defense agency), by July 1958 NASA had been formed, and by fall most of the civilian space effort had been transferred from ARPA to the new agency, including the Saturn and Centaur rocket development and the TIROS weather satellite program.

There was pressure on the military space side as well. The services had never fully accepted that a separate agency was needed for the military space program and noted that even at its peak ARPA had returned 80% of its programs back to the Air Force to manage [13]. Defense Secretary McElroy saw the issue the same way and on 18 September 1959 made the Air Force the executive agent for space, transferring the communication and early warning satellite programs to it.

Just 21 months after its creation as the central space agency for the United States, most of the ARPA portfolio of program, and much of its budget, had been transferred to other entities. It would have to find a new mission, not the last time a major portion of the agency would be split off to form a new agency. Many of the efforts that ARPA handled during its short period of control of the U.S. space program were concepts that were already in existence; ARPA established some order and the funding necessary. However, in losing existing, established technology programs to the services ARPA found its mission—to push the beyond the capabilities of today and bridge the gap to the future.

DARPA in Space: the 1980s and 1990s

Although the years in the 20th century following the early 1960s did not see frenetic space activity at DARPA, important programs and progress were nonetheless made. Significant DARPA space programs during the 1980s included the Global Low Orbiting Message Relay (GLOMR), manufactured by Defense Systems, Inc., now a part of Orbital Sciences Corporation. GLOMR was a store-and-forward system designed to relay data from remote Earth-based sensors. A small, basketball-sized 52-kg polyhedron, it was originally to be launched from shuttle mission STS-51B, but a battery issue forced a delay. It was eventually launched on mission STS-61A on 30 October 1985 and released into orbit two days later. (This was the second use of the shuttle "Get-Away-Special" or GAS payload approach.) It reentered the Earth's atmosphere 14 months later. The price tag on the satellite was a mere

$1 million, ushering in the idea of a "cheapsat" [17]. GLOMR was a "back to the future" program of sorts for DARPA, as it was the second store-and-forward relay satellite launched—SCORE being the first.

Significantly DARPA contributed to the development of what today can be considered the workhorses of small satellite launch, the Taurus and Pegasus launch vehicles built by Orbital Sciences. Both vehicles relied on similar solid rocket boosters. Taurus, a ground-launched system, was designed to be rapidly integrated and erected at a simply launch site. Pegasus was inspired by the idea that an air-launched system would provide enhanced flexibility as it steered clear of range impediments. (In addition, its performance was enhanced because of increased efficiencies at higher altitudes.)

DARPA's space activity in the 1980s and 1990s does appear at first glance to have been dwarfed by that which took place soon after Sputnik. But what is too overlooked is how other DARPA efforts in this time frame were integrated with newer space systems (with an ARPA/DARPA heritage) to provide unmatched advantage for the warfighter in the field during the last decade of the 20th century. Satellite-relayed UAV imagery, GPS-guided munitions, and fully networked satellite communications for command and control are examples.

DARPA'S SPACE MISSION ENTERS THE NEXT MILLENIUM

If the US is to avoid a 'space Pearl Harbor,' it needs to take seriously the possibility of an attack on US space systems.... The US is more dependent on space than any other nation. Yet the threat to the US and its allies in and from space does not command the attention it merits.

Space Commission Report, 2001

Thanks in part to the effort of early ARPA satellite and rocket pioneers, by 2000 space played a vital and enabling role in the nation's global military strategy and tactics and just as importantly was paramount to the American economy and safety of its citizens. Global communications, GPS, accurate weather forecasting, and freedom through knowledge of the adversary were all now possible because of space technology. But, with this capability came vulnerability—vulnerability not only in the form of catastrophic attack as made explicitly by the 2001 Space Commission, but vulnerability also in the form of technological surprise (like that of Sputnik) and the potential to fall behind in the technical supremacy needed to maintain a warfighting advantage. It was in this light that during his assignment to DARPA in mid-2001 that Anthony Tether was given the basic directive by then Secretary Donald Rumsfeld to get DARPA back into space. (Rumsfeld of course headed the 2001 Space Commission, which had published its findings early that year.) Tether focused DARPA's space activities into major focus areas, among them being space protection, space situational awareness, and access and infrastructure.

Two of the most dramatic programs that aimed to support the access and infrastructure domain are Falcon and Orbital Express. The Falcon program is "designed to vastly improve the U.S. capability to promptly reach orbit" [18]. This activity includes development of new hypersonic test vehicles that could bridge the gap between space and the atmosphere inhabited by aircraft. Falcon also focused efforts to develop new low-cost launch vehicles, including the SpaceX Falcon 1 launch vehicle. The initial first two tests were sponsored by DARPA, with the second launch successfully getting to space and providing important data for continued develop of this vehicle that could potentially open new markets to space users.

No other DARPA program in the recent past may have more profound impact on the future of space access and infrastructure than Orbital Express. Orbital Express was conceived in the late 1990s as a demonstration of autonomous robotic servicing and refueling of spacecraft. The Hubble repair missions conducted by space shuttle crews have demonstrated the undeniable value of satellite servicing. As in the case of Hubble, too often spacecraft that have required enormous resources to build and launch fail within weeks of placement on orbit. Likewise, as technologies rapidly evolve in the postinformation age, the ability to replace outdated sensors and computers onboard otherwise healthy satellites can bring increased utility to their users. Orbital Express was created with these thoughts in mind, but with the knowledge that human spaceflight repair missions—although very valuable—are very costly as well. By creating an autonomous capability for repair, costs could be vastly reduced, and mission turnaround time could be decreased. When considering Orbital Express, its developers evaluated historical space programs and concluded that a key life-limiting factor for many satellites is its store of propellant, used to provide for maneuvering and stationkeeping. The amount of propellant available is limited by launch-vehicle and spacecraft constraints on mass and volume. By allowing a means to refuel on orbit, satellites' operational lifetimes could be increased, or they could be allowed to maneuver more frequently. Other life-limiting elements include onboard batteries and computers, which can degrade (in the case of the former) or become obsolescent (in the case of the latter).

In 2000, the Boeing Company's Phantom Works was awarded a contract to develop the Orbital Express system. Phantom Works designed and built the autonomous servicer, ASTRO (Autonomous Space Transfer and Robotic Orbiter), while the client (service) spacecraft, NextSat was built by Ball Aerospace. ASTRO, which weighed in at 1000 kg (carrying propellant both for itself and NextSat), used two optical cameras, a laser rangefinder, a NASA-designed laser ranging and client pose determination system, and an infrared camera for rendezvous, proximity operations, and docking of NextSat. The smaller 225-kg (dry) NextSat was fitted only with targets and retroreflectors to assist ASTRO's sensors. A special docking mechanism allowed reliable mating of the two spacecraft, even at off-nominal docking angles. A fluid-transfer mechanism was developed to pump or pressure-feed hydrazine propellant from one vehicle to the other. A purpose-built robotic arm allowed for the berthing of NextSat and the removal and replacement of modular flight computers and batteries.

Launch of the Orbital Express (Fig. 16.9) took place on 8 March 2007 from Cape Canaveral. Over the next 135 days, the mission conducted 14 refueling operations, demonstrated six battery transfers and a flight computer changeout, and performed seven discrete autonomous rendezvous and docking operations from separation distances of as much as 400 km. Each operation would begin with ASTRO and NextSat in the mated configuration. ASTRO would then fire its thrusters and retreat along a preset trajectory to a target separation distance and then return and dock.

With Orbital Express, DARPA offered a new way of thinking about the design and operation of future space systems: not only can serviceable satellites offer unmatched capabilities, but also they provide decision makers and warfighters with the ability to change or modify these capabilities at any time

Fig. 16.9   An in-orbit view of NextSat from ASTRO during DARPA's Orbital Express mission. (See also color figure section at the back of the book.)

in their life cycle, as well as the ability to continue to perform the intended mission despite changes to the operating environment. These are the respective definitions of flexibility and robustness. Flexibility and robustness will be critical in a future filled with uncertain threats, uncertain technological development timeliness, uncertain budgets, and uncertain performance.

CLOSING THOUGHTS

It was a simple beep from Sputnik that surprised a nation and spurred the creation of a radically innovative agency called ARPA. Now DARPA, the legacy of the agency in space, is incredible. The roots of vital communication, weather, sensing, and navigation spacecraft are found at DARPA. Today new roots are being planted that might yield yet again unparalleled space capabilities. The events of the future are difficult to predict. DARPA has no crystal ball. But DARPA will remain committed to preventing the United States from being surprised by the future. As in the past, DARPA, in fact, will continue to harness the genius of innovative people to create technologies that will change and indeed shape the future.

REFERENCES

[1] York, H., *Making Weapons, Talking Peace*, Basic Books, New York, 1987, pp. 101, 117, 139, 220.

[2] Hughes, K., "Pioneering Efforts in Space," U.S. Army Missile Command Historical Office, Huntsville, AL, 1990.

[3] Day, D. A., Logsdon, J. M., and Latell, B., *Eye in the Sky: The Story of the CORONA Spy Satellites*, Smithsonian Inst. Press, Washington, D.C., 1999, pp. 25, 120–125.

[4] Brown, "Signal Corps Space Odyssey, A: Part II—Score and Beyond," *The Army Communicator*, Vol. 7, No. 1, Winter 1982, p. 60.

[5] Martin, D., Anderson, P., and Bartamain, L., "The History of Satellites," *Communication Satellites*, 5th ed., The Aerospace Press, Los Angeles, CA, 2006.

[6] Spires, D., "Beyond Horizons: A Half Century of Air Force Space Leadership," *Air Force Space Command*, 1998, p. 58.

[7] Taubman, P., *Secret Empire: Eisenhower, the CIA, and Hidden Story of America's Space Espionage*, Simon & Schuster, Upper Saddle River, NJ, 2004, p. 322.

[8] Reed, S., Van Atta, R., and Deitchman, S., "DARPA Technical Accomplishments: An Historical Review of Selected DARPA Projects," Inst. for Defense Analyses Paper, 1990, pp. 2-1–2-3, 2-5, 3-7, 4-2, 4-5, 5-2, 5-6, 11-1, 11-6, 11-7.

[9] Widger, W., *Meteorological Satellites*, Holt, Rinehart, and Winston, Philadelphia, 1966, p. 136.

[10] Ashby, J., "A Preliminary History of the Evolution of the TIROS Weather Satellite System," NASA, 1964, p. 10.

[11] Sobel, D., "A Brief History of Early Navigation," *Johns Hopkins APL Technical Digest*, Vol. 19, No. 1, 1998, pp. 11–13.

[12] Danchik, R. J., "An Overview of Transit Development," *John Hopkins APL Technical Digest*, Vol. 19, No. 1, 1998, pp. 18–26.

[13] Magnuson, S., "History of DARPA in Space," *50 Years of Bridging the Gap*, 2008, pp. 112, 114.

[14] Argo, H., "Satellite Verification of Arms Control Agreements," *Arms Control Verification*, Pergamon, Oxford, England, U.K., 1985, p. 292.

[15] Sloop, J., "Liquid Hydrogen as a Propulsion Fuel," NASA SP 4404, NASA Historical Series, 1978, p. 223.

[16] "History of Centaur," NASA Lewis Research Center, undated, p. 2.

[17] http://msl.jpl.nasa.gov/QuickLooks/glomrQL.html.

[18] DARPA Strategic Plan, DARPA, Arlington, VA, 2007.

Chapter 17

POTENTIAL SUCCESS STORIES IN THE FUTURE

D. K. SACHDEV*

After 15 landmark success stories by industry leaders, it is time now to look ahead and try to project what kind of success stories we could be celebrating in the coming years and decades. Some of these could be extensions or derivatives of what my coauthors have presented so well in the preceding chapters whereas others could be totally new, perhaps stimulated by the changing needs of the society, technological advances, or system architectures not yet developed or even thought of. Before we do so, it is helpful to set some benchmarks. We do so by revisiting a well-known vision for telecommunications.

VISION FOR THE FUTURE

In most fields, their leaders and strategists often project long-term visions that can act as goalposts for scientists and system developers for a length of time dependent on the pace of innovation in that particular field. Such visions might be relatively short lived or might stay the same for several decades. As we make progress, a stage eventually comes when it is time to reset the target success criteria in order to align them better with the accumulated results of technological advances and to respond more effectively to the ever-changing needs and demands of the society. We are probably at that the stage now for telecommunications in general and for satellite-based systems in particular.

In the 1950s, before the very first satellites were even conceived, the undisputed leader of telecommunications was AT&T and in particular its legendary Bell Laboratories, the home of many far-reaching advances and breakthroughs in several fields of science and engineering. Its leaders often articulated what was billed as the "ultimate vision for a telecommunication networks." Although the actual words used to vary at various times and with various

*President, SpaceTel Consultancy LLC, Vienna, Virginia; Adjunct Professor, Electrical Communication Engineering, George Mason University, Fairfax, Virginia.

speakers, the sense of such an ultimate vision for telecommunications used to go something like this:

> As soon as a child is born, he or she will be given a phone number; anybody calling that number should get either a ring or a busy tone. If there is a ring tone but the person does not answer, he or she is either not yet born or no longer alive.

I first became aware of the preceding vision in 1958 as a young man, only a couple of years out of college, starting my career in Indian Telecommunications. Influenced no doubt by the often dismal status of telecommunications around me, I never expected this vision to be realized in my lifetime because I considered it almost utopian. However, we live in an age when all development life spans are constantly getting compressed. After half a century of innovation and recent waves of deregulation and huge investments around the world, capabilities matching this vision do exist in many parts of the world, especially if we also include the now commonplace messaging services. Although this vision is still not global in nature, its feasibility has indeed been established. From the relatively narrow perspective of the satellite medium and this book, I hope the readers would agree that the success stories in the preceding chapters do substantiate that the satellite technology has contributed to the realization of this vision in a variety of ways, more often than not working side by side with other technologies and systems.

As we look toward the future, a pertinent question is: do we still work towards the same vision or is it time to make some adjustments or even radical changes? To answer this fundamental question, all we need to do is to look around us, study the evolving trends, and project them forward even by a modest amount. In my assessment, that once seemingly utopian vision is no longer adequate for the next 20 to 30 years. Several fundamental changes that are taking place in an overlapping fashion lead us to this conclusion. Two such changes—one technological and the other societal—are contributing the most.

The first is the Internet revolution that has transformed the way we interact with the rest of the world and its people and its myriad systems. Internet has become so pervasive and universal that it is inconceivable to conduct any business and go through our daily lives without it. And something that the younger generation does not always recognize, the backbone of Internet is the global telecommunication network of which satellites are an integral part.

The second need for change is more fundamental and societal in nature and is related to how the coming generations will prefer to live and work. The now well-known iPod generation of teenagers is expanding to other age groups, and the older "Blackberry" community is not far behind with its ever-present e-mails and content in various forms. Several age groups now prefer customized content rather than a broadcast one while the Blackberry community cannot live without the always connected push e-mail technology that downloads messages around the clock. If we combine both these trends in the marketplace, we can envision a world in the not-too-distant future where ubiquitous connectivity and *customized* content often *with mobility* would increasingly become the rule rather than an exception.

This is not all. A recent issue of *Economist* magazine [1] developed in depth the simple theme of "Always Connected Society" largely driven by the phenomenal progress in wireless technology. Although that by itself is not new, what is significant is that the future always connected societies will not be limited just to the billions of human beings but also will include a variety of simple and complex machines, all ready to interactively exchange data and perform programmed as well as intelligent functions, all in a seamless network.

If we combine all of these trends, we can see a whole set of new demands for the global telecommunication systems of the 21st century and beyond. Such a vision will include not only all living beings on the Earth (and maybe beyond) but also many machines and interactive objects of all kinds, often with no distinction between the two in terms of universality of connectivity. Obviously, different visionaries would express such a vision differently. For now we offer the following formulation:

> Information exchange, often customized, will be instantaneous, global and ubiquitous not only among human beings anywhere but also between and to inanimate objects and machines as needed without any constraints about location, time or mobility.

Given the increased emphasis on mobility and ubiquitous coverage, such a vision does provide new opportunities and challenges for the satellite medium. Such opportunities are especially enabled by high-power satellites and in some cases by the expanding processing capabilities in both space and on the ground. The benefits should manifest through affordable ubiquity and small portable and flexible user devices. However, as we aspire for such opportunities, we have to be cognizant of any inherent limitations of this medium. One such limitation that has often been challenging in the past is the finite amount of radio spectrum; the other is the inevitable satellite signal shadowing by man-made structures in urban areas.

With this draft vision in mind, we will now try to identify, and perhaps speculate, the kinds of success stories we will be reading and writing about in the coming decades. Although most of the earlier chapters have been written largely around different systems or operating entities, when we are projecting into the future it is more relevant to focus primarily on specific services and technologies that could be instrumental in enabling such stories to come into being. Wherever relevant, we will refer to the operating organizations as well.

DIRECT BROADCASTING

Although the ubiquity of satellites was recognized right at the very beginning of the satellite technology, practically ubiquity in terms of direct access to users became feasible only in the 1990s through a combination of high power satellites, signal compression, and ASIC technologies. The first such broadcast service was television broadcasting or direct broadcast services (DBS) followed after a few years by satellite radio.

TELEVISION BROADCASTING

Several chapters in this book have addressed the progress around the world in this service sector that currently account for over 100 million DBS users around the world (available on-line at http://www.sia.org/file/2008SSIR.pdf). The business models range from integrated entities with their own infrastructures providing full end-to-end services to those providing operating platforms for multiple independent service operators via long-term lease of high-power transponders. After several years of steady growth, several new trends have emerged recently. Among these trends are the following:

1) Newer markets in Asia such as China and India are demonstrating extremely high rates of adoption. Currently such rapidly growing markets are making the orbital arc in that part of the world one of the most congested.

2) High-definition (HDTV) and video-on-demand offerings are creating new demands for higher satellite capacity often garnering much higher revenues from existing customers. These trends are in turn leading to demand for new spacecraft with higher capabilities.

3) After decades of anticipation, the fiber-to-home capabilities are becoming operational in selected markets. Although their cost per household is still much higher, they are capable of creating market pressure on satellite-based markets, particularly through bundles services combining with broadband and telephone service.

While the broadcasting market is getting stronger, particularly with high-definition global coverage for special events like the Olympics, there are definite signs of growing customer preferences for some kind of customization through video-on-demand and interactive features more easily provided via traditional cable-based systems and the newer fiber-based systems.

Overall, this sector still has significant growth ahead both in terms of number of users as well as the volume and quality of the average content per user. Such demand will no doubt create new success stories in the future.

DIGITAL RADIO

Until the late 1990s, there was no satellite digital radio at all, except for some limited experiments. Part of the impediment was spectrum allocation, which was agreed to in 1992. The first operational application was the WorldSpace system in Africa (1998) and Asia (2000). This was soon followed by the two digital radio systems in the United States, Sirius and XM Radio. In addition to space and time diversity in the space segment, the U.S. systems also have terrestrial repeaters to fill in the gaps in satellite reception created by shadowing in urban areas. These two systems have demonstrated some of the highest adoption rates for any consumer equipment and have today between them over 18 million subscribers and counting. One of their strongest assets is that practically all of the major automobile manufacturers have adopted one system or the other as Original Equipment Manager (OEM) equipment, thus facilitating easy capture of new subscribers. The two systems have merged in August 2008 after a protracted review of their justification that was largely based on the growing competition from alternative media. Chapter 13 describes the XM system before this merger.

Another digital radio system is the MBSat being used by both Japan and South Korea for innovative multimedia services. The Korean system in particular has captured in a few short years several million subscribers. These numbers are impressive when we recall that the total population of South Korea is only 25 million with less than half in urban areas.

European digital radio systems have been under serious consideration for almost two decades. WorldSpace is in the process of extending its service in Africa to some of the European markets after installing terrestrial repeaters. There is still no dedicated digital radio system covering the whole continent. Current contenders for Europe are Ondas Multimedia and WorldSpace. Europe and its industries have contributed in many ways to the success of most of the current digital radio, including significant industrial contributions. However, it seems to have a much tougher time adopting them for its own operational use. Among the possible reasons for this situation are stronger and consistent resistance from terrestrial broadcasting systems with superior service and multiple languages and cultural barriers for continent-wide broadcasts. The relatively higher latitudes of prime areas and the need to cater to multiple languages also make the space segment investment higher than that for other regions.

What about future evolution of this technology? Will the future systems be similar in architecture and business plans or different? Is the competition serious enough to slow down new investments in this field?

One of the reasons satellite digital radio systems took longer to be built was the fierce opposition from local radio stations in the United States, and this is one the factors still slowing down this technology in Europe. The satellite-based systems have the benefit of superior digital technology in terms of quality and fidelity and also are truly ubiquitous throughout the coverage area unlike the limited domain of local radio stations. These dual benefits are the principal reasons for the fast uptake of this new technology.

However, this is beginning to change albeit slowly. After over a decade of development and standardization work, the local radio stations are finally taking the first step in beginning a systematic transition from the decades-old analog AM and FM techniques to digital broadcasts with their superior quality and resistance to all kinds of radio interference. This alternative, commercially known as HD radio in the United States, is finally getting adopted by the broadcasters, and the new receivers on the market are beginning to get positive customer response. It also has a clear advantage of being free. It does not offer anything close to ubiquity of satellite radio because the coverage is still limited to populated areas with the older AM and FM towers. However it has some not-too-well known demographic data to draw inspiration from. Although there is no doubt that the ubiquitous coverage of satellite radio is its strong selling point, the majority of its so-called "satellite radio" subscribers are in fact most of the time listening through terrestrial wireless repeaters whose signal is much stronger in urban areas where a majority of the subscribers live and work. Whether such subscribers would give up the assurance of country-wide ubiquity of reception through the satellite in exchange for free higher-quality reception only in urban and suburban areas is debatable. Whether HD radio or its equivalent systems elsewhere have the strength in the future to put a significant dent in the growth of digital radio will not be known for some years to come.

The other challenges to satellite radio quoted in the justification for the merger of the U.S. systems are the phenomenally successful iPod and similar products. Their attractions are many; high in the list is that the owner can customize the content to his/her taste and mood. This does require deliberate actions (and cost) in order to create customized content periodically. While iPod is certainly a serious challenge to packaged music on tape or CDs, its direct challenge to satellite radio is difficult to measure. However, it does highlight that as the access to interactive media becomes more pervasive, the one-way broadcast architecture of satellite radio can become a drag on future growth for this medium.

So in what form will we be writing the future success stories in this technology? I believe in the coming decade we will certainly celebrate additional systems similar to those in operation today. These will still have one-way broadcasts to wide market areas, possibly with multiple beams for different languages and interest groups. The receivers will become part of the automobile Telematics systems, better integrated with other services in the cars. Concurrently, we will see fascinating range of portable receivers popping up as part of all kinds of devices or even human attire thanks to progress in ASIC technology and flexible antennas.

If we now look at television and radio broadcasting together, we see some commonalties and differences. Whereas radio was the first ubiquitous mobile service enabled by the use of wireless technology for terrestrial repeaters,

DBS is to stationary locations via line-of-sight links Both are one-way systems and currently lack real-time interactivity. Technologically they are subsets of multimedia services and might one day combine into one integrated system. Finally, both have several success stories to come with or without new system features.

BROADBAND

When the work on this book was started almost two years ago, it was the editor's ardent wish that broadband satellite systems should also be in a separate chapter with their own success stories. However, a closer look at the status of satellite-based broadband systems at that time suggested such a chapter had to unfortunately to wait a while longer. Instead, several authors in this book have given due importance to such services in their respective domains. As we got closer to publication however, there has been substantial progress, and there is finally a sense of optimism in this industry for this sector, and it is evident that several success stories are around the corner.

Broadband, or by any other name, has traditionally meant connectivity to Internet. Starting small, it is growing exponentially. Thus, in 1999, the total global Internet traffic in the entire year was about two exabytes (one exabyte is a shade over one billion gigabytes). However, in 2007, according to a recent estimate it was at a level of about one exabyte per hour and growing. Currently, almost all of such traffic is carried by terrestrial means and only a small fraction by wireless and line-of-sight satellite links. Furthermore, as briefly discussed in the beginning of this chapter, the whole telecommunications is moving towards interactive exchange of information, thus widening the very definition of what has been called "broadband."

In this section, we will briefly discuss progress via line-of-sight satellite links or, as Mark Dankberg likes to call them, platform-based broadband traffic. Mobile interactive applications will be discussed in the next section under hybrid systems.

Satellite broadband started with Ku-band as an outgrowth of DBS but is now moving rapidly towards Ka-band at least in those areas that do not have excessive propagation impairments in this band. Multibeam Ka-band satellites have advantages over Ku-band versions in terms of lower user terminal cost and lower space segment costs per consumer through higher capacity with an equivalent investment. (See Chapter 15 for an interesting analysis on this topic.) This band also has significantly higher overall capacity in a given orbital arc because of the feasibility of closer spacing of spacecraft than with Ku-band. Although the industry is nowhere near exhausting the overall global orbital capacity with this band, the total throughput of orbital arcs accessible to populated landmasses can still capture a small fraction of the overall global broadband traffic.

The broadband satellite industry as a whole made its start through the globally backhauling of Internet traffic between areas without direct links to wideband fiber-optic networks. This role is still continuing primarily through telecommunication or distribution-type spacecraft.

In the late 1990s, successful DBS systems began to offer broadband services by leveraging their growing clientele and using part of the one-way high-power DBS satellites combined with return links either via phone lines or to separate transponders. Such systems had modest success because of their high costs. Among the factors driving up the costs included higher user equipment costs in part because of low antenna gain/noise temperature (G/T) of return link transponders and higher space segment costs per user. The net outcome of such system and cost implications was a kind of setback for the satellite medium, unfortunately about the same time that the cable industry had finally managed to solve its own technical issues for two-way links on cables and had moved to a common industry standard for subscriber equipment. As a result, the business community almost started to write off the satellite medium as one of the service options for the broadband revolution rapidly unfolding around the world.

MULTIBEAM BROADBAND SYSTEMS

Notwithstanding this discouragement, the satellite community regrouped and quickly identified what was needed at the system level. The answers came quickly: what was needed was an architecture that had much higher downlink capacity and also enabled less expensive user equipment through more efficient subscriber uplinks. The answer to both these imperatives was multibeam satellites with efficient and economic designs. Systems meeting these objectives have been built at both Ku-band and Ka-bands, and we will briefly summarize their status next. We will also summarize the progress being made with more efficient approaches using "traditional" Ku-band capacities around the world.

Relatively small countries in Eastern Asia have led the broadband arena in at least two ways. South Korea was the fastest to expand broadband access largely though with terrestrial (mainly ADSL) technologies and leads in the percentage of population having access to broadband, over 25%. The second country is Thailand where Thaicom deployed in 2005 a Ku/Ka-band satellite, IPSTAR. This satellite, now renamed Thaicom-4, provides nearly 45Gbit/s of both-way capacity in a single spacecraft. The subscriber access is at Ku-band, but unlike the DBS-derived systems in the United States and Europe, it has a large number of spot beams covering several countries in Asia. Only the feeder links are at Ka-band. The Ku-band beams cover a population of nearly two billion that is rapidly achieving a standard of living that can afford and demand broadband connectivity. The operating company, Thaicom, is progressively setting up national and regionally companies to provide gate-ways and customer service with user equipment tailored to each market. Is IPSTAR on way to become a success story in the near future? Most probably, yes. If indeed it does, it will be because it developed the system from scratch in order to match the market. The challenge for this system is that the intrinsic demand for broadband in this region is still somewhat in the future. Nevertheless, it is a system worth watching from multiple perspectives.

The WildBlue Ka-band Multi-Beam System was the first serious outcome of the modified approach just mentioned. The system operates via two Ka-band transparent or bent-pipe payloads from the same orbital location with coverage of North America. Whereas Anik-F2 is a multifrequency spacecraft (see Chapter 4 on Telesat), Wildblue-1 is an all Ka-band spacecraft. Together they share the Ka-band up- and downlink spectrum through com-mon 45 spot beams across North America. In terms of architecture, this pair meets the objectives just summarized. The market response is very positive, largely in areas not served by terrestrial systems. As the user equipment prices drop, this system should be able to compete with the much larger terrestrial counterparts in urban developed markets. If and when it does meet this challenge, it will certainly be a success story worth writing about, hopefully in the near future.

The Spaceway series of spacecraft are unique in several respects. They started out as highly efficient and flexible satellites targeting business users via spot and scanning beams. Three such spacecraft have been manufactured. However, shortly before the first of these spacecraft could be launched, the company changed hands, and the new owners did not see a future in satellite broadband given less than successful record of attempts until then with tradi-tional DBS satellites as just summarized. It was decided to quickly adapt the first two spacecraft for DBS, a much larger and established market with tre-mendous growth potential with HDTV, etc. The third spacecraft, Spaceway-3, however remained with the original owners, Hughes, and is now in orbit and has started broadband service in early 2008.

This series of spacecraft were quite costly, and at the time of their design there was no Ka-band competitor in the marketplace. Therefore it will be interesting to see the two systems, one a traditional multibeam bent-pipe system and another with high tech features such as scanning beams and onboard processing, compete in the marketplace. Certainly there is a room—and need—for both of these systems with the potential to become success stories in the coming years.

The success of the Anik F2/WildBlue spacecraft in the United States and similar Ku/Ka payloads in Europe has spawned significant new investment activity recently in this sector. Early 2008 saw the announcement of orders for two large-capacity broadband satellites, one by Viasat, the leading manu-facturers of ground equipment for WildBlue and other broadband systems around the world. The other "companion" order was by Eutelsat in Europe. Both systems are aiming for capacities of 100 Mbytes or higher in their respective markets, thus raising the possibility of more competitive offerings through lower space segment costs per user. A third such order by Hughes is expected before this book is published. All of these programs foreshadow a brighter future for this area compared to the status only a few years ago.

Ku-band Lease Capacity Broadband Services

As previously mentioned, the satellite-based broadband services started with traditional DBS satellites or their broad-coverage fixed satellite service (FSS) counterparts. They did not achieve much success. Nevertheless, thanks to continuing enhancements in system use and better ground systems, this sector is continuing to grow, not only in North America against the newer Ka-band systems, but particularly around the world where Ku-band capacity is the single largest resource available for lease. According to recent reports, some 900 transponders are currently in use for broadband services around the world. However, the total subscribers are still a small fraction of those carried by cable and ADSL.

In summary, Ka-band broadband systems have established an appropriate system approach for satellite broadband, a capability with potential to contribute substantially to the proposed vision for telecommunications. As the newer systems come into operation, and more experience is gathered, the capabilities are likely to become more competitive although the fiber-to-home bundled initiatives are in many ways shifting the competitive goal posts.

Lastly, although we have primarily talked about fixed applications, Ka- and Ku-band systems are also beginning to find niche applications for recreational vehicles, trains, and airplanes through tracking antennas—overall, an area replete with success stories in the near future.

Integrated Satellite-Wireless (Hybrid) Systems

We will now discuss a group of systems that could potentially lead to a whole set of success stories, not just a single one. This is also an area that appears closest to the draft vision proposed earlier in this chapter. First, here is some background.

Satellites are the most ubiquitous communication tool invented by man, although in reality it took nearly 50 years of progress and growth before true consumer ubiquity in terms of reasonable size user equipment could be realized. However, the microwave frequencies used in satellites require line of sight between the transmitter and user. Any soft or hard blockage of such a visibility leads to degradation of quality or loss of communications.

During the last 20 years or so, terrestrial wireless communication has grown out of the category of old-fashioned and often discarded communication tools to one of the hottest connectivity technology with mobility. Under favorable circumstances, wireless frequencies can also "bend" to a certain extent, a factor critical to the phenomenal success of cellular mobile phones. However, the range of wireless reception is limited because of much faster signal attenuation as a result of absorption by ground and man-made structures.

As the satellite technology reached the maturity where direct consumer access could be provided, a whole host of services began to emerge. Initially these were limited to static installations with permanent line of sight to the satellite such as those for DBS and VSATs. However, for services requiring mobile access such as first-generation satellite radio, it became obvious that satellites alone could not satisfy the quality of service (QoS) for a mobile user as a result of too frequent quality degradations or complete loss of service because of signal blockages by natural obstructions or buildings, etc. The solution came from wireless systems, and the first systems to adopt commercially what we are calling in this section a "hybrid" architecture were the two satellite radio systems in the United States. As already discussed and explained in Chapter 13 on XM Radio, these systems use terrestrial repeaters in areas of high blockages. Such terrestrial repeaters rebroadcast the satellite signal in their respective coverage areas through what are frequently referred to as singe-frequency networks. It is fair to say that without this symbiosis of satellites and wireless in a hybrid system, digital radio would not have been a success story today.

The preceding "hybrid" concept is now being extended to other systems. The first one in advanced implementation phase is the group of mobile systems, with one notable difference. In digital radio systems, the one-way content is broadcast continuously throughout their coverage area; therefore, the associated terrestrial repeaters in urban areas need to use dedicated frequency band segments. However, in multibeam mobile systems, each beam (that can cover typically a whole urban area) uses only a fraction of the total allocated spectrum for the system. It is therefore feasible with some guard bands and regions to dynamically use the remaining segments of the total allocation for terrestrial repeaters in the same area. In other words, no additional spectrum is needed for the terrestrial segment. We will discuss a few such systems next.

NEXT-GENERATION SATELLITE MOBILE SYSTEMS

In this book, we have included two success stories on mobile satellite systems, both for geostationary satellite systems. The first, Inmarsat, is a global network that has steadily expanded over the decades but has generally addressed professional users on land and seas. The second story is on the Thuraya geostationary direct-to-user geostationary satellite mobile system. Both of these systems are satellite-only systems and operate very well as long as reasonably clear line of sight is maintained between the satellite and the user equipment. This often translates to either keeping the equipment stationary while in use or the user moving to an area with lower blockages. Notwithstanding these limitations, both networks have been successful and are expanding steadily. In addition to these geostationary systems, there are, of course, the two well-known low Earth orbit (LEO) mobile systems, Iridium and Globalstar, just discussed. These LEO systems also have the same concern regarding blockages as their geostationary counterparts.

This concept of dynamically sharing the same spectrum between satellite and terrestrial components of one integrated system was approved by the FCC in early 2005. The terrestrial component of such systems is designated as ATC (auxiliary terrestrial component). The EEU has recently adopted a similar approach but has named it CGC (complementary ground component).

Such hybrid mobile systems will compete head on in the terrestrial markets with large well-established cellular operators. Therefore, their user equipment has to be comparable in features and size with that of competition. This requirement is leading to the development of some of the largest commercial spacecraft with large reflectors.

An attractive feature of these systems is their ability to either direct higher capabilities in one particular area or alternatively to switch to a broadcast mode over the coverage area as required.

Currently, one such system, ICO, has deployed one such spacecraft over the United States in early 2008. Other systems planning similar launches in the near term are Terrestar and MSV. Current mobile systems such as Inmarsat, Globalstar, and Iridium have also shown interest in the ATC for their respective space segments. The European Union is likely to award spectrum to selected parties sometime in 2009.

Will such systems be the success stories of the near future? Very likely, yes. Such an achievement will be particularly important for the satellite industry as a whole given the rather mixed success record in the truly mobile field so far. More significantly, such systems come close to meeting the vision of mobility and ubiquity developed earlier in this chapter. If demand for such systems grows, the limited spectrum currently allocated could become a constraint. This might suggest extending such concepts to other band segments as well.

SENSOR-BASED SYSTEMS

During the last decade or so, tremendous advances have been made in all kinds of sensors for biological, industrial, and security purposes. At the center of this progress are advanced miniature sensor modules often combining the microelectromechanical systems (MEMS) and wireless technologies in an innovative way. When interconnected in an appropriately configured wireless network, such sensors can gather and transmit a whole range of information that until recently required manual data collection by qualified personnel. The first set of major advances in wireless-based sensors was made through a DARPA-sponsored program, now popularly known as Smart Dust. It realized for the first time the miniature sensor technology with tiny wireless transreceivers with built-in batteries all in a size comparable with medium-size dust particles, hence the project's name. This technology has since then expanded and branched off and is finding applications for all kinds of human, industrial, and security needs.

One general area where there is a real potential for a combination of satellite and wireless technologies is for wide-area security systems. Typical examples currently receiving serious attention and funding include border security, infrastructure surveillance, tracking of containers on high seas, and first responders' networks. Although the objectives and detailed implementations differ quite a bit, the architectures for most of these systems consist of a number of sensor networks in specific areas or infrastructures (often unattended) interconnected with a central network center via secure satellite links. Such networks can be configured for simple monitoring or also for remote actuations of controls through properly controlled sensors. Although cost and reliability considerations are relevant, the primary emphasis is on speed and survivability under any disaster scenarios. When properly designed and implemented, such systems can have a greater survivability than landline systems.

The potential of this technology is still under development, and a number of young entrepreneurs are pushing the capabilities in different directions such as size, battery capacity, ability to self-diagnose any malfunctions, and operation in flexible and ad hoc network configurations—almost definitely an area with many success stories in the future.

NEXT-GENERATION LEO SYSTEMS

Just before the phenomenal success and growth of wireless cellular systems started in earnest, two major satellite mobile systems were conceived and deployed in the second half of the 1990s. These are, of course, the Iridium and Globalstar LEO constellations. These global systems were the very first serious entries by the satellite industry in the portable mobile communication arena. Unfortunately, both of these systems filed for bankruptcy soon after going into operation. The net result was that commercial LEO systems of practically any kind were considered a taboo both by the professional as well as by the financial communities.

That is until now. The financial community during early 2007 supported the order for a second-generation 45-satellite constellation for Globalstar, and the Iridium system also announced that it will follow suit in a few years when its first-generation satellites are approaching their in-orbit lifetimes. How do we explain this complete turnaround? The Iridium system had its "near-death" experience when it was close to being deorbited in order to save the expenditure on the ground for maintaining its barely used 66 satellites in orbit. Under the then geopolitical environment, a fortunate turn of events was that the U.S. military found this derelict constellation of immediate use for data applications in remote areas. The successor company, with just $25 million investment in a bankrupt constellation with an original investment of over $5 billion, found this opportunity good enough to continue operations rather than to liquidate its assets. The rest, as they say, is history. Such data applications have grown since then and have stimulated interest from other niche markets in remote areas where other systems have a hard time providing service. In this regard, it is perhaps worthy of note that very recently one of my students, with personal involvement in one of these systems, remarked that had Iridium held on financially until that fateful day of 11 September 2001, it might not have gone bankrupt—an interesting "what-if" analysis indeed.

For telecommunications, the fundamental issue not yet fully resolved is still the same, namely, is there a role for commercial LEO systems as an alternative architecture for future satellite systems beyond the limited government applications? The main attraction of such orbits is still the much shorter time delay compared to that for geostationary orbit (GEO) architectures. The first-generation LEO systems had their premature financial collapse primarily because of rapid evolution of cellular systems and unattractive user equipment. While the electronic industry continues to shrink the handset size while adding newer feature, such advances are benefiting cellular systems even more. In addition, as just discussed, the geostationary mobile systems are making a strong return to provide another challenge to LEO telecommunication systems.

Should we expect future success stories here? The kind of services being nurtured today by the first-generation Iridium and Globalstar are certainly large enough to attract funding for the replacement constellations. These are likely to be success stories of the future.

POSITIONING AND NAVIGATION

As Chapter 12 highlights, GPS has been a phenomenal success that continues to grow by the day. Not only was it a totally new service, but also it had the advantage of starting right from the outset as a global system, thanks to the adoption of a medium Earth orbit (MEO) architecture by the U.S. military, which continues to provide free and universal access for all transmissions that are open for public use.

Unlike other applications that tend to be self-contained businesses, GPS is different as it provides a basic universal facility of highly accurate reference frequencies that can be combined with a seemingly unending range of end-user applications with their own business plans and objectives. Thus, the one set of GPS signals entering the car for navigation could also be enabling other location-sensitive services and systems.

Can we identify future success stories in this category? First of all, the U.S. system described in this book is continually being upgraded for newer services and better accuracies. One of innovations underway is the addition of intersatellite links in order to substantially reduce the time needed to periodically synchronize all of the spacecraft. In addition to improvements in the spacecraft constellation proper, perhaps this system is contributing to many current and future success stories in the application arena and will continue to do so in the future.

In parallel with the evolution of the U.S. system, several other similar systems are being operated, built, or planned by other countries or groups of countries. Notable examples currently know are the European system (Galileo), Russian system (Glonass), Chinese system (Beidou and Compass), Japanese (QZSS) system, and the Indian system. The Glonass is not really new but is being considerably expanded. All of these have one common objective and many additional specific missions. The common objective is to ensure that as they integrate the GPS system more and more into their public and commercial applications, they would like to have another alternative if for any reason—technical or political—the access to the GPS system is not possible. In addition, in many cases concurrent operation with the GPS is expected to provide additional enhancements including higher accuracy, availability, etc. Overall, given the amount of talent and innovation being devoted to this field, there is no doubt that there will be future success stories as well, not always as linear extrapolations of the past.

SPACECRAFT TECHNOLOGIES AND CONCEPTS

The spacecraft technology continues to advance in almost all relevant areas. Currently, the drivers towards larger spacecraft include direct broadcasting and geomobile spacecraft. Electric propulsion, after decades of development, is accumulating in-orbit fault-free periods to a point that within a few years, one can envision spacecraft relying on this technology exclusively without having bulky chemical backups. Electric propulsion for transfer orbit itself is some ways away, but here also recent science missions hold the prospect of substantial launch mass savings. If such savings are transferred to higher payload capabilities, we could see more competitive offerings. One area that is still waiting for a definitive solution is higher efficiency solid-state power amplifiers.

Spacecraft subsystem technology advances are generally enablers of success stories in the operational and business sense. However, there is one notable area that maybe an exception. We will discuss this briefly now.

FRACTIONATED SPACECRAFT

As discussed in the overview (Chapter 1) and in other several chapters, system designers for quite some time now have been exploring means of progressively building capacity in space for business reasons or technological limitations. The business drivers can be to invest progressively rather than right at the outset as a hedge against market uncertainties or limited availability of spectrum in the initial stages. Often the need to occupy all of the spectrum allocated at a particular location can sometimes force the operator to build right at the outset a spacecraft much bigger than the demand forecast might warrant.

A well-established approach to meet the objectives of staggered investment matching the demand profile in time is to progressively collocate multiple spacecraft at the same location. Among several operators adopting such an approach, SES has perhaps been the most ardent follower of this principle in Europe. Each such spacecraft in such a "cluster" is autonomous and complete in all of its functions and is managed directly from the ground.

At the other extreme is the principle of hard assembly in space. The leading example of such an approach is the giant International Space Station (ISS) that has been progressively built over a period of almost 20 years via successive shuttle launches. Such an assembly has so far involved considerable human interaction for hard docking and assembly. Automatic docking has been successfully used for unmanned cargo vehicles like the Progress and more recently with the automatic transfer vehicle (ATV).

Owen Brown and his colleagues at the Defense Advanced Research Projects Agency (DARPA) have been addressing this subject from a novel and different perspective and have come up with some very interesting solutions and architectures [2]. They approach the subject from the perspective of responsive space, which they define as "the capability of the space systems to respond rapidly to uncertainty" [2]. They go on to define different types of uncertainties to include technical, environmental, demand, and even funding. After several years of study and reviews with experts in industry and academia, they have developed the concept of F6, which stands for future, fast, flexible, free-flying fractionated spacecraft. In early 2008 contracts were awarded to industrial teams to develop the key enablers for the fractionated spacecraft [3].

The most notable departure from traditional thinking in the F6 is how the functionalities are subdivided or "fractionated," borrowing a term from the art world. Instead of each module being a small but complete spacecraft, it is instead responsible for a selected functionality for the complete cluster. All of the modules are "free-flying" as they are only connected by wireless links using standardized software. Thus, one module could be responsible for the TT&C for the complete cluster while another one might generate power and distribute it wirelessly over short distances. With such a "virtual satellite," a replacement module could be launched near the cluster and commanded to "log" itself into the network without any complicated rendezvous, docking, or robotic service.

Like many other disruptive approaches from DARPA (e.g., see Chapter 16), F6 could also change the nature of the space industry at least for certain centrally controlled objectives. It has several potential benefits: for example, selective upgrade of technology and change in overall capacity with demand and softening the shock of launch failures compared to large monolithic structures. Could this be a really successful story in the future? Yes, it will be definitely worth watching its progress.

SPACE ELEVATOR

And finally, we will end this chapter and the book with a story that in fact started in the 19th century and may come to fruition in the mid-21st century.

As we all know, whenever we watch a spectacular launch of a rocket, we are so fascinated by its roar and fireworks that we tend to ignore that it is probably one of the most inefficient tools known to man. Even one of the most advanced rockets today, Ariane 5 ECA, weighs 700 tons, but it can only deliver 9.8 tons in transfer orbit. After subtracting the amount of the onboard fuel required for reaching the geostationary orbit, the delivered mass in that orbit would be at best 7 tons, representing a net efficiency of 1%. Apart from the straight economics, a more alarming part is that the bulk of this 99% wasted mass is either the fuel burnt during ascent thus creating pollution, or metallic structures that either end up in the sea or, worse, increase the amount of the debris circulating the Earth.

Alternative concepts of reaching beyond the Earth had begun appearing as early as 1895 when a Russian scientist, Konstantin Tsiolkovsky, published a concept to go into space inspired by the Eiffel Tower. Several others followed him. In 1979 Arthur Clarke popularized this concept through his science fiction novel, *Fountain of Paradise*.

Among the several alternatives being considered, the tethered approach has been developed much more than others. Essentially, this envisions a rope that is tethered to a structure about 100,000 km above the Earth with its center of gravity around the geostationary orbit. Such a rope could in principle be used to carry spacecraft up to the geostationary orbit (and beyond) if it had its own powered propulsion system, much like an everyday elevator. Such a system if realized in actual operations in principle would have no wastage, would have no pollution or debris, and would be far more efficient compared to the rocket technology today and perhaps forever.

Can this concept be realized? Yes, but there are still many challenges that may take years to conquer. However, the most difficult of these could finally have a viable solution. This was to find a material for the long rope that would have much higher tensile strength/density ratio than the best steel. The technology for such a material is nanotubes; however, building a rope over 100,000 km long is still a distant target. There is consensus now that the elevators or climbers, as they have come to be named, would be powered by laser beams from the ground. The physical process of building such an elevator is by no means easy; the preferred concept is to roll out the cable from a very large "deployment spacecraft" launched by conventional means. The most promising prognosis is that the lead time estimates have dropped from 300 years to 20 to 30 years and the expected costs from some $40B to less than $10B.

It would indeed be a fitting tribute to the memory of Arthur Clarke if a space elevator were to actually be built by 2045, which is one hundred years after the concept of geostationary orbit was first presented by him. It would be even more memorable if the ground anchor for such a space elevator is near Sri Lanka, which was the site of the fictional elevator in *Fountain of Paradise*!

REFERENCES

[1] "When Everything Connects," *The Economist*, 26 April 2007, Economics.com.

[2] Brown, O., and Eremenko, P., "Fractionated Space Architectures: A Vision for Responsive Space," AIAA 4th Responsive Space Conference, RS4-2006-1002, April 2006.

[3] Brown, O., and Eremenko, P., "Application of Value-Centric Design to Space Architectures: The Case of Fractionated Spacecraft," AIAA Paper 2008-7869, Sept. 2008.

INDEX

CONTRIBUTING AUTHORS

Al Mazrooei, Ali Saeed has been the Chief Technology Officer of Thuraya Telecommunications since 2002. He joined Thuraya in 1998 to supervise the design phase of the ground segment and since then has held several senior positions. Al Mazrooei started his career in 1992 with Emirates Telecommunications Corporations, Etisalat, in building and operating the UAE GSM network. Then, he began working with the Dubai government in 1996 to support IT systems and networks.

Berretta, Giuliano is Chairman and Chief Executive Officer of Eutelsat Communications. He joined Eutelsat in 1990 as its first Commercial Director and was Director General from 1999 to July 2001 to steer the transformation of the assets and activities of the organisation into a private company. Prior to Eutelsat, Berretta worked at the European Space Agency's (ESA's) headquarters in Paris and its technical centre (ESTEC) in the Netherlands where he was instrumental in the design of Europe's pioneering communications satellite programs. He began his career in Italian industry (Telettra, Selenia) in the field of military and civilian radio links and television broadcasting.

Bhaskaranarayana, A. is the Indian Space Research Organization's (ISRO's) Scientific Secretary and Director, Satellite Communication Programmes and Frequency Management, Program Director-INSAT and Edusat and Telemedicine. He received the Distinguished Achievement Award for his contributions to the Aryabhata project, National Research Development Corporation (NRDC) award for Innovative Hardware Development, and ASI Outstanding Achievement Award for his contribution to spacecraft and related systems. He is a graduate of Indian Institute of Technology Madras and Fellow of The Institution of Electronics and Telecommunication Engineers (IETE).

Brown, Owen is a Program Manager in the Tactical Technology Office at the Defense Advanced Research Projects Agency (DARPA), where he conceives, develops, and manages radically innovative space systems for national defense. He is the creator and manager of the System F6 fractionated spacecraft program. He previously provided support to DARPA as a consultant, where he helped lead a team of technical experts on the RASCAL space launch program. He was a spacecraft engineer for seven years at Space Systems/Loral in Palo Alto, California. He served for six years on active duty in the U.S. Navy as a nuclear submarine officer. He holds a BS in Engineering Science from Loyola College in Baltimore and an MS and PhD in Aeronautical and Astronautical Engineering from Stanford University.

Covens, Lloyd is an international journalist who has written and researched the global direct-to-home and advanced consumer technologies (HDTV, CDMA, VoiP) since the 1970s. Covens is an award-winning TV/radio broadcaster, a national speaker/panelist and a founding member/past board member of the Society of Satellite Professionals International. In Colorado, Covens has been an active local community planner, worked on regional transportation infrastructure improvements, and now oversees a bio-fuels start-up firm growing non-food products for alternative energy deployment. Lloyd received his MA from the University of New Mexico and is a candidate for his doctorate in business on Sustainable Energy at Argosy University, Denver.

Dankberg, Mark is a cofounder of ViaSat, Inc. and its Chief Executive Officer and Chairman of the company since its inception in 1986. As a startup, ViaSat was selected to the Inc. 500 list of fastest growing private companies three times. Dankberg has been recognized as an Entrepreneur of the Year in San Diego in 2000, as the Satellite Industry Executive of the Year in 2003, and received the American Institute of Aeronautics and Astronautics (AIAA) Aerospace International Communications Award in 2008. He holds a number of patents in communications and satellite networking technologies. Dankberg earned BSEE and MEE degrees from Rice University and is a member of the Rice University Electrical and Computer Engineering Hall of Fame.

Dealy, John F. is President of The Dealy Strategy Group, LLC. For the past 26 years, he has provided strategic advice and negotiating/operational support to a number of satellite communications companies building and launching next generation commercial systems. He served as a Senior Advisor to the CEO of XM Satellite Radio from 1997–2008. He was Distinguished Professor at Georgetown University School of Business from 1982–1998 and previously held executive positions at Maryland Health Care Product Development Corporation, American Satellite Corporation, Space Communications, Inc., and Fairchild Industries, Inc. He was an Attorney-Advisor in the Office of the Secretary of the Air Force from 1964–1967. Dealy is a graduate of Fordham College and New York University School of Law.

Dest, Leonard R. is President and Chief Executive Officer of RD AMROSS, a joint venture of Pratt & Whitney Rocketdyne and NPO Energomash, which supplies RD-180 rocket engines for the Atlas V launch vehicle. He is the founder of Duckwood Advisors, an international marketing and business development management consulting firm. His professional career includes positions at Fairchild (now Orbital Sciences), RCA Americom (now SES Americom), Comsat, Intelsat, Hughes Space and Communications (now Boeing Satellite Systems), and Lockheed Martin International Launch Services (ILS). In 2000 Dest was awarded an *Aviation Week* Laurel Award for Space in recognition of creative launch services business arrangements.

Feltes, Yves, Vice President Media Relations at SES and SES-AMERICOM-NEW SKIES, has been with the SES Group since 1989. As ASTRA press spokesman and then SES spokesman he accompanied the company in its phenomenal growth from a single satellite operation into the world's premier satellite operator with a global fleet of 40 spacecraft. Prior to joining SES, Yves Feltes worked as a journalist with a special interest in the field of media. He studied political sciences, mass media, and journalism at the University of Vienna in Austria.

Ghais, Ahmad F. has participated in U.S. space programs starting in 1961. He has been involved in Inmarsat matters since it was first conceived in 1973, and served as its Director, Engineering and Operations in London during 1983–1993. He retired from that position in 1994 to continue his career as an independent consultant, specializing in international satellite telecommunications, particularly where technology, policy, and business converge. He also served as president of the Mobile Satellite Users Association (MSUA). He is an alumnus of NASA, Comsat, Intelsat, and Inmarsat. Ghais received his education in telecoms at MIT.

Kennedy, Fred is a Lieutenant Colonel in the U.S. Air Force. Currently a student at the U.S. Army War College, he was recently a Program Manager at DARPA, where he led the successful launch and demonstration of the Orbital Express mission. He also started several programs that aim to develop advanced propulsion and power capabilities for space systems. He has held several technical positions while serving in the Air Force, including assignments to the National Reconnaissance Office and the Air Force Phillips Laboratory. He holds an SB and SM from MIT in Aeronautics and Astronautics, an MA from the George Washington University, and a PhD in Electrical Engineering from the University of Surrey.

Kowalik, Harry retired as Vice President of Space Systems at Telesat in 1993. In 1960 the Canadian government began the Alouette satellite program in which Kowalik worked as a design engineer. In 1969 he joined the newly created Telesat Canada to pioneer geostationary orbit technology. He led an innovative team in the development of the world's first orbit control system for geostationary domestic communications satellites, for which he received the McCurdy Award in 1980. He has a BS in Engineering Physics and a MS in Applied Science. He now lives in Ottawa, Canada.

Kullman, Conny led Intelsat through its privatization in 2001 and its metamorphosis into a highly competitive operator. He was the driving force behind commercial, operational, and engineering changes as Chief Executive Officer from 1998 to 2005. In 2003, Kullman led the acquisition of Loral's North American satellites, providing access to the U.S. market and completing Intelsat's global system. In 2004, Kullman recognized the opportunity to obtain premium shareholder returns, by completing an LBO process resulting in the acquisition of Intelsat by private equity funds at a very competitive price. Kullman capped his career in 2006 as Chairman with the completion of Intelsat's acquisition of PanAmSat. He is an inductee to the Society of Satellite Professionals International (SSPI) Satellite Hall of Fame.

McDonald, Keith D. is Chairman and Technical Director of NavtechGPS and President of Navtech Consulting. He was Scientific Director of the U.S. Department of Defense Navigation Satellite Program during the initial four years of the development of the Global Positioning System (GPS) and has remained involved continuously. He also served as Chairman of the DoD Navigation Satellite Management Office that established the operating parameters for the system. He was a member of the National Academy of Science/National Research Council Committee on the Future of GPS, served as President of the U.S. Institute of Navigation in 1990–1991 and as President of the International Association of Institutes of Navigation from 1997–2000. He is a Fellow of the Royal Institute of Navigation of the United Kingdom and of the U.S. Institute of Navigation.

Morris, Adrian J. has been Executive Vice President of Hughes Network Systems since early 2006. Prior to that, he has been Senior Vice President of Engineering since 1996 and worked on the first generation Thuraya system from 1997 onwards. His career began at Hughes in 1982 as a hardware designer, and he has held a variety of technical and management positions throughout his career. Prior to joining Hughes, he worked for Ferranti Electronics and Electro Optics Division for nearly 5 years. He holds a number of patents in digital communications and has authored several published papers.

Morris received a BS from Trinity College Dublin, and an MS in Digital Techniques from Heriot Watt University, Edinburgh.

Nagai, Yutaka is Director of the Board and Senior Executive Vice President of SKY Perfect JSAT Corporation (SPJ) and is also Board Director of SKY Perfect JSAT Holdings Inc., the parent company of SPJ. He started his career at Nippon Telegraph & Telephone Public Corporation in 1971 and was engaged in research and development of its satellite communications system. In 1986, he joined Japan Communication Satellite Company Inc. Then he developed his career from satellite operation to satellite engineering, business planning, and business development. He received a BE from Tokyo University.

Pulliam, Wade is a Program Manager in the Tactical Technology Office, DARPA, where he pursues high-altitude, long-endurance aircraft technologies, including the Vulture and Rapid Eye UAV programs. Prior to joining DARPA, he served at the Homeland Security Advanced Projects Research Agency and in the Future Combat Systems Technology Office. An entrepreneur, Pulliam was an early employee of Luna Innovations, a fiber optic medical and sensor company, and a cofounder and President of Fortis Technologies, an active material research and development firm. He is a holder of two patents and a PhD from Virginia Tech.

Rao, U. R., former Chairman of ISRO, has made original contributions to the development and application of space technology in India. Starting with Aryabhata, India's first satellite, Rao was responsible for the development of INSAT and IRS series of communication and remote sensing satellites and ASLV and PSLV launch vehicles. He was Chairman of the U.N. Committee on Peaceful Uses of Outer Space during 1997–2000 and President of UNISPACE-III Conference. *Space News* Magazine lists Rao among the top 10 International Space Personalities. Rao has received numerous national and international awards including the Life Time Achievement Award and Padma Bhushan, the second highest civilian award granted by the government of India.

Rosen, Harold A. has earned worldwide recognition for his pioneering work in the field of communications satellites and is widely recognized as "the father of the geostationary satellite." He is a recipient of the 1995 National Academy of Engineering's Draper Prize, the 1990 Arthur C. Clarke Award, the 1985 National Medal of Technology, the 1985 Communications and Computing Prize from NEC, the 1982 Alexander Graham Bell Medal, and the 1976 Ericsson International Prize in Communications. In 2003, he was inducted into the National Inventors Hall of Fame, and has received numerous other awards and honors. A holder of over 75 space-related patents, he is a Fellow of the IEEE and the AIAA. He is a Distinguished Alumnus of Caltech, from which he received his PhD in electrical engineering (with a minor in aeronautics). He now consults for Boeing in the design of new satellite systems.

Schaeffler, Jimmy is the Chairman and CSO of The Carmel Group, which has, since 1995, provided consulting, research, analytical, publishing, and conference organizing services to a broad clientele base. Schaeffler has recently authored the 275-page industry standard text *Digital Signage—Software, Advertising and Displays: A Primer For Understanding The Business.* His insights and views are reported on a regular basis in such media as NPR, *Investor's Business Daily, Time Magazine,* and *The Wall Street Journal.* Schaeffler is a frequent speaker and moderator at telecommunications industry events, such as CES, NAB, NCTA, and IBC. Schaeffler is a graduate of the University of California, Berkeley, and has a DJ from the University of Pacific's McGeorge School of Law.

SUPPORTING MATERIALS

Many of the topics introduced in this book are discussed in more detail in other AIAA publications. For a complete listing of titles in the Library of Flight, as well as other AIAA publications, please visit http://www.aiaa.org.

Fig. 3.2   The team in front of the road-show airplane, minutes before we got the IS-804 news.

Fig. 4.4   The release of Anik C3 from STS 5 Columbia.

Fig. 4.9   The Anik F2 in Ka-band.

Fig. 4.10   The Earth station at Eureka.

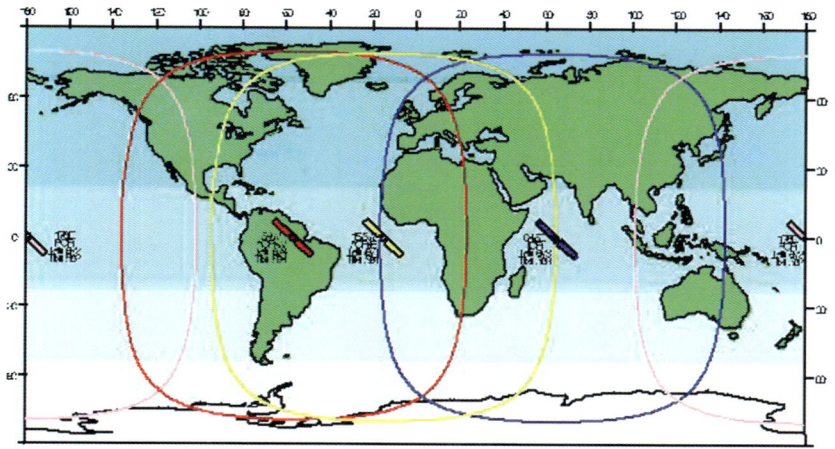

Fig. 5.1   Three or four ocean regions provide global coverage.

Fig. 6.4   Thuraya phone users on Mount Everest.

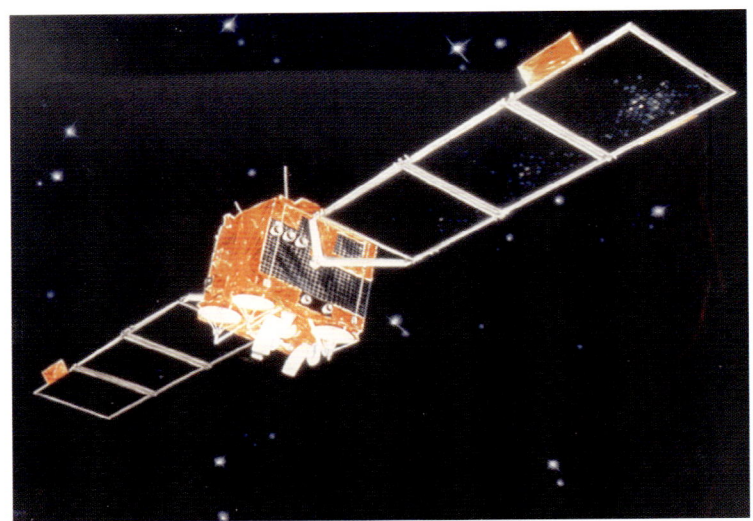

Fig. 7.1   The EUTELSAT 1 satellite.

Fig. 9.3   The first Japanese satellite CS-1 (Sakura) [copyright held by Japan Aerospace Exploration Agency (JAXA)].

Fig. 9.5 The experimental geostationary communication satellite ECS-1 (Ayame 1) [copyright held by Japan Aerospace Exploration Agency (JAXA)].

Fig. 9.11 The N-STAR spacecraft.

Fig. 10.2   The three-axis-stabilized, geostationary communication satellite APPLE.

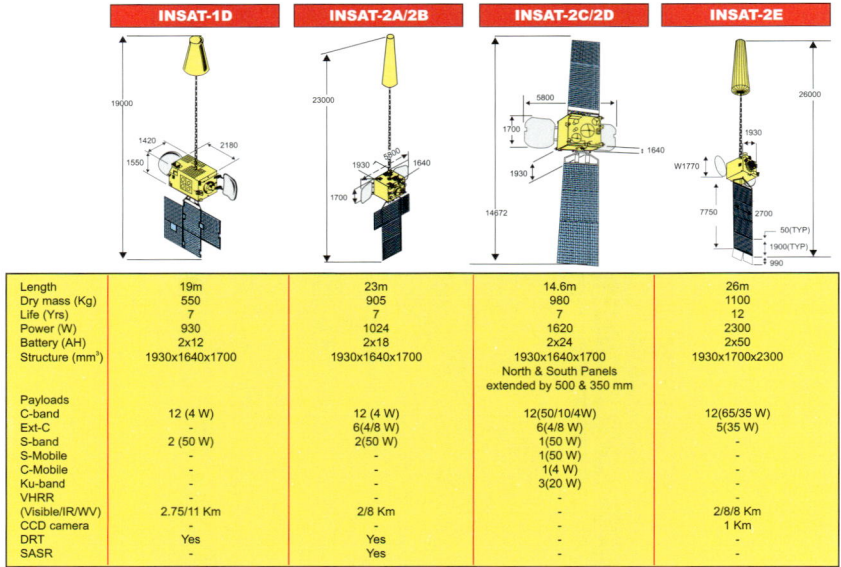

	INSAT-1D	INSAT-2A/2B	INSAT-2C/2D	INSAT-2E
Length	19m	23m	14.6m	26m
Dry mass (Kg)	550	905	980	1100
Life (Yrs)	7	7	7	12
Power (W)	930	1024	1620	2300
Battery (AH)	2x12	2x18	2x24	2x50
Structure (mm³)	1930x1640x1700	1930x1640x1700	1930x1640x1700 North & South Panels extended by 500 & 350 mm	1930x1700x2300
Payloads				
C-band	12 (4 W)	12 (4 W)	12(50/10/4W)	12(65/35 W)
Ext-C	-	6(4/8 W)	6(4/8 W)	5(35 W)
S-band	2 (50 W)	2(50W)	1(50 W)	-
S-Mobile	-	-	1(50 W)	-
C-Mobile	-	-	1(4 W)	-
Ku-band	-	-	3(20 W)	-
VHRR	-	-	-	-
(Visible/IR/WV)	2.75/11 Km	2/8 Km	-	2/8/8 Km
CCD camera	-	-	-	1 Km
DRT	Yes	Yes	-	-
SASR	-	Yes	-	-

Fig. 10.7   INSAT-1–INSAT-2 comparison is given.

Payload	EIRP	G/T
5 Ku -Regional beam	53 dBW	7 dB/K
1 Ku - National beam	50 dBW	3 dB/K
6 Ext-C-National beam	37 dBW	-1 dB/K

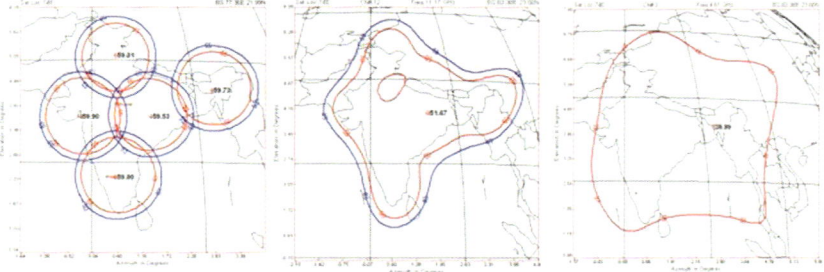

Fig. 10.9   The Edusat satellite.

Fig. 12.3   The NNSS Transit satellite on the Oscar spacecraft.

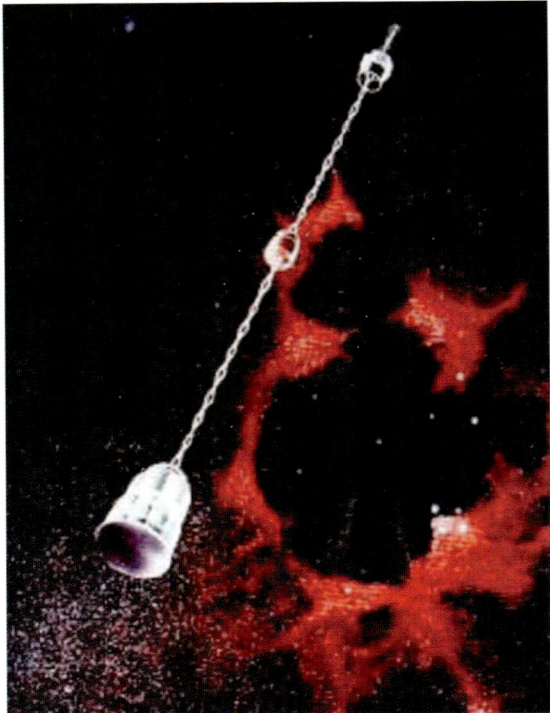

Fig. 12.4   The Transit TRIAD satellite configuration in orbit (courtesy of Johns Hopkins University Applied Physics Laboratory).

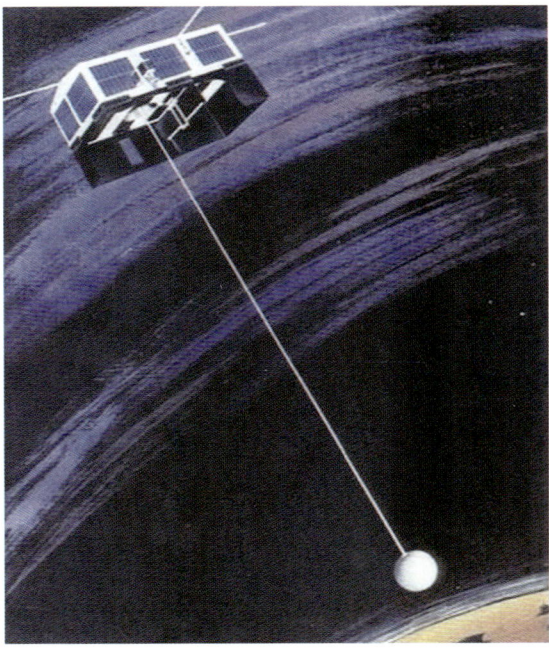

Fig. 12.5   The NRL Timation satellite (courtesy of U.S. Navy Naval Research Laboratory).

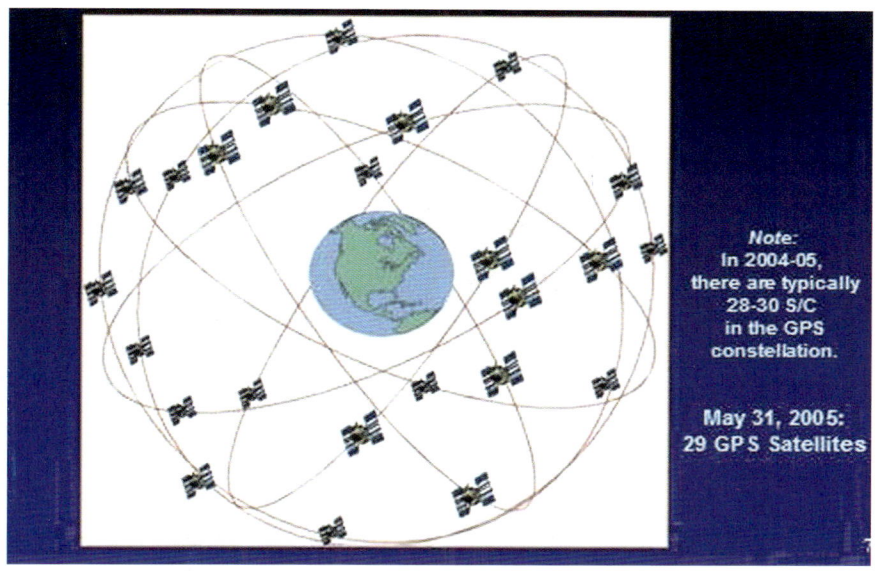

Fig. 12.7   The Navstar GPS operational constellation (24-satellite baseline configuration).

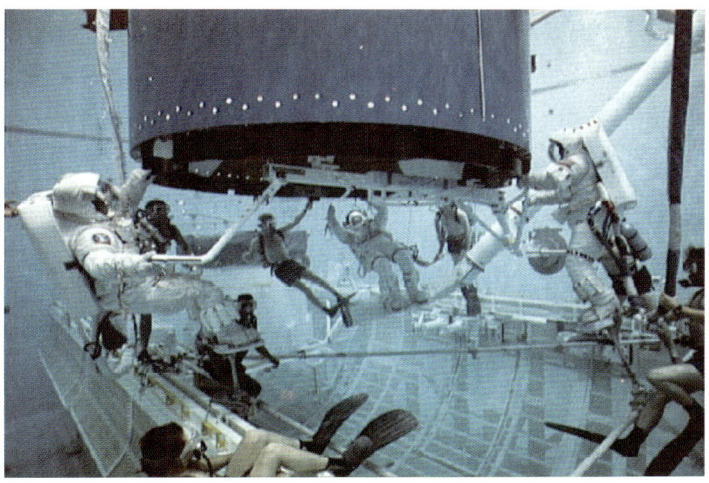

Fig. 14.2   Engineers work underwater alongside reboost mission astronauts.

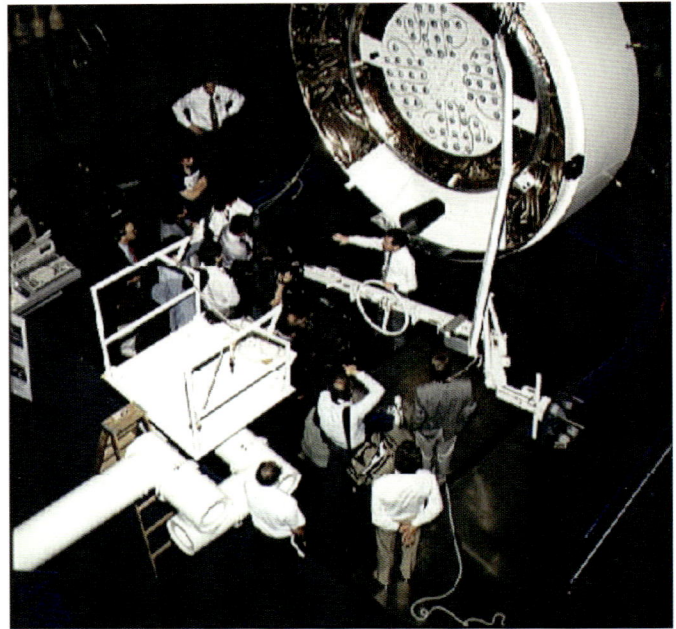

Fig. 14.3   The capture bar is designed by NASA.

Akers   Brandenstein   Chilton   Hieb   Melnick   Thornton   Thuot

Fig. 14.5   The STS-49 crew.

Fig. 14.6   Intelsat VI (F-3) satellite is shown prior to rendezvous with Space Shuttle Endeavour.

Fig. 14.8   EVA Mission Specialist Thuot standing on the RMS end effector with the Intelsat capture bar.

Fig. 14.9   STS-49 crew captures Intelsat VI above OV-105's payload bay during EVA.

Fig. 14.10   Three mission specialists perform a manual capture.

Fig. 14.11   Intelsat VI F-3 separation and ejection from the STS cargo bay.

Fig. 14.12   Intelsat VI F-3 drifts in space after servicing.

Fig. 15.1   Intelsat Standard A Earth station (courtesy of Globecomm, Inc.).

Fig. 16.1   The Soviet launch of the world's first artificial satellite, Sputnik I provided the impetus for many changes in scientific, educational, and military policy in the United States, including the formation of ARPA (downloaded from http://nssdc.gsfc.nasa.gov/planetary/image/sputnik_asm.jpg).

Fig. 16.4   The first weather satellite program, TIROS.

Fig. 16.6   VELA 5 satellite: a presidentially directed effort, the ARPA VELA HOTEL satellites, part of the larger VELA program, provided the United States with the means to detect atmospheric and space nuclear detonations, a key verification capability in support of the Limited Test Ban Treaty (downloaded from http://imagine.gsfc.nasa.gov/Images/vela5b/vela5b_5sm.gif).

Fig. 16.7 The Atlas Centaur launch system. ARPA funded the development of the LH$_2$/ LOX Centaur upper stage, as a high-performance stage necessary to lift heavy payloads. Combined with the Atlas and Titan boosters, Centaur was the upper stage for launches of probes to Mercury, Venus, Mars, Jupiter, Saturn, Uranus, and Neptune. Centaur continues to serve as the Untied States' most powerful upper stage (downloaded from http:// exploration.grc.nasa.gov/education/rocket/gallery/atlas/AtlasCentaur.jpg).

Fig. 16.9 An in-orbit view of NextSat from ASTRO during DARPA's Orbital Express mission.